Third Edition

Ecosystems and Human Health

Toxicology and Environmental Hazards

Third Edition

Ecosystems and Human Health

Toxicology and Environmental Hazards

Richard B. Philp

CRC Press
Taylor & Francis Group
Boca Raton London New York

CRC Press is an imprint of the
Taylor & Francis Group, an **informa** business

CRC Press
Taylor & Francis Group
6000 Broken Sound Parkway NW, Suite 300
Boca Raton, FL 33487-2742

© 2013 by Taylor & Francis Group, LLC
CRC Press is an imprint of Taylor & Francis Group, an Informa business

No claim to original U.S. Government works

Printed on acid-free paper
Version Date: 20130125

International Standard Book Number-13: 978-1-4665-6721-4 (Hardback)

Library of Congress Cataloging-in-Publication Data

Philp, Richard B.
    Ecosystems and human health : toxicology and environmental hazards / Richard
B. Philp. -- Third edition.
      pages cm
    Includes bibliographical references and index.
    ISBN 978-1-4665-6721-4 (hardback)
    1. Environmental toxicology. 2. Environmental health. I. Title.

RA1226.P48 2013
615.9'02--dc23   2012050915

Visit the Taylor & Francis Web site at
http://www.taylorandfrancis.com

and the CRC Press Web site at
http://www.crcpress.com

Contents

Author...xix

1. **Principles of Pharmacology and Toxicology**..................................1
   Introduction...1
   Pharmacokinetics..3
      Absorption..4
      Distribution..5
      Biotransformation...7
      Elimination..10
   Pharmacodynamics..12
      Ligand Binding and Receptors...12
      Biological Variation and Data Manipulation.............................14
      Dose Response...15
      Probit Analysis...17
      Cumulative Effects..19
      Factors Influencing Responses to Xenobiotics..........................21
         Age...21
         Body Composition...22
         Sex...22
         Genetic Factors...22
         Presence of Pathology...25
         Xenobiotic Interactions..25
   Some Toxicological Considerations..27
      Acute versus Chronic Toxicity..27
      Acute Toxicity..27
         Peripheral Neurotoxins...28
         Central Neurotoxins...28
         Inhibitors of Oxidative Phosphorylation............................28
         Uncoupling Agents...28
         Inhibitors of Intermediary Metabolism..............................29
         Chronic Toxicity...29
   Mutagenesis and Carcinogenesis..29
      Introduction...29
      Genetics of Carcinogenesis..31
         Oncogene...31
         Proto-Oncogene..31
         Tumor Suppressor Genes..31
         Growth Factor Receptors..32
         Hormone Receptors..33
         Drug Resistance Genes..33

Antisense Genes...34
Genetic Predisposition to Cancer 34
Epigenetic Mechanisms of Carcinogenesis...............................34
Viruses and Cancer..35
Models of Carcinogenesis...37
Model 1 37
Model 2 38
Model 3 38
Model 4 38
Model 5 39
Model 6 39
Stages of Chemically Induced Carcinogenesis........................39
Initiation 39
Promotion...39
Progression...40
DNA and Cell Repair 41
Response of Tissues to Chemical Insult 41
DNA Repair 41
Cell Repair and Regeneration in Toxic Reactions 42
Fetal Toxicology 42
Teratogenesis...42
Transplacental Carcinogenesis...44
Population and Pollution 45
Review Questions 46
Answers..50
Further Reading...51

2. **Risk Analysis and Public Perceptions of Risk**.....................53
Introduction...53
Assessment of Toxicity versus Risk..54
Predicting Risk: Workplace versus the Environment.................55
Acute Exposures...55
Chronic Exposures..55
Very-Low-Level, Long-Term Exposures..............................55
Carcinogenesis...56
Risk Assessment and Carcinogenesis 56
Sources of Error in Predicting Cancer Risks.........................59
Portal-of-Entry Effects 59
Age Effects...61
Exposure to Co-Carcinogens and Promoters.......................61
Species Differences 61
Extrapolation of Animal Data to Humans 63
Hormesis..63
Natural versus Anthropogenic Carcinogens 64
Reliability of Tests of Carcinogenesis.................................65

Environmental Monitoring 65
Setting Safe Limits in the Workplace...66
Some Important Definitions..67
Environmental Risks: Problems with Assessment
and Public Perceptions...68
  Psychological Impact of Real and Potential
  Environmental Risks..69
  Voluntary Risk Acceptance versus Imposed Risks................70
  Costs of Risk Avoidance...71
Some Examples of Major Industrial Accidents and Environmental
Chemical Exposures with Human Health Implications.........72
  Nuclear Accidents 72
  Formaldehyde..73
  Dioxin (TCDD)..74
Some Legal Aspects of Risk 75
  De Minimis Concept..75
  Delaney Amendment 76
Statistical Problems with Risk Assessment..............................77
Risk Management 78
Precautionary Principle 79
Case Study 1 80
Case Study 2 80
Review Questions 81
Answers..84
Further Reading 85

3. Water and Soil Pollution 87
Introduction..87
Factors Affecting Toxicants in Water 88
  Exchange of Toxicants in an Ecosystem...................................88
  Factors (Modifiers) Affecting Uptake of Toxicants
  from the Environment..89
    Abiotic Modifiers..89
    Biotic Modifiers...91
  Invasive Species..92
Some Important Definitions..94
Toxicity Testing in Marine and Aquatic Species 94
Water Quality 95
  Sources of Pollution 96
  Some Major Water Pollutants 98
  Chemical Classification of Pesticides 99
Health Hazards of Pesticides and Related Chemicals.............99
  Chlorinated Hydrocarbons...99
  Chlorophenoxy Acid Herbicides...100
  Organophosphates (Organophosphorus Insecticides) 100

Carbamates 101
  Bhopal Disaster.. 101
Acidity and Toxic Metals 102
Chemical Hazards from Waste Disposal 104
  Love Canal Story 105
    Problems with Love Canal Studies.. 107
Toxicants in the Great Lakes: Implications for
Human Health and Wildlife 108
  Evidence of Adverse Effects on Human Health 110
  Evidence for Adverse Effects in Wildlife.. 111
Global Warming and Water Levels in the Great Lakes....................... 111
Marine Environment.. 112
  Sources of Marine Pollution 113
    Nonpoint Sources of Pollution 113
    Point Sources of Pollution 116
Biological Hazards in Drinking Water 118
Walkerton Water Crisis.. 119
Review Questions 122
Answers... 124
Further Reading 125

4. Airborne Hazards... 129
Introduction.. 129
Types of Air Pollution 129
  Gaseous Pollutants.. 129
  Particulates.. 130
  Smog... 130
Sources of Air Pollution 130
Atmospheric Distribution of Pollutants 131
  Movement in the Troposphere 132
  Movement in the Stratosphere 132
  Water and Soil Transport of Air Pollutants.................................... 132
Types of Pollutants... 133
  Gaseous Pollutants.. 133
  Particulate Pollutants... 134
Health Effects of Air Pollution... 134
  Acute Effects 134
  Chronic Effects 135
Adverse Effects of Aerial Spraying 136
  Light-Brown Apple Moth... 137
  Painted Apple Moth and the Asian Gypsy Moth........................... 138
  Other Incidents... 139
  Spraying with Conventional Insecticides....................................... 140

Air Pollution in the Workplace 141
    Asbestos.. 142
    Silicosis 143
    Pyrolysis of Plastics 143
    Dust.. 143
    Methane... 143
    CO and NO_2 144
    Multiple Chemical Sensitivity.. 144
Chemical Impact of Pollutants on the Environment 149
    Sulfur Dioxide and Acid Rain.. 149
    Chemistry of Ozone... 150
    Chlorine... 151
Climate Change.. 152
    Global Warming Debate.. 152
    Chemistry of Climate Change 154
        Water 154
        Carbon Dioxide... 154
        Methane.. 155
        Sulfur Dioxide... 156
        Motor Vehicle Exhaust... 156
    Subtle Greenhouse Effects 157
    Global Cooling: New Ice Age? 157
Natural Factors and Climate Change 158
Remedies.. 159
Case Study 3 160
Case Study 4 161
Case Study 5 161
Case Study 6 162
Case Study 7 162
Case Study 8 162
Review Questions 163
Answers.. 165
Further Reading 165

**5. Halogenated Hydrocarbons and Halogenated Aromatic
   Hydrocarbons** 171
Introduction.. 171
Early Examples of Toxicity from Halogenated Hydrocarbons 171
Physicochemical Characteristics and Classes
of Halogenated Hydrocarbons... 172
    Antibacterial Disinfectants... 172
    Herbicides.. 173

Dioxin (TCDD) Toxicity.. 173
    Hepatotoxicity.. 174
    Porphyria... 174
    Chloracne... 174
    Carcinogenicity.. 175
    Neurotoxicity 177
    Reproductive Toxicity 177
    Metabolic Disturbances.. 177
Role of the Aryl Hydrocarbon Receptor (AhR)
and Enzyme Induction.. 177
Paraquat... 179
Insecticides.. 179
Industrial and Commercial Chemicals... 180
    Biphenyls 180
    Toxicity... 180
    Pharmacokinetics and Metabolism 180
    Biodegradation.. 181
Accidental Human Exposures 181
Problem of Disposal... 182
Solvents.. 182
    Toxicity... 182
    Mechanism of Toxicity 183
    Trihalomethanes 183
Case Study 9 184
Case Study 10.. 185
Review Questions 185
Answers.. 187
Further Reading 188

6. **Toxicity of Metals** 189
Introduction.. 189
Lead.. 190
    Toxicokinetics of Lead 191
    Cellular Toxicity of Lead... 191
    Fetal Toxicity 192
    Treatment 192
Mercury 193
    Elemental Mercury Toxicity 194
    Inorganic Mercurial Salts.. 194
    Organic Mercurials... 194
    Mechanism of Mercury Toxicity.. 195
    Treatment of Mercury Poisoning... 196
        The Grassy Narrows Story.. 196
Cadmium 198
    Cadmium Toxicokinetics 198

Cadmium Toxicity..199
    Treatment...200
Arsenic...200
    Toxicokinetics of Arsenicals 201
    Toxicity of Arsenicals 201
    Treatment 202
    Environmental Effects of Arsenic...202
Chromium..202
Other Metals..203
    Aluminum...203
    Manganese 203
    Uranium 204
    Antimony 204
    Nutritional Elements 204
    Metallothioneins 204
Carcinogenicity of Metals...205
Unusual Sources of Heavy Metal Exposure 205
Case Study 11..206
Case Study 12 207
Case Study 13..207
Review Questions 208
Answers...210
Further Reading..210

7. **Organic Solvents and Related Chemicals** 213
Introduction...213
Classes of Solvents...213
    Aliphatic Hydrocarbons...213
    Halogenated Aliphatic Hydrocarbons 214
    Aliphatic Alcohols...215
    Glycols and Glycol Ethers 216
    Aromatic Hydrocarbons 217
Solvent-Related Cancer in the Workplace 219
    Benzene 219
    Bis(Chloromethyl) Ether...219
    Dimethylformamide and Glycol Ethers....................................219
    Ethylene Oxide (CH_2CH_2O) 220
Factors Influencing the Risk of a Toxic Reaction...........................221
Nonoccupational Exposures to Solvents 221
Case Study 14..222
Case Study 15..222
Case Study 16..223
Review Questions 223
Answers...225
Further Reading..225

8. Food Additives, Drug Residues, and Food Contaminants................227

Food Additives...227

Food and Drug Regulations...227

Some Types of Food Additives..228

Artificial Food Colors..230

Banned or Restricted Artificial Food Colors..............................232

Emulsifiers...232

Preservatives and Antioxidants...233

Artificial Sweeteners...235

Flavor Enhancers...237

Drug Residues...239

Antibiotics and Drug Resistance...239

Infectious Drug Resistance...240

Infectious Diseases...244

Allergy..245

Hormones as Growth Promotants in Livestock...............................245

Diethylstilbestrol...245

Bovine Growth Hormone..248

Other Hormonal Growth Promotants..249

Natural Toxicants and Carcinogens in Human Foods......................250

Some Natural Toxicants...250

Favism...250

Toxic Oil Syndrome..251

Eosinophilia-Myalgia Syndrome..252

Herbal Remedies..252

Natural Carcinogens in Foods...253

Case Study 17..254

Review Questions...255

Answers..257

Further Reading..258

9. Pesticides..263

Introduction...263

Classes of Insecticides...265

Organochlorines (Chlorinated Hydrocarbons)................................265

Organophosphorus Insecticides..267

Carbamate Insecticides...267

Botanical Insecticides...268

Herbicides..269

Chlorphenoxy Compounds..269

Dinitrophenols...269

Bipyridyls..270

Carbamate Herbicides...270

Triazines...271

Fungicides 271
   Dicarboximides 271
   Newer Biological Control Methods 271
Government Regulation of Pesticides 272
Problems Associated with Pesticides 273
   Development of Resistance 273
   Multiple Pesticide Resistance 274
   Nonspecificity 275
   Environmental Contamination 275
Balancing the Risks and Benefits 276
Toxicity of Pesticides for Humans 277
Case Study 18 278
Case Study 19 279
Review Questions 279
Answers 281
Further Reading 281

10. **Mycotoxins and Other Toxins from Unicellular Organisms** 283
Introduction 283
Some Health Problems Associated with Mycotoxins 284
   Ergotism 284
   Aleukia 285
Some Specific Mycotoxins 285
   Aflatoxins 285
   Fumonisins 287
   Ochratoxins 288
   Patulin 289
   Fusarium Species 290
      Zearalenone 291
      Vomitoxin (Deoxynivalenol or DON) 291
      Other Tricothecenes 293
Economic Impact of Mycotoxins 294
Detoxification of Grains 294
   Harvesting and Milling 294
   Chemical Treatments 295
   Binding Agents 295
   Other Techniques 295
Other Toxins in Unicellular Members of the Plant Kingdom 296
Review Questions 296
Answers 299
Further Reading 300

11. **Animal and Plant Poisons** 303
Introduction 303
Toxic and Venomous Animals 304

Toxic and Venomous Marine Animals 304
Scale Fish Toxins 304
    Ciguatoxin 304
    Tetrodotoxin 305
    Scombroid Toxins 306
    Ichthyotoxin 306
Red Tide Dinoflagellate Toxicity for Higher Species 307
Shellfish Toxins 307
    Saxitoxins 307
    Brevetoxins 308
    Domoic Acid 308
    Okadaic Acid 308
    Azaspiracid Toxin 308
    Yessotoxin 309
    Palytoxin 309
Stinging Fish Venoms 309
Mollusk Venoms 310
    Conotoxins 310
Coelenterate Toxins 311
Echinoderm Venoms 312
Freshwater Algae 312
Toxic and Venomous Land Animals 313
    Venomous Snakes 313
    Snake Venoms 317
    First Aid 318
Venomous Arthropods 319
Toxic Plants and Mushrooms 321
    Introduction 321
    Vesicants 321
    Cardiac Glycosides 321
    Astringents and Gastrointestinal Irritants (Pyrogallol Tannins) 322
    Autonomic Agents 322
    Dissolvers of Microtubles 323
    Phorbol Esters (for Example, Phorbol Myristate Acetate, PMA) 323
    Cyanogenic Glycosides 323
        Detoxification of Hydrogen Cyanide 324
    Convulsants 324
    Used in Research and Treatment 325
Case Study 20 326
Case Study 21 327
Case Study 22 327
Case Study 23 328
Case Study 24 328
Case Study 25 329
Case Study 26 329

Review Questions 329
Answers... 331
Further Reading 331

12. Environmental Hormone Disrupters 335
Introduction... 335
Lake Apopka Incident.. 336
Brief Review of the Physiology of Estrogens and Androgens 336
Disruption of Endocrine Function 337
  Mechanisms.. 337
  Methods of Testing for Hormone Disruption...................... 337
  Some Examples of Xenoestrogen Interactions with E2 Receptors
  In Vitro or Effects in *In Vivo* Tests 338
  Some Effects of Xenoestrogens on the Male
  Reproductive System 339
  Modulation of Hormone Activity through Effects
  on the Ah Receptor 340
    Estrogen/Androgen Effects 340
    Effects on Thyroid Function 340
Plastic-Associated Chemicals... 341
Phytoestrogens... 341
Results of Human Studies on Xenoestrogens......................... 342
  Males.. 342
  Females 343
Effects of Xenoestrogens and Phytoestrogens
in Livestock and Wildlife.. 345
Problems in Interpreting and Extrapolating Results
to the Human Setting.. 346
Case Study 27 347
Review Questions 348
Answers... 349
Further Reading 350

13. Radiation Hazards.. 353
Introduction... 353
Sources and Types of Radiation.. 354
  Sources.. 354
    Natural Sources of Radiation 354
    Man-Made Sources of Radiation.. 354
    Cause of Radiation 354
  Types of Radioactive Energy Resulting from Nuclear Decay 355
Measurement of Radiation 355
  Measures of Energy 355
  Measures of Damage 356

Some Major Nuclear Disasters of Historic
and Current Importance 357
   Hiroshima 357
      Hiroshima Update 357
   Three Mile Island 358
   Hanford Release 359
   Chernobyl 359
   Fukushima, Japan 361
Radon Gas: The Natural Radiation 362
Tissue Sensitivity to Radiation 364
Microwaves 365
Cell Phone Use and Brain Tumors 366
Ultraviolet Radiation 368
   Medical Uses of UV Radiation 369
Extra-Low Frequency Electromagnetic Radiation 369
Irradiation of Foodstuffs 372
Irradiation of Insect Pests 373
Case Study 28 373
Review Questions 374
Answers 375
Further Reading 376

14. **Gaia and Chaos: How Things Are Connected** 379
Gaia Hypothesis 379
Chaos Theory 381
Other Examples of Interconnected Systems 382
   Vicious Circle 382
   Domino Effects of Global Warming 384
   Feedback Loop 385
Food Production and the Environment 386
   Meat versus Grain 386
   Genetically Modified Plant Foods 388
The Environment and Cancer 391
Further Reading 391

15. **Case Study Reviews** 393
Case Study 1 393
Case Study 2 393
Case Study 3 394
Case Study 4 394
Case Study 5 394
Case Study 6 395
Case Study 7 395
Case Study 8 395
Case Study 9 396

Case Study 10.. 396
Case Study 11.. 397
Case Study 12 397
Case Study 13.. 398
Case Study 14.. 398
Case Studies 15 and 16 399
Case Study 17.. 400
Case Study 18.. 400
Case Study 19.. 401
Case Study 20 401
Case Study 21.. 402
Case Study 22 402
Case Study 23 403
Case Study 24.. 403
Case Study 25 403
Case Study 26 404
Case Study 27 404
Case Study 28 405
References 405

Index 407

Author

Richard Philp is an emeritus professor and former chair of the Department of Pharmacology and Toxicology at The University of Western Ontario. After spending five years as a country veterinarian, he returned to Western and earned his PhD in pharmacology. This was followed by a postdoctoral year at the Royal College of Surgeons of England as the Canadian Defence Research Board Fellow in Aviation Medicine. Returning to Western, he conducted research in the physiology of deep sea diving, investigations into arterial thrombosis, and, for the last 15 years of his 40-year career, taught a course on environmental toxicology and authored two textbooks on the subject. He has conducted studies on heavy metal pollution of the waters of the Gulf of Mexico and was a member of the Thames-Sydenham Regional Sourcewater Protection Committee. He has served as a consultant in environmental toxicology to citizens' groups and to Santa Cruz County, California. He has also published more than 100 papers in peer-reviewed journals and 6 books, mostly on environment-related subjects.

1

Principles of Pharmacology and Toxicology

> The right dose differentiates a poison and a remedy.
>
> **Paracelsus, 1493–1541**

Introduction

The past century has seen a tremendous expansion in the number of synthetic chemicals employed by humankind as materials, drugs, preservatives for foods and other products, pesticides, cleaning agents, and even weapons of war. The American Chemical Society maintains a chemical registry. Since 1907, it has recorded 33 million organic and inorganic substances and 58 million sequences as of 2008. About 4000 new chemicals are added each day. A study by the Danish Environmental Protection Agency calculated that 13.4% of them possess acute toxicity, 2.5% reproductive toxicity, 3.9% are mutagens, 1.8% carcinogens, and 3.5% are dangerous to the aquatic environment. Four thousand chemicals are used as medicinals and at least 1200 more as household products. Add to this the numerous natural substances, both inorganic and organic, that possess toxic potential, and it is little wonder that the public expresses concern and even, sometimes, panic about the harmful effects these agents may exert on their health and on the environment. Tens of thousands of these agents have never been subjected to a thorough toxicity testing.

According to the Danish study, thousands of chemicals are potential carcinogens but the number that has been confirmed to be human carcinogens is much smaller. About 500 chemicals have been evaluated for carcinogenic potential. Some 44 have been designated as possible human carcinogens on the basis of evidence, either limited or conclusive, obtained from human studies. Of these, 37 were tested positive for carcinogenicity in animal tests and were later shown to be carcinogens for humans. There are, however, numerous other agents that have been shown to be carcinogenic in rodents but that are yet to be identified as human carcinogens. This creates significant problems regarding the legislative and regulatory decisions that need to be made about their use. Some of the areas of uncertainty that surround the extrapolation of data from the animal setting to the human one will be discussed in the following chapter. The process of extrapolation requires input from many different disciplines that may include engineering, physics, biology, chemistry, pathology, pharmacology, physiology, public health, immunology, epidemiology, biostatistics, and occupational health. The field of toxicology thus depends on all of these, but perhaps draws most heavily on pharmacology, biochemistry, and pathology. It is the identification of the degree of risk to which individuals or groups are exposed in a given set of circumstances that directs all of this activity.

Other forms of toxicity, hepatotoxicity, nephrotoxicity, and neural toxicity, for example, may be more important in acute exposures that might occur in the industrial setting. Reproductive and fetal toxicity has been demonstrated frequently experimentally, but their significance for the general population exposed to low levels of toxicants in the environment remains unclear.

The (U.S.) Agency for Toxic Substances and Disease Registry, and The (U.S.) Environmental Protection Agency jointly maintain a priority list of toxic substances. The "top 15" in the 2011 list are arsenic, lead, metallic mercury, vinyl chloride, polychlorinated biphenyls (PCBs), benzene, cadmium, benzo[a]pyrene, polycyclic aromatic hydrocarbons (PAHs), benzo[b]fluoranthene, chloroform, aroclor 1260, P'P'-DDT, aroclor 1254, and dibenz[a,h] anthracene. The complete list can be viewed on the Internet at www.atsdr. ctv.gov "Priority List of Hazardous Substances."

Considerable difficulty attends efforts to extrapolate the results of toxicity tests in experimental animals to humans exposed to very low levels in their environment, especially with regard to the risk of cancer. Current legislation requires testing in two species with sufficient numbers for reliable statistical analysis. Rats and mice are generally used, as hamsters are resistant to many carcinogens and primates are too expensive and, in the case of some species, too environmentally threatened. For statistical purposes, cancer includes all tumors whether benign or malignant. A 2 year carcinogen study employing two species cost, in 2010, at least $2,000,000 plus the costs of 1 year for preparation, one for analysis (pathology, etc.), and one for documentation and statistics. Since it is not practical to test every chemical, several factors need to be considered in selecting test chemicals. These include the frequency and severity of observed effects, the extent to which the chemical is used, its persistence in the environment (examples of persistent chemicals include chlorinated hydrocarbons), and whether transformations to more toxic agents occur.

Heavy metals, the by-products of most mining and ore extraction processes, are examples of ubiquitous toxicants with almost infinite half-lives. Mercury (Hg), for example, is present in all canned tuna at about 5 parts/million (ppm), mostly from natural sources. Aquatic bacteria can transform mercury to methylmercury. This has a different toxicity profile. Cadmium (Cd) enters the environment at about 7000 ton/year and is concentrated by livestock because they recycle it in feces used for fertilizer. It is then passed on to forage grasses. Radioactive isotopes of cesium and iodine entered the food chain after Chernobyl.

The estimation of the degree of risk associated with the presence of a potentially toxic substance in the environment is the basis for all decisions relating to the legislative controls over that chemical, including its industrial use and eventual disposal. Pharmacological/toxicological principles are essential for understanding the processes involved in toxicity testing.

Pharmacology may be defined as the science of drugs. It includes a study of their sources (*materia medica*); their actions in the living animal organism (pharmacodynamics); the manner in which they are absorbed, moved around in the body, and excreted (pharmacokinetics); their use in medicine (therapeutics); and their harmful effects (toxicology). In this context, a drug is any substance used as a medicine but pharmacology generally includes the study of substances of abuse and, in the broadest sense, deals with the interactions of *xenobiotics* (literally, substances foreign to living organisms) whether they be natural or man-made (anthropogenic), therapeutic or toxic. In this sense, toxicology can be considered to be a branch of pharmacology. Xenobiotics may also be exploited as research tools to reveal mechanisms underlying physiological processes.

Toxicology is the study of the harmful effects of xenobiotics on living organisms, the mechanisms underlying those effects, and the conditions under which they are likely to occur.

Environmental toxicology is the study of the effects of incidental or accidental exposure of organisms, including human beings (the focus of this text), to toxins in the environment, i.e., air, water, and food. While the greatest concern today centers on pollutants of human origin, it should not be forgotten that toxic substances, including carcinogens, abound in nature. The subject of environmental toxicology embraces the study of the causes, conditions, environmental impact, and means of controlling pollutants in the environment. It may also be extended to include the environment of the workplace (industrial hygiene). The related term *ecotoxicology* deals with the harmful effects of chemicals, usually of anthropogenic origin, on ecosystems.

Economic toxicology is the study of chemicals that are developed expressly for the purpose of improving economic gain by selectively eliminating a species (insecticides and herbicides), improving health and productivity (drugs), preserving foodstuffs (food additives), or for the manufacture of a marketable product (industrial solvents, cleaning agents, etc.).

Forensic toxicology refers to the medico-legal aspects of the harmful effects of drugs and poisons administered or taken deliberately or accidentally. Detection of xenobiotics in tissues and fluids and in, or on, objects is an important aspect of this field as is the preparation of evidence for submission in court.

Pharmacokinetics

There has been a trend recently to attempt to separate toxicology from pharmacology by the use of such terms as *toxicokinetics, toxicodynamics,* etc. The distinction is largely semantic. The principles governing the absorption, distribution, metabolism, and excretion of a xenobiotic are the same regardless of whether it is used as a therapeutic agent or is a toxin. Throughout this text, *pharmacological* can be taken also to represent *toxicological*.

The response of organisms to drugs and chemicals is governed by natural laws. One of these is the *Law of Mass Action*, which dictates that, in the absence of a transport system, chemicals in solution will move from an area of high concentration to one of low concentration. If a semipermeable membrane is interposed between these areas, the chemical will move across it, assuming the chemical can penetrate the membrane. In reality, molecules wander randomly across the barrier, but the frequency of transfers will be greater from the area of high concentration to that of the low one until equilibrium is established. Cell walls and other biological membranes function as semipermeable membranes, and The Law of Mass Action influences the uptake of most drugs and toxicants by living organisms. The concentration of a toxicant in the environment (water, air, soil) is thus an important determinant affecting its uptake. Transport mechanisms are dealt with under "Absorption" and "Distribution" sections.

Partition coefficient is the ratio of a chemical's relative solubility in two different phases. The ratio of solubility in oil (often n-octanol) to that in water is frequently used to predict the distribution of a xenobiotic between the aqueous and lipid phases in the body.

Absorption

Whether or not a xenobiotic is toxic, and how that toxicity is manifested, depends largely on how the body deals with it. Substances that are not absorbed from the gastrointestinal tract have no systemic toxicity. This fact allows barium to be used as an x-ray contrast medium, despite barium's toxicity by other routes of administration. The selective toxicity of most insecticides depends solely on a greater ability to penetrate the chitin of the insect's exoskeleton than to penetrate human skin. A substance that is not readily excreted by the body (usually through the kidneys or in the feces) will accumulate to toxic levels.

The primary routes of absorption for toxicants are the skin, the lungs, and the gastrointestinal tract. The latter two are important for the population at large but the skin may be a very significant site in certain industrial settings. The site of absorption, more commonly called the *portal of entry* in toxicology, can have a significant influence on the toxicity of a substance.

Larger molecules require a degree of lipid solubility to cross biological barriers since cell membranes consist of a fluid phospholipid matrix with embedded proteins that may penetrate partway or all the way through the membrane. Factors that influence the lipophilicity of a chemical will therefore affect its absorption. Many chemicals are weak acids or bases that may exist in an ionized (polar) or a nonionized (nonpolar) state with equilibrium established between them; for example,

$$\underset{\text{Nonpolar}}{R-H} \rightarrow \underset{\text{Polar}}{R^-} + H^+$$

The polar form is water soluble whereas the nonpolar form is lipid soluble. The pH will influence the equilibrium and, hence, the amount of the lipid-soluble form that is available for absorption. The *dissociation constant* (pK_a) of a substance is defined as that pH at which 50% of it will exist in each state. Weakly acidic drugs are shifted to the nonpolar state in an acid medium and to the polar state in an alkaline medium. The reverse is true for weakly alkaline drugs. The pH of the stomach and upper small bowel is acidic (pH 2–4); therefore, acidic chemicals will be absorbed here. Alkaline substances tend to be absorbed in the lower small bowel and the upper colon that are more alkaline, whereas the descending colon becomes acidic again.

Lipid solubility is not essential for the passage of all molecules across membranes. There is the bulk transfer of water across the cell membrane that can carry very small (less than 200 Da) water-soluble molecules with it. Metallic ions such as calcium, sodium, and potassium, as well as chlorine, can pass through special channels, some of which are regulated by the transmembrane potential (voltage-regulated) and others by specific receptors (receptor-activated). Specialized exchangers also exist, for example, the sodium pump.

Active transport is an energy-consuming process by which a substance may be moved against a concentration gradient. Active transport is important in the kidney and the liver. In addition to energy consumption, it is also characterized by saturability, selectivity for specific chemical configurations, and the ability to move substances against an electrochemical gradient. *Facilitated diffusion* is similar except that no energy is consumed, and it cannot occur against an electrochemical gradient.

Pinocytosis is a process whereby a segment of the plasma membrane of a cell invaginates to form a sack in which extracellular fluid and colloidal particles can be taken into the cell by pinching off the "mouth" of the sack. This is an important mechanism by which the mucosal cells of the intestinal tract take up nutrients and some drugs and chemicals.

Distribution

Once absorbed, the agent may be distributed throughout various compartments in the body. Serum albumin possesses many nonspecific binding sites for xenobiotics, especially weakly acidic ones, and it therefore becomes a transport system for many substances. The balance between dissociated (polar) and undissociated (nonpolar) states also affects the distribution of a chemical as well, since pH changes from the extracellular fluid (pH 7) to the plasma (pH 7.4). The partition coefficient of a substance also influences its distribution, determining, for example, the extent to which it will be sequestered in fat. Highly lipid-soluble substances will be sequestered in body fat, where they may remain for long periods. Everyone has DDT and its metabolites dissolved in their fat. The amount varies with their age and location.

The use of DDT in North America was drastically reduced in the 1970s and a complete ban was legislated in Canada in 1990. Substances like DDT, which are sequestered in fat, may be released during periods of starvation, extreme dieting, as a result of illness, and even during lactation, when lipids are transferred to milk. The released toxicant may reach concentrations at target sites sufficient to cause a toxic response. Figure 1.1 illustrates these relationships among storage fat, blood, and target organ.

The rate of distribution of a substance is a function of the rate of blood flow through the tissues (tissue perfusion). Highly vascular organs will accumulate it first; organs that are poorly perfused will accumulate it last. The substance is thus distributed initially on the basis of tissue perfusion, then, as equilibrium states are reached, it will redistribute on the basis of its solubility. Following the intravenous injection of a chemical with a high partition coefficient, equilibrium will be established instantly with the kidney

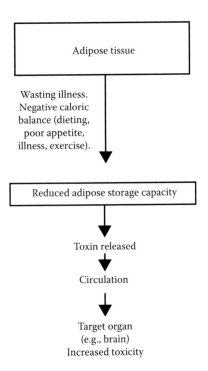

FIGURE 1.1
Disposition of lipid-soluble chemicals in adipose tissue and blood and the effect of severe weight loss.

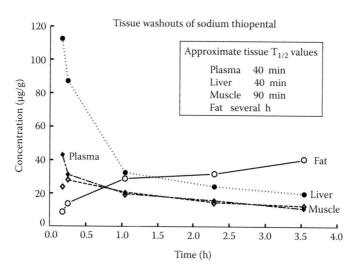

FIGURE 1.2
Tissue $T_{1/2}$ values for sodium thiopental, a highly lipid-soluble, ultrashort-acting barbiturate anesthetic.

and liver because of their high vascularity, almost as quickly with the brain, with muscle in about 30 min, and with fat in about 3 h. The membranes surrounding the brain and separating it from its blood vessels constitute the *blood–brain barrier* that generally will pass only quite lipid-soluble agents such as all anesthetics.

Thus, tissue perfusion and partition coefficient may play important roles in determining the onset and termination of either a therapeutic or a toxic response. Sodium thiopental, an ultrashort-acting barbiturate, is used for anesthetic induction. The rate of biotransformation is so slow as to have little effect on recovery. The drug readily penetrates the blood–brain barrier because of its high lipid solubility and the brain, which is richly perfused, rapidly takes it up and anesthesia ensues. This effect is terminated because the drug is redistributed to other tissues, including depot fat, which is poorly perfused. New equilibria are established among blood, brain, and other tissues so that, while initial recovery is rapid, a state of sedation may persist for several hours. In Figure 1.2, the effects of perfusion and partition coefficient on $T_{1/2}$ s of thiopental in different tissues are shown.

Biotransformation

Biotransformations of xenobiotics are classified as either *Phase I reactions* or *Phase II reactions*.

Phase I chemical reactions, also known as nonsynthetic biotransformations, convert a lipophilic (fat soluble) substance to a more polar, and hence more water-soluble, substance. This metabolite is excreted more readily by

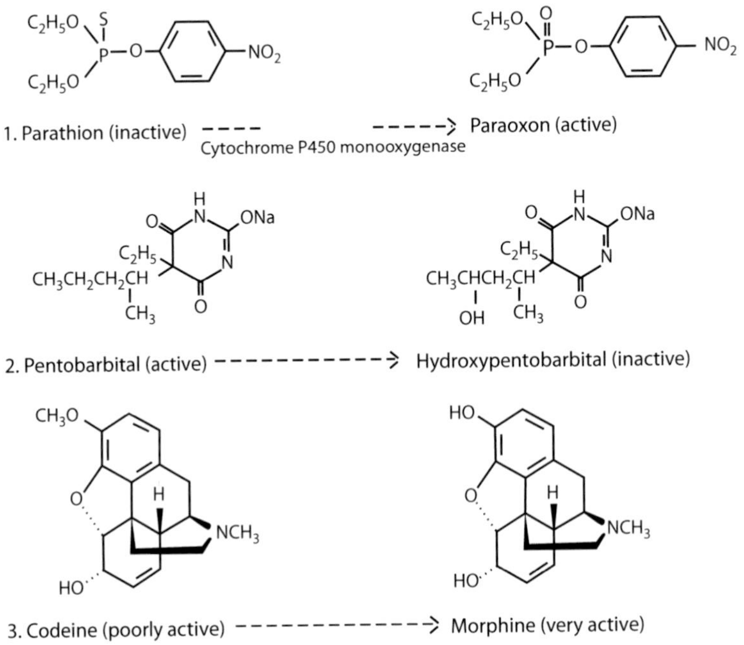

1. Parathion (inactive) $-\,-\,-\,-$ $-\,-\,-\,-\,\rightarrow$ Paraoxon (active)
Cytochrome P450 monooxygenase

2. Pentobarbital (active) $-\,-\,-\,-\,-\,-\,-\,-\,-\,-\rightarrow$ Hydroxypentobarbital (inactive)

3. Codeine (poorly active) $-\,-\,-\,-\,-\,-\,-\,-\,-\,-\rightarrow$ Morphine (very active)

FIGURE 1.3
Some examples of Phase I reactions. The product may be more active or less active than the parent compound or it may be inactive.

the kidneys than the parent compound, but it usually retains significant bioactivity. It may be more active, or less active, than the parent substance. If the parent chemical is nontoxic but the metabolite is toxic, this is a toxication reaction. A drug that requires biotransformation to become active is referred to as *pro-drug*. Figure 1.3 shows some examples of Phase I reactions and their consequences.

Phase I chemical reactions include oxidation, reduction, and hydrolysis and generally unmask or introduce a functional (reactive) group such as $-NH_2$, $-OH$, $-SH$, COOH. The oxidation reactions are listed in the next paragraph dealing with the cytochrome P-450 enzyme group. Hydrolysis of esters and amides also occurs. Reduction reactions may involve azo ($RN = NR$) or nitro (RNO_2) groups.

Many oxidation reactions are under the control of a group of mixed-function oxidases for which members of the cytochrome P450 (CYP450) group serve as a catalyst. CYP450 enzymes are widely distributed in nature and are involved in the biotransformation of a multitude of xenobiotics as well as numerous naturally occurring substances. They are located primarily in the smooth endoplasmic reticulum (SER) of hepatic cells, but they exist in many tissues as well as many species, including single-celled organisms. The CYP450 monooxygenases have tremendous substrate versatility, being able to oxidize lipophilic xenobiotics plus fatty acids, fat-soluble vitamins, and various hormones. This is partly because there are at least 20 variants of the enzyme (isozymes) and because each is capable of accepting many substrates. CYPs 1, 2, and 3 are isozymes especially involved in xenobiotic transformations. The reactions they catalyze include aromatic and side-chain hydroxylation; N-, O-, and S-dealkylation; N-oxidation; N-hydroxylation; sulfoxidation; deamination; dehalogenation; and desulfuration. There are likely 1000 or more CYP450 isozymes in nature with perhaps 50 that have significance in mammals.

It should be noted that pro-carcinogens are converted to carcinogens by Phase I reactions. Examples of this include benzo[a]pyrene, the fungal toxin aflatoxin B_1, and the synthetic estrogen diethylstilbestrol. This process often involves the formation of an epoxide compound, as it does in the three examples given. An epoxide has the chemical configuration shown in Figure 1.4, making it highly nucleophilic and chemically reactive. Many epoxides are carcinogens. Figure 1.4 shows this chemical transformation for stilbestrol and benzo[a]pyrene, which is an example of a polyaromatic hydrocarbon (PAH). Many of these are carcinogenic and are environmental pollutants. Other enzymes called epoxide hydrolases may detoxify the epoxides.

Phase II reactions are conjugation (synthetic) reactions that render the agent not only more water soluble, but biologically inactive, with a very few exceptions. A common conjugation reaction is with glucuronic acid. Conjugation also occurs with sulfuric acid, acetic acid, glycine, and glutathione. Many Phase I metabolites are still too lipophilic (fat soluble) to be excreted by the kidneys and are subjected to Phase II conjugation. All chemicals need not be subjected first to Phase I transformations. Many, if they possess the necessary functional groups (e.g., –OH, –NH₂) are conjugated directly.

An important concept for understanding toxication and detoxication of xenobiotics is *enzyme induction*. Hepatic enzymes of the SER can be stimulated

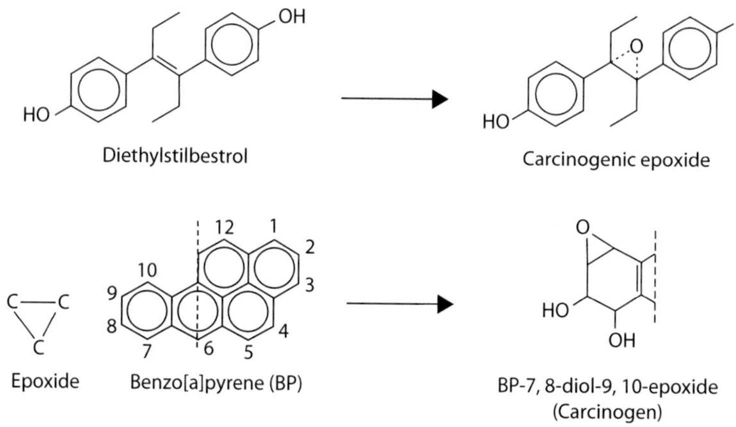

FIGURE 1.4
Examples of epoxide formation to potentially carcinogenic metabolites.

to a higher level of activity by many highly lipophilic agents. Because these enzymes are nonspecific, this has consequences for many other agents transformed by the same enzymes. Induction is accomplished by the increased synthesis of more enzymes, so the SER actually increases in density. The result may be increased detoxication of a chemical, or the increased synthesis of a toxic metabolite. Cigarette smoke contains many inducers and may increase the breakdown of many drugs (theophylline, phenacetin, etc.) but conversely it may act through this mechanism as a promoter or as a co-carcinogen.

Elimination

Every secretory or excretory site in the body is potentially a route of elimination for xenobiotics. Thus, they may be excreted in saliva, sweat, milk, tears, bile, mucus, feces, and urine. Of these, the most significant site is urine, followed by feces and bile. The kidney (Figure 1.5) is the principal organ for

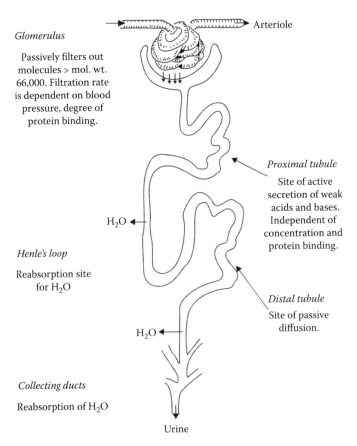

FIGURE 1.5
The Nephron is the basic renal unit.

the elimination of natural waste metabolites. Most of these are toxic if they exceed normal levels. The kidney also is the main organ for maintaining fluid and electrolyte balance. It is therefore not surprising that the kidney also is the main site of elimination of xenobiotics, including drugs. Although it constitutes only 0.4% of total body weight, it takes 24% of the cardiac output. It is a highly efficient filter of blood.

The basic physiological unit of the kidney is the nephron (see Figure 1.5), which is composed of the glomerulus (a tightly wound bundle of blood vessels) and the tubule, which is closed at the glomerular end to provide a semipermeable membrane. The tubule is composed of several segments with different functions. Molecules with a molecular weight less than 66,000 Da are passed through the glomerulus. They may be reabsorbed further down the tubule and even re-secreted. This occurs with uric acid, which is completely passed through the filter, 98% reabsorbed, and further secreted. The pH of urine will determine the degree of dissociation of acids and bases and hence influence their movement across the reabsorption sites. Manipulating urine pH is a method of accelerating the elimination of some xenobiotics. For example, increasing urinary pH from 6 to 8 can increase the excretion of salicylate by almost 1 order of magnitude.

Passive diffusion across the distal tubule depends on the degree of ionization in the plasma and extracellular fluid as only the lipid-soluble form will be diffused. The concentration gradient also thus is an important rate-limiting factor. Very-water-soluble agents are passed through the glomerulus if they are small enough, and this is the reason why most biotransformations result in increased water solubility. Other substances are actively secreted (an energy-consuming process) at tubular sites (see Figure 1.5).

It should be noted that the lungs are a very important site of elimination for volatile substances including solvents, alcohols, and volatile and gaseous anesthetics. These can, in fact, be smelled on the breath, which can be an important first aid procedure to determine the cause of unconsciousness or stupor. Ketoacidosis in diabetics also can be detected by the acetone-like, or fruity, odor on the breath. Young diabetics have been suspected of glue sniffing when brought to an emergency department in a stupor or coma because of this fact.

Many drugs and chemicals are excreted into the bile. These tend to be polar agents, both cationic and anionic, the latter including glucuronide conjugates. Nonselective active transport systems, similar to those in the kidneys, are involved in the excretory processes. Once they enter the small intestine, these chemical metabolites may be excreted in the feces or reabsorbed back into the bloodstream. Enzymatic hydrolysis of glucuronide conjugates favors a return to the more lipid-soluble state and hence reabsorption. Purgatives may sometimes be used to facilitate the elimination of chemicals from the bowels.

The excretion of xenobiotics in mother's milk may not be an important route of elimination, but it can have significance for toxicity in the infant.

The chloracne rash associated with the now-obsolete bromide sedatives appears to be related to the secretion of this halogen in sweat. It is distributed in the body like chloride ion.

Extensive batteries of enzymes in the body may render the chemical nontoxic (detoxication), more water soluble, and hence more easily excreted, or they may activate it to a toxic form (toxication). The liver is the primary site of xenobiotic biotransformation in the mammalian body, but it is by no means the only one. Indeed, significant biotransformation can occur at the portal of entry. The chemical pathways are often the same. The response of the body to chemical insult also depends on the mitotic activity of the target tissue. Rapidly dividing tissues allow little time for repair to occur before cell division, so that the chance of a mutation is increased. Moreover, tissues that regenerate poorly are vulnerable to permanent damage by toxicants.

Pharmacodynamics

Ligand Binding and Receptors

Since only the molecules that are free in solution contribute to the concentration gradient, their binding to tissue components, or their chemical alteration by tissue enzymes will contribute to the maintenance of the gradient. The nature and strength of the chemical bond determines how easily the xenobiotic will dissociate when the concentration gradient is reversed. Drugs interact with specific sites (receptors) on proteins such as plasma membrane proteins, cytosolic enzymes, membranes on cell organelles, and, in some cases, nucleic acids (e.g., certain antineoplastic drugs). Membrane receptors and enzymes have molecular configurations that will react only with certain molecules in a kind of "lock-and-key" manner. Ease of reversibility is an important characteristic for most drugs, so that as concentration of the free substance falls, the drug comes off the receptor and its effect is terminated. This is often expressed by the following equation:

$$\text{Drug (D)} + \text{Receptor (R)} \longleftrightarrow \text{DR complex} \longleftrightarrow \text{Response}$$

The magnitude of the response is determined by the number (percentage) of receptors occupied at any given time. Neither the drug nor the receptor is altered by the reaction, which is defined as *pharmacodynamic*. Drugs are generally classified as being agonists, partial agonists, or antagonists, and receptors can exist in an active or an inactive state. An agonist is a drug that promotes the active state because it has a high affinity for receptors in that state. A drug with modest affinity for the active state of the receptor will be a partial agonist and a drug with equal affinity for both the active and the inactive states will be a competitive antagonist; meaning that its antagonism can be overcome by a sufficient concentration of an agonist. An antagonist that binds so strongly to the active receptor that it cannot be reversed is a noncompetitive antagonist.

In many cases, drugs and toxicants interact with receptors that normally accept physiological ligands such as neurotransmitters, hormones, ions, and nutritional elements. The proteins of cell surface receptors may penetrate to the interior of the cell in the case of ion channels and exchangers, or they may connect with other proteins in the membrane to transduce signals. Many neurotransmitters operate through a family of receptors that share the property of connecting to a protein having seven, membrane-spanning peptide chains. These "G" proteins (G for guanosine triphosphate or GTP) are transducers that interact with enzymes such as adenlycyclase or phospholipase C to initiate intracellular second messengers. G proteins may be inhibitory (G_i), stimulatory (G_s), or operate through other, unidentified, mechanisms (G_o). The neurotransmitters noradrenaline, acetylcholine, dopamine, serotonin, histamine, gamma-aminobutyric acid (GABA), glycine, and glutamic acid have been shown to act through G-protein receptors. Many centrally acting drugs work through these receptors.

Steroid receptors also exist. These are soluble cytosolic receptors that bind to the steroid after it diffuses into the cell and carry it to the nucleus. Opioid receptors in the CNS accept the endogenous peptide endorphins and encephalins. These receptors are the site of action of the narcotic analgesics.

Any receptor is a potential target for a toxicant interaction. A special case is the aryl hydrocarbon, or Ah, receptor. This cytosolic receptor binds to aromatic hydrocarbons like dioxins and it is believed that it is involved in their toxicity. No natural ligand for this receptor has yet been identified in mammals. This subject is discussed in detail in the chapter on halogenated hydrocarbons.

The chemical bond with the target receptor can involve covalent bonds, as well as noncovalent bonds including ionic, hydrogen, and van der Waal's forces. If the xenobiotic interacts irreversibly with a component of a cell, the effect may be long-lasting. Indeed, irreversibility of effect is an important characteristic of many toxicants (organophosphorus insecticides are examples of irreversible inhibitors of the enzyme acetylcholinesterase). If a chemical reacts irreversibly with DNA, a mutation may result in carcinogenesis or teratogenesis. This effect is sometimes described as "hit-and-run" because it is unrelated to any measurable concentration of the agent in the serum (see the following text).

Irreversibility of binding does not always mean irreversibility of effect. The drug acetylsalicylic acid (aspirin) is an irreversible inhibitor of the enzyme cyclo-oxygenase, which accounts for many of its pharmacological actions. Provided that exposure to aspirin is terminated, the effect declines as new enzyme is synthesized.

Biological Variation and Data Manipulation

Within any given population of organisms, there will be some that will respond to a drug or toxicant at the lowest concentration, and others that only respond at the very highest concentration, whereas most subjects will be grouped around the mean response. This is true of all organisms, including human beings and single-celled ones. It is even true of populations of like cells (liver cells, kidney cells, blood cells) within the body, and may partly explain why some cells may become malignant while others do not. It is the existence of biological variation that necessitates the use of large populations of test subjects and the development of mathematical treatments of data to permit the comparison of different populations of test subjects. If the responses of the species in question are grouped symmetrically about the mean response, a "normal" or Gaussian distribution curve is obtained (Figure 1.6). In this case, 68.3% of the population will fall within ±1 standard deviation (SD) of the arithmetic mean, 95.5% between ±2 SDs, and 99.7% between ±3 SDs. A datum point lying outside these limits is assumed not to belong to the test population. Sometimes the population is skewed, however, with more subjects falling on one side of the mean than on the other. Factors accounting for variability could include differences in the rate and degree of uptake, distribution, biotransformation, and excretion and even the nature and number of binding sites and receptors for the agent. These factors may be under genetic control, or they could be due to environmental differences in such things as temperature, nutrition, disease, the presence of other xenobiotics including medications, and so on. They also tend to vary with age and sex.

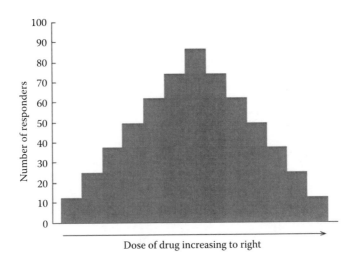

FIGURE 1.6

Normal (Gaussian) distribution of a population responding to different drug doses. A theoretical normal or "Gaussian" dose response distribution curve for a drug.

Dose Response

A population distribution in response to a drug or chemical applies even to like cells within the same organism or cell culture, because some cells will be more aged than others or may be defective in some way. Therefore, it is impossible from a single dose of a drug or toxicant to draw any conclusions about its potency, since it is not known whether one is recruiting only the most sensitive cells, or nearly all of them. For this reason, it is necessary to construct a dose–response curve whether testing a new drug for its effective dose or a chemical for its toxicity. Typically, the response rises rapidly once the threshold is exceeded, and then flattens out as fewer and fewer cells remain to be recruited (Figure 1.6). This type of response is a graded or *dose-dependent response*. There is another type of response that can be described as yes/no or all-or-nothing, and this is a *quantal response*. Lethality is an example. For graded responses, it is important to establish standard points of comparison, since comparing a dose at the low end of the response for one chemical, with one at the top end for another is not statistically reliable. The point usually chosen is that dose which produces 50% of the maximum effect, the effective dose (or concentration) 50% (ED_{50} or EC_{50}). "Dose" is used when the test substance is administered individually to the test subjects, and "concentration" when it is added to the surrounding medium, such as the water in an aquarium or the fluid bathing an isolated tissue in an organ bath.

A quantal response can be converted to a graded one by using several test groups, each receiving a different dose of the agent being tested. The percent of animals showing the expected response can then be plotted against dose. Thus, one can calculate the dose that, on average, will kill 50% of the test animals (lethal dose 50%, LD_{50}). If the response is a toxic one (liver necrosis, kidney failure), the value is the toxic dose 50% (TD_{50}). Variations on this approach include the LD_{10}, TD_{10}, etc. The LD_1 is also called the *Minimum Lethal Dose*. Values such as the LD_1 and the LD_{10} are replacing the LD_{50} in many jurisdictions.

In attempting to compare responses to two different chemicals, it is useful to perform a mathematical manipulation on the data so that differences or similarities in the shapes of the dose–response curves are more obvious. This involves plotting the logarithm of the dose against the response and this converts the exponential curve shown in Figure 1.7 to the sigmoidal one (S-shaped, Figure 1.8).

It is now much easier to interpolate to the EC_{50} (since the selected doses might not have included it) and to compare these points. Using the log of the dose tends to overcome the fact that large increases in dose result in small increases in response on the right side of the curve, whereas small increases in dose result in large increases in response on the left side of the curve. Thus, the midpoint, the EC (or TD)$_{50}$, provides the greatest statistical reliability. Parallel slopes of curves suggest similar mechanisms of action, and comparisons based on molar concentrations provide information on relative potencies. Toxicity comparisons may be done by calculating the Therapeutic Index (TI) if the

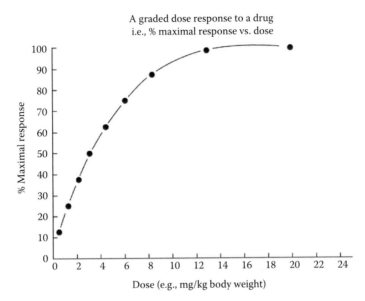

FIGURE 1.7
Typical, graded dose–response curve plotted arithmetically.

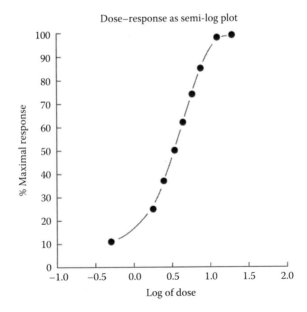

FIGURE 1.8
Semilogarithmic plot (response vs. the log of the dose) of the dose–response curve shown in Figure 1.7.

substance is a therapeutic agent. This is the LD_{50}/ED_{50}; the higher the number, the safer the agent. Other estimates of safety, more appropriate to toxicity studies of nontherapeutic agents, involve comparisons of the TD_1 and the EC_{99}.

It is important to note that all toxicity tests contain a temporal factor in that the determination of toxic effects is conducted at a specific time after exposure. Acute toxicity studies generally involve determinations made 72 h after a single high dose, whereas long-term toxicity requires multiple exposures with measurements made at least 28 days later. These studies are defined by government regulations in jurisdictions where there is a legal requirement for testing new chemicals.

Another value that is frequently used is the *NOEL or NOAEL, the No Observable (Adverse) Effect Level*. The NOEL includes effects, such as minor weight loss, that are not considered to be adverse. These values are applicable only to that species in which the test was conducted. Extrapolation to other species will require dosage adjustment.

Probit Analysis

It is often desirable to compare the toxicity of one xenobiotic to that of another. This information may help to determine whether a substance used commercially or industrially can be replaced with a safer one, or whether a metabolite of a parent compound is more or less toxic than the compound itself. For this purpose, probit analysis is often used. When a toxic reaction is expressed as the number of experimental animals in a group displaying that reaction (e.g., kidney failure), the percent of a group responding to a given dose or exposure can be expressed as units of deviation from the mean. These are called normal equivalent deviations (NEDs). The NED for the group in which there were 50% responders would be zero, since it lies right on the mean. A NED of +1 corresponds to 84.1% responders. NEDs are positive or negative relative to the mean, so a value of 5 is added to each to make them all positive. The result is called a probit (for probability unit). Table 1.1 shows the equivalent probits and NEDs for given percent responses.

TABLE 1.1

Conversion of Percent Responders to Probit Units

Percent Responders	NED[a]	Probit
0.1	−3	2
2.3	−2	3
15.9	−1	4
50.0	0	5
84.1	+1	6
97.7	+2	7
99.9	+3	8

[a] NED, normal equivalent deviation.

TABLE 1.2

Lethality in Fathead Minnows for Two Toxic
Chemicals

Lethality (%)	Fluorine (mg/L)	Naphthalene (mg/L)
10	25.0	0.5
20	50.0	1.0
60	100.0	2.0
84.1	200.0	4.0

When quantal data are plotted as probit units against the log of the dose, a straight line results, regardless of whether the original data were distributed normally or were skewed. The method, in fact, assumes that the data were distributed normally. It is now easier to compare the quantal data for two different xenobiotics exhibiting the same toxic manifestation (or their lethality). These concepts apply equally to toxicological studies in mammals and in nonmammalian species. The following example illustrates these concepts using hypothetical toxicity data (Table 1.2) for two toxicants tested in fathead minnows (0.25–0.5 g). Each test group consisted of 100 fish. Values listed are mg/L concentration in water. Tables are available for conversion to probits.

When using aquatic or marine organisms for toxicity studies, it is important to remember that they are exposed to a given concentration of the test substance continuously, but they may not take it up instantly or even rapidly. A consistent time of exposure must therefore be incorporated into the experimental design. Figures 1.9 through 1.11 illustrate arithmetic, semilogarithmic, and probit plots for these data.

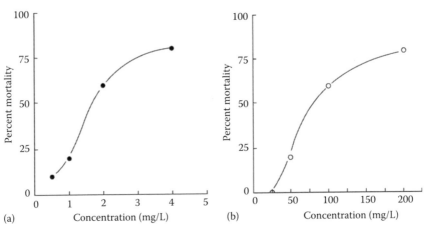

FIGURE 1.9
Arithmetic plot of the lethality data shown in Table 1.2 for (a) naphthalene and (b) fluorine.

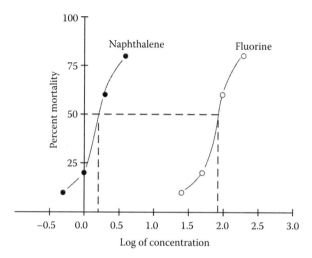

FIGURE 1.10

Semilogarithmic plot of % mortality vs. log of concentration. Semilog plot of data shown in Table 1.2 and in Figure 1.9.

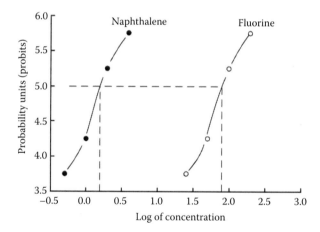

FIGURE 1.11

Probit plot of mortality for naphthalene and fuorine. Probits vs. log of concentration. Probit plot of the data shown in Table 1.2 and Figures 1.9 and 1.10.

Cumulative Effects

It may be that the manifestation of toxicity does not occur until the individual has been exposed continuously or repetitively (as with repeated, daily injections) for a prolonged period, perhaps days or weeks. This generally occurs with agents that are metabolized or eliminated very slowly, so that the rate of intake exceeds slightly the rate of detoxification and the drug slowly accumulates until a toxic level is reached. It is analogous to filling a bathtub with a faulty drain. The tub will fill, but only slowly because the water is running out almost as fast as it comes in. This involves the concept of *biological half-life or* $T_{1/2}$. The plasma $T_{1/2}$ of a drug or chemical is the time required for the plasma concentration to fall 50%. It is important to note that in most cases, this value is a constant for a given xenobiotic, i.e., it remains the same regardless of the initial level of the chemical. This is because the rates (pl) of biotransformation and excretion are usually concentration driven (the Law of Mass Action again), increasing or decreasing according to the plasma level. Biotransformations are enzymatic processes that generally obey first-order kinetics; i.e., the conversion rate is dependent upon the initial concentration of substrate. $T_{1/2}$ values may also be established for other tissue compartments in the body. If the exposure interval greatly exceeds the $T_{1/2}$ of a substance, it may be virtually eliminated between exposures. It requires about 5 × the plasma $T_{1/2}$ to achieve virtual elimination. If the exposure interval is equal to or less than the $T_{1/2}$, and the dose is constant, plasma steady-state concentration will be achieved in five plasma $T_{1/2}$. The xenobiotic may be sequestered in organs and tissues. If no detoxification or elimination occurs, of course, the chemical will accumulate significantly with each exposure. Cumulative effects can occur as the result of repeated exposures even though no detectable levels of the toxicant accumulate, as when genetic damage is induced by carcinogens. Figure 1.12 illustrates the relationship of frequency of exposure and $T_{1/2}$

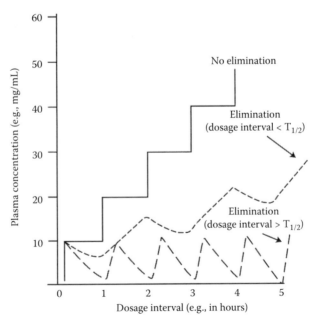

FIGURE 1.12
Influence of frequency of exposure (or dose) on tissue concentration of a chemical.

with respect to clearance from tissues. These factors influence whether a toxic reaction is defined as *acute* (within 48 h), *subacute* (in 7–9 days), *subchronic* (±90 days), or *chronic* (>90 days).

Factors Influencing Responses to Xenobiotics

It should be evident that anything that influences the absorption, distribution, metabolism, or excretion of a xenobiotic will affect its toxicity. Many such factors exist.

Age

In general, biotransformation and excretion are less efficient at the extremes of life. Although drug metabolizing enzymes are detectable at mid-gestation, they do not become fully developed until 6–12 months of age. Thus, the toxicity of many substances is higher in neonates. An example is the antibiotic chloramphenicol, which can accumulate to toxic levels because the glucuronide-conjugase enzyme is lacking. Renal function also is underdeveloped so that excretion of drugs is impaired. The $T_{1/2}$ for insulin clearance is 100 min in infants younger than 6 months of age versus 67 min in adults.

Body composition also differs with age. Total body water is 70%–75% of body weight in neonates versus 50%–55% in adults. Extracellular fluid is 40% of body weight in newborns versus 20% in adults. There is, thus, a greater fluid volume for dilution of water-soluble drugs in these infants. Their basal metabolic rate, moreover, is higher than in adults.

Other differences include greater permeability of skin, which has resulted in toxicity due to absorption of the hospital germicide hexachlorophene. Gastric pH is higher and gastric emptying is prolonged so that heavy metal absorption is increased. The intestinal flora will differ, and biotransformation by microbes will be different as well. In later childhood, these situations may be reversed because systems are functioning at peak efficiency.

After age 75, these same systems may have slowed down significantly compared to a 30 year old. Renal function and respiratory tidal volume are down, drug metabolizing enzymes are less efficient, and even the number of drug receptors may be lower. Body composition also changes. The ratio of fat to lean body mass increases with age and total body water is down. Cardiac output is reduced and perfusion is lower. Plasma albumin content is down. The toxicity of water-soluble toxicants like alcohol will be increased because it will be more concentrated. The concentration of the free component of albumin-bound agents will be increased because fewer albumin-binding sites are available. Biotransformation and excretion of xenobiotics will be impaired. The following are some examples of function at age 75 expressed as a percent of function at age 30 (Table 1.3).

TABLE 1.3

Function at Age 75 as a Percent
of Function at Age 30

Nerve conduction velocity	90%
Basal metabolic rate	84%
Cardiac output	70%
Glomerular filtration	69%
Respiratory function	43%

Body Composition

Body composition also varies considerably among normal individuals independent of age and that the factors discussed earlier with respect to fat and water content will be in play here as well.

Sex

Men and women may differ in response to xenobiotics. In part this is due to differences in body size, fat content, and basal metabolic rate. In one study, the $T_{1/2}$ of antipyrine (an old and toxic analgesic) was 30% longer in young men than in young women. Differences in response to sex steroids obviously occur. Pregnancy is a special situation involving great changes in the metabolism, body composition, and fluid content of the mother. Placental transfer of many agents occurs and this may put the fetus at risk (see the following). Sex-related differences also occur in experimental animals. Cessation of respiration has been shown to occur more frequently in female than in male rats after barbiturate anesthesia.

Genetic Factors

For the majority of the population, biotransformations are controlled by multigenetic determinants. This is the basis of the continuous variation in response as reflected in the characteristic, normal population distribution curve. In some cases, however, a single gene locus may be responsible for altering the metabolism of a substance. This occurs in a subset of the population and affects a particular enzyme. The result is a discontinuous, bimodal distribution that reflects two, overlapping normal distribution curves.

In Figure 1.13, representative response distributions are shown for acetylsalicylic acid (ASA, aspirin) as an example of the multigene type of control and isoniazid as the single gene type, in this case for an acetylating enzyme.

Pharmacogenetics is the subdiscipline of pharmacology that deals with this phenomenon. Over 100 examples have been identified of genetic differences in the biotransformation of drugs and chemicals. One of special concern for toxicology is the enzyme N-acetyltransferase. It acetylates and detoxifies many drugs and chemicals including the aryl amines that

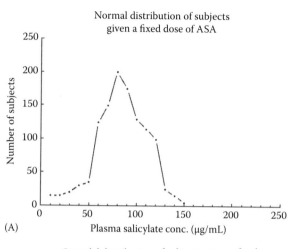

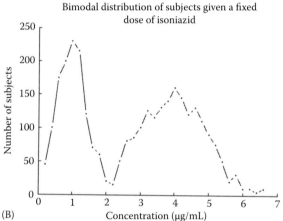

FIGURE 1.13

Hypothetical dose–response distribution curves for a drug influenced by many genes (A) and one influenced by a single gene affecting drug biotransformation (B).

are potential carcinogens, as well as many drugs including isoniazid, an antitubercular agent, and sulfa drugs. Slow acetylation is the dominant pattern in most Scandinavians, Jews, and North African Caucasians. Fast acetylation predominates in Inuit and Japanese. Similar patterns have been found in other species including rabbits. Slow and fast oxidative metabolism has also been reported. About 9% of North Americans are slow metabolizers. In one study, a group of dye workers who were exposed to N-substituted aryl compounds were surveyed for the occurrence of in situ carcinoma of the bladder.

Those with the disease were predominantly slow acetylators, suggesting that slow acetylators probably accumulated a carcinogenic agent normally detoxified by N-acetylation.

Thiopurine S-methyltransferase (TPMT) also is subject to genetic poly-morphism. This enzyme is responsible for S-methylation of the antineoplas-tic drugs azathioprine and 6-mercaptopurine. About 88.6% of humans have high, 11.1% intermediate, and 0.3% low or undetectable levels of enzyme activity. Individuals with high levels may be poor responders to cancer che-motherapy unless dosage is adequate, while those with low or undetectable levels are in danger of developing complete suppression of bone marrow function unless, again, dosage is adjusted downward. Other aromatic and heterocyclic sulfhydryl compounds are methylated by this class of enzyme, and there is little doubt that toxicity can be affected by genetic differences in enzyme activity. Similar genetic polymorphism has been demonstrated for the drugs debrisoquine and metoprolol, which are detoxified by hydroxyl-ation. About 10% of male blacks develop hemolytic anemia when exposed to the antimalarial drug primaquine.

Pharmacogenomics is the science of using genotyping to identify indi-viduals in advance who may pose therapeutic problems because of a lack or excess of detoxifying enzyme activity. It is now technically possible to do this through the use of modern techniques in molecular biology, nota-bly DNA genotyping by the use of appropriate primers and the Polymerase Chain Reaction (PCR) method. Such technology may also be applied to the workplace to identify persons who are at excessive risk of a toxic reaction from a chemical in the work environment, or perhaps individuals who have already incurred DNA damage because of past exposure to a mutagen. These people could be excluded from high-risk areas (for them) or denied employment. Paradoxically, this possibility has already raised concerns in some union quarters that individuals might be denied their right to earn a living on biochemical grounds. In the future, employers might have to bal-ance union concerns against the possibility of future litigation by workers made ill by their jobs. There is certainly no doubt that new ethical dilemmas will arise out of this technology.

Genetic factors also determine the emergence of strains of organisms resis-tant to normally toxic agents. There are many examples, including resistance of mosquitoes to DDT, rats to warfarin, malarial parasites to many drugs, cancer cells to anticancer drugs, and bacteria to antibiotics. In all cases, sus-ceptible cells are killed off, leaving resistant mutants to proliferate. Mutations are occurring all the time. In bacterial populations, it has been estimated that a mutation imparting a degree of resistance to a drug occurs once in every 10^9 cell divisions. Unless that particular drug is present, however, the mutant strain will have no selection advantage and it will be overwhelmed by non-mutant cells. If the drug is present, however, it will have a distinct survival advantage and it will become the dominant form.

Genetic factors also may influence the response to a toxicant at the target site as well as at the site of biotransformation. Inherited disorders that ren-der individuals more susceptible to drug-induced hemolytic anemia include glucose-6 phosphate dehydrogenase (G-6-PD) deficiency and sickle cell anemia.

A host of drugs, including antimalarials and sulfonamides will induce a hemolytic attack in these people. There are also several inherited disorders of hemoglobin synthesis that act similarly. Many of the drugs involved are oxidizing, aromatic nitro compounds and nitrates. As noted earlier, altered gene coding plays a major role in carcinogenesis.

Presence of Pathology

Given that the liver and the kidneys are the major organs of detoxication, it follows that any serious impairment of their function will have a significant impact on the toxicity of xenobiotics. This has been observed in fatty necrosis of the liver, hepatitis, and cirrhosis. These will impact mostly on highly lipid-soluble agents requiring biotransformation to more water-soluble forms. Pharmacologically, this is seen with CNS depressants including the tranquillizers Librium and Valium. Kidney dysfunction is reflected mainly on water-soluble agents and their elimination may be greatly impaired by renal disease. Water-soluble antibiotics such as gentamicin have a greatly prolonged $T_{1/2}$ in the presence of renal disease. Cardiovascular disease may affect tissue perfusion and the delivery of the xenobiotic to, or conversely, its removal from, its target site. Pulmonary disease will also affect the transfer of volatile agents across the alveolar membrane. In meningitis, the presence of inflammation compromises the integrity of the blood–brain barrier, and substances that would normally be excluded may reach significant concentrations in the spinal fluid. This fact makes some antibiotics (e.g., penicillin) useful to treat meningococcal meningitis even though they normally do not penetrate the barrier.

One of the most serious consequences of preexisting pathology may simply be the fact that if an organ has already lost much of its function, further damage by a toxicant may destroy it completely and create a life-threatening situation. This is one of the main reasons why the elderly are more vulnerable to toxic effects of drugs and chemicals.

Xenobiotic Interactions

The effect that one drug may have on the action of another, collectively known as drug interactions (or drug–drug interactions), is an important aspect of clinical pharmacology. The effects of two drugs given together may be *additive* if they induce the same response (even through different mechanisms), *synergistic* if the total response of their combined effect is greater than the predicted sum of their individual effects, or *antagonistic* if one drug diminishes or prevents the effect of the other.

Mechanisms involved in drug interactions include altered absorption from the gastrointestinal tract, altered excretion (renal, biliary, respiratory), competition for receptors (antagonism), summation of pharmacological effects, and altered biotransformation. Drug interactions are usually associated with drug adverse reactions, but they may also be exploited. All antidotal remedies are based on drug or chemical interactions. Examples of antidotes include the use of atropine to treat organophosphorus poisoning, metal chelators for treating lead and other heavy metal poisoning, the use of activated charcoal to bind toxicants in the gastrointestinal tract and prevent their absorption, emetics to induce vomiting and the removal of toxicants, naloxone to reverse the respiratory depression of opiates, and N-acetylcysteine to treat acetaminophen poisoning.

When more than two substances are present, the possibility for a drug interaction is much greater. This is of special concern in the area of environmental toxicology because of the multiplicity of xenobiotics that may be present in water, food, and air. Most of the attention has centered on the possibility that the presence of one substance may increase the concentration of carcinogens from other sources by affecting their metabolism through enzyme induction. We have already discussed the effect of cigarette smoke on hepatic enzymes. Many volatile solvents also are enzyme inducers, and chronic exposure to low levels of these, as in the industrial setting, can lead to increased enzyme activity. It has been shown experimentally that bedding rodents on soft-wood shavings induces hepatic microsomal enzyme activity because of the volatile terpenes given off. Workers in soft-wood sawmills could experience similar effects. Literally hundreds of industrial chemicals have been shown to be enzyme inducers, including benzo[a]pyrene and 3-methylcholanthrene. Others include insecticides, (DDT, aldrin, dieldrin, lindane, chlordane), PCBs, polybrominated biphenyls (PBBs), dioxin, and drugs such as phenobarbital and other barbiturates, steroids, and others. Even though a substance is itself not toxic at a particular exposure level, it may influence the toxicity of other agents. The enzymes usually involved are the CYP450 monooxygenases, but conjugating enzymes also may be induced. Food may contain natural enzyme inducers. Potato contains "a-solanin," and tomato "tomatin," both of which are steroidal alkaloids. The bioflavonoids are fairly potent inducers. They are found in species of *Brassica* including Brussels sprouts, where they have been shown to affect the metabolism of some drugs. Rutin is a bioflavonoid found in buckwheat in fairly large amounts. While it is difficult to identify situations where these various inducers have influenced toxicity (with the exception of cigarette smoke), their ubiquity illustrates how different individuals or groups may display different sensitivities to the same toxic exposure on different occasions. More recently, it has been shown that fruit juices, especially grapefruit juice, can have a significant impact on the metabolism of some drugs. Naringin is a flavonoid in grapefruit juice and hesperidin, a flavonoid in orange juice. Both have been shown to inhibit enteric CYP3A4, which is a barrier protein for the absorption of many drugs. They also inhibit the organic anion-transporting polypeptide (OTAP) 1A2 and are known to interfere with the absorption of many drugs. The list of food–drug interactions grows longer year by year.

Organic solvents have been shown to influence ethanol toxicity. Toluene depresses alcohol dehydrogenase and prolongs the ethanol $T_{1/2}$. Ethanol itself may increase the liver and CNS toxicity of CCl_4, trichloroethylene, and others.

Diet can also affect the toxicity of substances in other ways. A toxicant may be adsorbed to dietary components that reduce its absorption. The ability of dietary calcium to reduce lead toxicity is well documented. Lead and calcium appear to compete for the same absorption site on the intestinal mucosa so that a diet high in calcium will reduce lead absorption. A diet high in fiber will shorten the transit time of the gastrointestinal tract so there is less time available for absorption. This may be one reason why high fiber diet is associated with a lower incidence of colon cancer.

Some Toxicological Considerations

Acute versus Chronic Toxicity

Acute and chronic toxicity for a single agent may be quite different and one is not a reliable predictor of the other. For example, acute benzene intoxication involves CNS disturbances such as excitation, confusion, stupor, and convulsions. Chronic toxicity includes depression of the bone marrow and a reduction of all circulating blood cells (pancytopenia) and benzene is carcinogenic in experimental animals. Chronic carbon monoxide (CO) poisoning is experienced by heavy smokers who may suffer from headache, dizziness, and shortness of breath. Acute CO poisoning affects firemen and fire victims who may become comatose. Numerous other examples exist. A major area illustrating this is that of tumor formation. A short-term exposure to a substance may elicit acute toxicity without significant risk of carcinogenesis, whereas long-term exposure to very low levels may not result in any toxic manifestation but could induce tumor formation. Dioxin (TCDD) is a prime example of this. Acute exposure causes the skin rash known as chloracne, whereas long-term exposure may be carcinogenic.

A detailed discussion of toxic mechanisms is beyond the scope of this chapter. More detailed descriptions are given with the discussions of specific chemicals and groups of chemicals. What follows is a brief overview to illustrate the role of target organs and systems in toxic reactions.

Acute Toxicity

Acute toxicity generally refers to effects that occur following a 24–72 h exposure to a single or multiple doses of a toxicant. Effects are usually observed within a few days. If the agent is rapidly absorbed, the effect may be immediate. The CNS is very vulnerable to acute toxicity from very-lipid-soluble agents.

Peripheral Neurotoxins

Organophosphorus and carbamate pesticides inhibit acetylcholinesterase (AChE) so that acetylcholine (ACh) accumulates and overstimulation of receptors occurs. ACh is a neurotransmitter in both the central and peripheral nervous systems. There are many naturally occurring neurotoxins, including tubocurarine, which blocks nerve transmission to voluntary muscle; botulinum toxin (from *Clostridium botulinum*), which prevents the release of ACh from nerve endings; and tetrodotoxin (from the puffer fish), which paralyses nerves by blocking sodium channels. Belladonna alkaloids (atropine and scopolamine) from nightshade are muscarinic blockers with peripheral and central effects and muscarine from mushrooms is a muscarinic stimulant. Nerve gases are also irreversible AchE inhibitors. Many other examples exist (see Chapter 11).

Central Neurotoxins

The inhalation of many volatile organic solvents and petroleum distillates can act like anesthetics and may cause unconsciousness. This has occurred in industrial accidents and in substance abuse (e.g., gasoline sniffing). Ethyl, methyl, and isopropyl alcohols are CNS depressants.

Inhibitors of Oxidative Phosphorylation

Cyanide (CN) in the form of cyanogenic glycosides (e.g., amygdalin) is present in many plant components including almonds, the pits of cherries, apples and peaches, plums, apricots, and wild chokecherries. Human poisonings have occurred from consuming too many of these seeds, and livestock are often poisoned from eating chokecherry bushes. Cyanide binds to heme to prevent electron transfer. Cyanide may be present in metal ores, in some pesticides, and in metal polishes. The tragic accident at Bhopal, India, which killed over 2000 people, was due to the release of 40 tons of methyl isocyanate from an American Cyanamid plant. Azide and hydrogen sulfide act like CN.

Carbon monoxide (CO) combines with hemoglobin and cellular cytochromes and prevents association with O_2. It thus causes cell hypoxia and also interferes with O_2 transport by red blood cells.

Uncoupling Agents

Many agents act as uncouplers, preventing the phosphorylation of ADP to ATP, the high energy phosphate. Uncouplers include the herbicide 2,4-D, halogenated phenols, nitrophenols, and arsenate. Most contain an aromatic ring structure. O_2 consumption and heat production are increased without an increase in available energy.

Inhibitors of Intermediary Metabolism

Certain fluoroacetate compounds of natural origin are used professionally as rat poisons. They inhibit the citric acid cycle to deplete available energy stores. The heart and the CNS are the organs of toxicity.

Chronic Toxicity

Exposure to some toxicants must occur over days, weeks, or months before signs of toxicity appear. Heavy metals tend to act in this manner. Several outbreaks of methylmercury poisoning have followed this pattern. Mercury (Hg) from industrial discharges may be converted to methylmercury by microorganisms in the water. Monomethylmercury is CH_3Hg and dimethylmercury is $(CH_3)_2Hg$. These accumulate up the food chain to concentrate in fish and shellfish that may be consumed as food. As the toxicant accumulates in the tissues, severe neurological disorders occur. This happened at Minamata Bay in Japan, and on the Grassy Narrows reserve in Northern Ontario. Infants born to exposed mothers may suffer from a cerebral palsy-like syndrome.

Cadmium (Cd), used in nickel–cadmium batteries, in electroplating, and in pigments, may accumulate in workers and cause kidney damage. Similar cases in the general population have been reported in Japan, from eating rice and other grains grown in soil contaminated with industrial wastes. Carbon tetrachloride (CCl_4) was used extensively in the dry-cleaning industry before it was discovered that is caused hepatic necrosis because it was activated to a free radical by CYP-450-dependent monooxygenase.

Mutagenesis and Carcinogenesis

Introduction

Sir Percival Pott was a senior surgeon at St. Bartholomew's Hospital in London, England, in the latter half of the eighteenth century. He is widely credited with being the first person to make a connection between exposure to a foreign substance and the development of a cancer. In 1775, he noted a high incidence of scrotal cancer, subsequently shown to be squamous cell carcinoma, in chimney sweeps. The common term for this condition, soot wart, suggests that workers in the trade had already made the connection. Boys as young as four were employed as chimney sweeps, often pressed into service by unscrupulous adults. In 1788, the Chimney Sweepers Act was passed by the British Parliament, becoming the first legislation designed to eliminate child labor, at least in this trade.

Cancer will afflict one in four North Americans at some point in their life and will be responsible for the death of about one in five of them. While chemicals are responsible for many cancers (this is the cause that creates the most concern among the populace), viruses and radiation, either natural or from anthropogenic sources, are also important causes. Cancer can be described as an aggressive and inappropriate growth of a cell type in the body. If the cell type remains true to its origin and can always be identified as that cell type, the tumor is generally benign and does not metastasize beyond the primary site. It can still become excessively large and interfere with normal bodily function if not removed. If the cancer cells lose their defining characteristics and become undifferentiated, the tumor is malignant and will most likely metastasize to distant sites if it is not treated successfully by surgery and/or antineoplastic drugs.

A characteristic of chemically induced carcinogenesis is that it is believed usually to involve decades-long exposure to very low levels of carcinogens, making predictions from animal studies very difficult. This is clearly illustrated by the association of cigarette smoking and lung cancer, which is discussed in Chapter 4. An exception is the process of chromothripsis (see Model 6 given in the following). A wide range of chemically diverse, natural and synthetic chemicals may induce alterations in DNA that, depending on the nature of the defect and the timing of its occurrence, can cause a neoplasm, a heritable change (*mutation*), or a developmental birth defect. It should be noted that birth defects (*teratogenesis*) may result from chemical interference with many other cell processes not involving altered DNA, such as interference with essential substrates and precursors, impaired mitosis, enzyme inhibition, altered membrane characteristics, etc. Most anticancer drugs are teratogenic for the same reason they are antineoplastic. Mutations are not always harmful, and they provide the necessary genetic diversity for natural selection to occur, but they can also be responsible for fertility disorders, hereditary diseases, cancers, and malformations. Three types of genetic abnormalities may be induced.

Chemically induced carcinogenesis differs from other manifestations of toxicity in that the usual exposure–response relationship does not generally apply. The xenobiotic or its metabolite attaches irreversibly by binding to an essential component of the living cell to disrupt normal function. The binding sites are usually on macromolecules such as nucleotides, nucleosides, regulatory proteins, RNA, and DNA. The effects may be cumulative, and they cannot be related to blood or tissue levels at the time of their manifestation. There may be a considerable delay from the initial exposure to the emergence of toxicity. This kind of reaction is involved in mutagenic and carcinogenic effects, and also causes a type of anemia, called aplastic anemia, in which the capacity of the bone marrow to produce blood cells is permanently destroyed. Some older drugs such as chloramphenicol and phenylbutazone have caused aplastic anemia in a very small number of patients; perhaps one in 40,000 exposed persons. The solvent benzene, a carcinogen, has also been associated with aplastic anemia. These low frequency toxicities are difficult to identify and necessitate careful risk/benefit analysis when therapeutic interventions are contemplated.

Our understanding of the carcinogenic process has been greatly enhanced by the development of DNA technology and the mapping of the human genome. Nonetheless, it is a highly complex, even confusing, process that is in a continual state of flux. What follows is an attempt to organize and simplify the story. It should not be viewed as a fully detailed, complete description of the process.

Genetics of Carcinogenesis

Oncogene

This is a gene that predisposes a normal cell to become a malignant one. There are scores of known human oncogenes with more being discovered all the time. Oncogenes can be turned on by external factors such as radiation, viruses, chemicals that damage DNA throughout life, or by inherited factors actors. Unlike normal genes, they cannot be turned off but remain in a state of constant activity. The first oncogene to be identified was a component of a cancer-causing virus in chickens, the Rous sarcoma, named after the discoverer. Labeled the *src* (sarc) oncogene, it has been identified in humans and is associated with colon, liver, lung, breast, and pancreatic cancer. It has since been shown that viral oncogenes like src are not part of the normal genetic complement of the virus but are acquired from normal cells by the retrovirus. Retroviruses may cause the excessive production of an oncogene, induce its expression in an inappropriate cell, or mutate the coding of a gene during transduction. Human tumor oncogenes have been identified as alleles of the *ras* family of proto-oncogenes. Others include *fos, jun, and myc*. The *myc* oncogene was first identified as a viral one and is important in human tumorigenesis as a growth factor stimulant. Cellular oncogenes are designated by the prefix "c" (e.g., c-myc, c-abl). Viral oncogenes are designated by the prefix "v." The genetic mutations BRCA1 and BRCA2 are predisposing risk factors for breast cancer.

Proto-Oncogene

A proto-oncogene is the normal counterpart of an oncogene. Proto-oncogenes are normal constituents of the cell that exist in every cell component (membrane, cytoplasm, nucleus, etc.) and that are involved in every signal transduction cascade. They modulate cell function, growth, and proliferation. When acted upon by external influences they are converted to oncogenes, resulting in uncontrolled cell growth. If a proto-oncogene undergoes a somatic mutation, regulation of growth in that cell is lost and cancer can occur. This can be triggered by a number of molecular mechanisms.

Tumor Suppressor Genes

These recessive genes are part of the normal human genome and are capable of preventing the onset of cancer even in the face of the molecular events noted earlier. They are sometimes referred to as "anti-oncogenes." They were discovered as a result of the observation that when normal cells are fused with tumor cells in culture, suppression of the tumor cells resulted. Inactivation of these suppresser genes may be an essential step in tumorigenesis, at least in some cases, and prevention of this inactivation by inserting normal copies of tumor suppresser genes, or by mimicking their function pharmacologically, may lead to new therapies. Tumor suppressor genes (TSGs) may be responsible for the existence of "cancer families." If an inherited defect in one copy acquires a matching defect in the other copy through retroviral or chemical damage, or from the other parent, cancer may result. Much has been learned about TSGs from the study of a juvenile eye cancer, retinoblastoma. This is caused by a mutation in the Rb gene located on chromosome 13. It normally suppresses the tumor in its dominant phenotype. The mutation must be present in both alleles for the tumor to develop. The retinoblastoma story is rather complex and it serves to illustrate the complicated nature of carcinogenesis. The product of the normal Rb gene interacts with a protein, E2F, a nuclear transcription factor involved in cellular replication functions during the S phase of the cell cycle. The interaction with the Rb product prevents this function, inhibiting mitosis. The mutant form of Rb cannot interact with E2F, so cell growth is not inhibited. The retinoblastoma story is discussed further under the "Two-hit" model.

Another TSG is *p53* located on chromosome 17. It is a protein containing 393 amino acids, and a substitution of a single one can lead to loss of function. Both recessive and dominant mutations can occur. Other TSGs include Rb, involved in retinoblastoma and osteosarcoma and APC, involved in colon cancer. E-cadherin, a Ca^{++}-dependent transmembrane protein involved in epithelial cell–cell interactions, has been suggested as a candidate TSG.

Growth Factor Receptors

Many tumors may express genes for receptors that are receptive to agents that promote the aggressive growth of the tumor. The epidermal growth factor receptor is one such. It is known by many names such as EGFR, ErbB-1, and, in humans, HER-1. It is one of a family of receptors known as the ErB family. Others are HER2/c-neu (ErbB2), HER 3 (ErbB-3), and HER 4 (ErbB-4). These receptors bind to specific ligands like epidermal growth factor and transforming growth factor α (TGFα). The result may be accelerated tumor growth through uncontrolled cell division.

Many tumors have receptor HER2 including breast cancer, colon cancer, lung, and likely squamous cell carcinoma of the head and neck, and probably many others. New drugs that alter the receptor affinity for HER2 can be useful in suppressing tumor growth. These new drugs include trastuzumab (Herceptin®) and cetuximab (Erbitux®).

Hormone Receptors

The presence of cytoplasmic receptors for estrogen and progesterone is a critical factor in breast cancer and determines the nature of the therapy. Although it may seem paradoxical, patients whose primary tumor is receptor-positive generally have a more favorable course than those whose tumors are receptor-negative. This is because it is possible to manipulate the hormonal environment with receptor-blocking drugs. Tamoxifen is an estrogen receptor modulating agent that is a first-line drug in treating patients with estrogen receptor-positive breast cancer.

A difficult form of breast cancer to treat is the so-called triple "negative" breast cancer. This cancer lacks receptors for estrogen, progestin, and the epidermal growth factor (receptor HER2). They are thus not responsive to treatment with receptor blocking agents like tamoxifen for estrogen receptors or Herceptin for HER2 receptors but the oncologist must rely on other treatments like radiation and conventional chemotherapy.

Hormonal manipulation is also employed in the treatment of metastasized carcinoma of the prostate gland in men. These tumors have cytoplasmic androgen receptors that are responsive to androgens like testosterone (which is actually converted to dihydrotestosterone) that promote proliferation of the tumor. Antiandrogen therapy is used to retard this. Older techniques included removal of the testes and the use of the synthetic estrogen diethylstilbestrol. Flutamide and bicalutamide are androgen receptor antagonists. A family of drugs that includes leuprolide (Lupron) and goserelin (Zoladex) are analogs of gonadotropin-releasing hormone (GnRH). There is a complex feedback loop controlling the synthesis of androgens and estrogens. The hormone GnRH (also known as LHRH for luteinizing hormone-releasing hormone) is synthesized and released from the hypothalamus in a pulsatile fashion. Low frequency pulses stimulate the release of follicle-stimulating hormone (FSH) and high frequency pulses the release of luteinizing hormone (LH), both from the anterior pituitary gland. Synthesis and release of FSH and LH are also affected by the levels of circulating estrogens and androgens in a negative feedback manner.

Analogs of GnRH occupy GnRH receptors on the anterior pituitary, initially causing increases of FSH and LH until the receptors become desensitized and levels of FSH and LH begin to fall, as do levels of androgens. Circulating levels can reach those associated with castration and the name chemical castration has been applied to this hormonally based approach.

Endocrine-disrupting chemicals, both natural and synthetic, may also play a role in the development of mammary tumors. These will be discussed in Chapter 12.

Drug Resistance Genes

It is well known that cancer cells can develop resistance to antineoplastic drugs just as bacteria become resistant to antibiotics and rats to warfarin.

The T790M mutation and MET oncogene impart resistance to drugs targeting the EGF receptors. Natural EGFR inhibitors also exist, one being a potato carboxypeptidase inhibitor. A multidrug-resistant gene, MDR-1, encodes for a P-glycoprotein that pumps cytotoxic, antineoplastic drugs out of tumor cells.

Antisense Genes

An even more recent development in the genetic manipulation of tumor cells is the synthesis of "antisense" sequences. These oligonucleotides are short (15–25 bases), single-stranded, DNA sequences that have been altered so that they target a specific mRNA sequence, resulting in a defective DNA segment in a highly specific manner. Thus, if a specific segment (sense) that plays a key role in tumorigenesis can be identified, the appropriate antisense sequence can be formulated and inserted to interfere in the process. This advance holds great promise for the future treatment of cancer.

The World Health Organization's International Agency for Research on Cancer maintains an excellent website with evaluations on the current carcinogenicity rating of hundreds of chemicals (http://www.iarc.fr/).

Genetic Predisposition to Cancer

Just how much cancer is due to environmental factors and how much is due to a genetic predisposition has been a matter of considerable debate. Recently, a study conducted by the Department of Medical Epidemiology at the Karolinska Institute has shed some light on this question. They combined data from 44,788 pairs of twins listed in the Swedish, Danish, and Finnish twin registries to examine the risk of cancer for persons with a twin with cancer at one of 28 anatomical sites. Statistically significant evidence of effects of heritable factors was observed for several cancers. Analysis indicated that genetic predisposition accounted for 27% of the risk for breast cancer, 35% of the risk for colorectal cancer, and 42% of the risk for prostate cancer. Overall, the risk of being diagnosed with any of these cancers by age 75 ranged from 11% to 18% for a person whose identical twin had that cancer and from 3% to 9% for a person whose fraternal twin had that cancer. The authors concluded that environmental factors were by far the greatest contributors to the risk of getting cancer.

Epigenetic Mechanisms of Carcinogenesis

Although it must be stated that no malignancy progresses without some of the genetic factors discussed earlier, carcinogens that function through non-DNA-reactive processes have been identified in recent years. They do not induce mutation in cell assays nor do they induce direct DNA damage in target organs. Precise mechanisms by which these agents operate have not been established but changes in gene expression and cell growth (proliferation) seem to be central to the process. Some may be tumor promoters that cause proliferation of initiated cells. Phorbol esters of plant origin seem to act this way. Cytotoxic agents, by killing large number of cells, trigger the production and release of growth factors, and the frequency of mutations is thus also increased. Some of these may be malignant cell types. Chloroform, by inducing liver necrosis and hyperplasia, can induce tumors in this way. Phenobarbital, conversely, can induce mitosis without cytotoxicity by mechanisms that are not clear. Rapid cell division can lead to error-prone repair and hence mutation. This relates to the Ames hypothesis discussed later. Under the influence of mitogenic factors, a mutant cell may proliferate and become the dominant cell type.

Another mechanism that may operate for nongenotoxic carcinogens is interference with normal apoptosis. Apoptosis, or programmed cell death, is a normal process by which senescent cells or cells with massive DNA damage are removed. Cancer results from an imbalance between cell growth and cell death with resulting unchecked proliferation. Several tumor promoters have been shown to inhibit normal apoptosis. The process of apoptosis involves oncogenes, TSGs, and cell cycle regulatory genes, and interference with these can lead to cell proliferation and neoplasia. Obviously there are genetic influences at work here as well.

Many of these processes may be facilitated by interference with cell–cell communication across gap junctions. Gap junctions are areas of cell–cell close contact containing channels that allow the passage of ions and small water-soluble molecules. Signal-transducing substances such as calcium, inositol trisphosphate, and cyclic AMP are transmitted in this way. Gap junction communication has been shown to be decreased in the presence of some epigenetic carcinogens.

Induction of cytochrome P450 may result in the formation of reactive intermediates that can act as promoters, often through the formation of free radicals.

Viruses and Cancer

The story of the Rous chicken sarcoma virus and its role in oncogene formation has been dealt with earlier. The list of viruses that can play a role in carcinogenesis is growing quite long and no doubt will continue to do so. The list includes Epstein–Barr virus (EBV), human papilloma virus (HPV), hepatitis C virus (HCV), hepatitis B virus (HBV), human herpes virus 8 (HHV-8), Human T-cell leukemia virus type 1 (HTVL-1).

The Rous sarcoma virus is a retrovirus. Retroviruses are a major factor in the conversion of proto-oncogenes to oncogenes. These viruses contain RNA rather than DNA and redirect host cell to produce viral DNA through the process of reverse transcription. This involves an enzyme, "reverse transcriptase." Proto-oncogenes may be mutated to oncogenes during this process. Over 30 human oncogenes have been identified to date. In general, tumor viruses have the ability to integrate into the host genome and express their viral oncogenic proteins. Prior exposure to EBV and HPV has become a recognized risk factor for many cancers.

Viruses are also being employed as therapeutic agents in cancer therapy. Viruses can be reprogrammed into oncolytic vectors. Several modifications are employed. Targeting introduces multiple layers of cancer specificity, improving safety and efficiency. Arming refers to the expression of pro-drug convertases and cytokines. Shielding refers to coating the virus with polymers and the use of different envelopes (capsids) to protect the virus from the host's immune response. While gene therapy has many potential therapeutic uses, over half of currently approved gene therapies relate to the treatment of cancers. The following lists ways in which such therapy is employed.

Inactivation of oncogenes: Oncogenic proteins are associated with many malignancies and blocking their expression would be a desirable target in cancer therapy. The adenoviral E1A interferes with the transcription of the oncogenic protein erb-2, common in breast and ovarian cancer.

Augmenting TSGs: Mutations in the numerous TSGs are the factor in the development of a neoplasm. Using adenoviral vectors some success has been obtained by introducing the TSG p53. BRCA1 and BRCA2 (standing for breast cancer susceptibility genes 1 and 2) are TSGs. A mutation of one of these genes is a significant risk factor for breast and ovarian cancer and these mutations can be heritable. A viral vector has been used to introduce normal BRCA1 into ovarian cancer in an attempt to induce suppression of the tumor with some success. Some cases fail because the mutant gene can be dominant and prevent the normal gene from acting.

Induced cell death: The inclusion into a virus of a gene that promotes the conversion of an antineoplastic pro-drug to its active form can, when given in combination with that drug, cause lethal concentrations of the drug to accumulate in the tumor cell. This is accomplished by using a pro-drug-converting enzyme such as cytosine deaminase for 5-fluorocytidine, or deoxycytidine kinase for fludarabine and 2-chlorodeoxyadenosine.

Chemoprotection: There is a protein known as P-glycoprotein that is capable of transporting or "pumping" numerous antineoplastic drugs out of tumor cells, thus preventing the drugs from reaching lethal concentrations. This form of multidrug resistance (MDR) can be a thorny problem for cancer chemotherapy. The MDR-1 gene encodes for the P-glycoprotein. In a departure from the usual gene therapy approach, the MDR-1 gene in a retroviral vector is being studied to protect normal bone marrow cells from the cytotoxic effects of antineoplastic agents used to treat tumors elsewhere. Because of their high rate of cell division, bone marrow cells are very vulnerable to cytotoxic effects.

Viral-mediated oncolysis: Some viruses, notably adenovirus and HSV-1, are capable of lysing tumor cells that they infect. When the tumor cells lyse, the virus is released to infect adjacent cells. Using such viruses in combination with other therapies, including other gene therapies, is showing promise for cancer treatment.

Immunomodulation: Tumor cells generally have impaired immune function. Cytokines may enhance immunity against cancer cells. Genetically engineered tumor cells have been shown experimentally to be less able to establish tumors *in vivo*. Infecting tumors with these cytokine-producing genes may increase their vulnerability to the host immune system, especially if used in conjunction with immunostimulatory agents that do not affect tumor immunity.

The search for new ways to exploit knowledge of tumor genetics for therapeutic gain continues and will doubtlessly lead to novel and hopefully successful approaches.

Models of Carcinogenesis

An interesting article by Vineis et al. discusses the changes in our understanding of carcinogenesis and how it has evolved over time. They note that six alterations in cell physiology are required to initiate and perpetuate malignant growth. These are as follows: (1) self-sufficiently in growth signals; (2) insensitivity to inhibitory (antigrowth) signals; (3) avoidance of programmed cell death (apoptosis); (4) unlimited replicative potential; (5) sustained angiogenesis (development of new blood vessels); (6) tissue invasion and metastasis. They also propose several models of carcinogenesis. They chose the term model rather than theory because there is experimental and mathematical evidence in support of them. While conceding that their approach is largely artificial they find it useful to promote discussion and understanding.

Model 1

The "mutational" model places emphasis on point mutation and chemical carcinogens. Mutation is the main change that leads to a malignant phenotype. The mutation becomes fixed so that the progeny of the mutated cell is abnormal. (Spontaneous natural mutations also occur, and this is the basis of natural selection. If a mutation provides a survival advantage to the individual it may become a dominant feature of the species. If it is disadvantageous it likely will not survive.) Point mutation, also called gene-locus mutation, involves the alteration in some way of a small number of base pairs of nucleic acids. This may involve deletion, addition, or the substitution of an incorrect base pair. Strictly speaking, the term mutagenesis refers exclusively to this type of DNA alteration.

The discovery of the carcinogenicity of tobacco smoke and of PAHs contributed to the formulation of this model. Research into the identification of reaction products between chemicals like PAHs and macromolecules like DNA adducts contributed to the development of the model as did the study of induced mutations in bacteria. Viral research also played a role (see src oncogene discussed earlier).

Bruce Ames, in his 80s (as this is written), is Professor of Biochemistry and Molecular Biology at the University of California in Berkeley. He is well known for the development of the Ames test, a bacterial culture test widely used for the identification of mutagenic chemicals. It is widely recognized that the frequency of mutagenesis increases with the frequency of cell divisions (mitogenesis) and thus the likelihood of a neoplastic mutation also increases as well. Ames postulated that the standard animal tests for detecting chemical carcinogens, that employ very high doses to elicit a measurable response, were producing an inordinately high rate of false positives. Very high doses of many, if not all, chemicals are cytotoxic, and when large numbers of cell are killed it stimulates mitosis. The development of cancer, therefore, could be a nonspecific response to the high rate of mutagenesis. Ames also postulated that endogenous mutations were a major contributor to the aging process. Prof. Ames' website may be viewed at http://www.bruceames.org. Ames incurred the wrath of environmentalists and some fellow scientists when he declared that our obsession with eliminating trace levels of chemicals, notably pesticides, from our diet was misplaced. He based his premise on the knowledge that we consume hundreds, if not thousands, of natural pesticides that have never been tested for carcinogenicity or mutagenicity and the few synthetic ones to which we are exposed are inconsequential by comparison. While his theory remains largely untested, it constitutes an important reminder that our natural environment is not without risks and that we are in possession of protective mechanisms that have evolved against them, as is discussed later. Some perspective is often required in judging relative risks whether from anthropogenic or natural sources.

Model 2

The "genome instability" model evolved from research on familial cancers. The study of retinoblastoma was the basis for the "two-hit" hypothesis of cancer development and the (then) theory of TSGs. The discovery of Rb1, the TSG that leads to retinoblastoma when both copies of the gene are mutated, confirmed the theory. In this case, one hit was the inherited predisposition for the tumor and the second hit was the deactivation of the TSG. A double hit could also involve growth promotion of the tumor cells combined with deactivation of a TSG. A related, recently described model called chromothripsis is discussed later.

Model 3

The "nongenotoxic" model is characterized by emphasis on nongenotoxic effects of cancer risk factors that do not act through DNA alterations. These would include such risk factors as diet, obesity, hormones, and insulin resistance.

Model 4

The "Darwinian" model applies Darwin's theory of natural selection at the cellular level. Cells with a greater propensity to divide may have a selection advantage and may through other processes mutate to become malignant cells.

Model 5

Vineis and his group note that more recent models put emphasis on the microenvironment surrounding the cancer cells that may have been involved in creating favorable conditions for their development.

Model 6

A new model has been added since the paper by Vineis' group. Chromothripsis is the name applied to a new form of carcinogenesis. In January of 2011, a paper appeared in the journal *Cell* describing how a catastrophic event could shatter a chromosome into myriad pieces, which might then be reassembled incorrectly, producing numerous mutants that could conspire to cause a multihit form of carcinogenesis. The conventional theory that cancer always required many years to develop was thus also shattered as this form of cancer could appear without needing any preexisting genetic predisposition or concomitant risk factors. What might cause such a catastrophic event remains largely a mystery, although radiation is thought to be one possibility. It has been observed that bone cancer sometimes develops suddenly in persons exposed to radiotherapy. Two to three percent of all cancers and 25% of bone cancers appear to result from such a chromosome-shattering event. It seems highly likely that the story of carcinogenesis is not yet finished.

Stages of Chemically Induced Carcinogenesis

The induction of a cancer is a complex, multistage process involving interactions among the carcinogen, environmental, and endogenous factors. Three stages are generally recognized: (1) initiation, (2) promotion, and (3) progression. It should be noted that the description of these stages is largely one of observed behavior rather than any genetic or biochemical characteristic.

Initiation

The induction of a mutation by an electrophilic chemical or metabolite (a genotoxic carcinogen) that binds to DNA is believed to be the initial step. DNA thus altered is called an adduct. The heritable characteristic is not expressed. Chemicals that are capable of inducing tumors after a single exposure are sometimes called complete carcinogens (see also Model 6). This is usually seen in animal studies. PAHs such as dimethylbenz[a]anthracene and 3-methylcholanthrene are initiators.

Promotion

Promoters are agents that increase the number of tumors, increase their growth rate, or decrease the latency period. They do not bind to DNA and are, therefore, called epigenetic carcinogens. Exposure to the initiator must occur first. Promoters do not themselves generally cause tumors, although there is evidence that some promoters are merely weak carcinogens that act synergistically with other carcinogens. Co-carcinogens are agents that, when present just before, or together with, a carcinogen, result in significantly higher tumor yields in experimental animals some substances may be both promoters and co-carcinogens (e.g., phorbol esters). Nutritional factors (e.g., saturated fat intake), hormones, trauma, and viruses may act as co-carcinogens.

Progression

This refers to the natural history of the disease. At this stage there are irreversible changes resulting from mutation in one or more genes controlling key aspects of cell regulation. This instability of the karyotype increases with tumor growth. Some chemical carcinogens act at this stage, including arsenic salts, benzene, and asbestos. Figure 1.14 summarizes the steps involved in carcinogenesis.

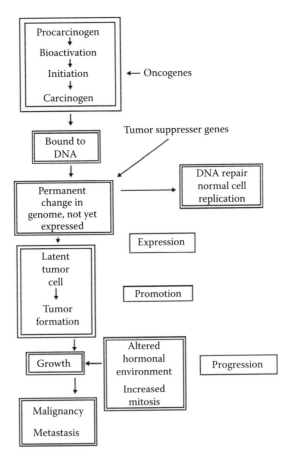

FIGURE 1.14
Stages in the development of a disease or malignancy.

Given the large number of chemicals that are carcinogenic, and the number coming onstream daily, testing for carcinogenicity is of paramount importance and of some difficulty. This subject will be dealt with in Chapter 2.

DNA and Cell Repair

Response of Tissues to Chemical Insult

Tissue regeneration occurs after destruction of some cells. This may lead to the release of growth factors that stimulate cell division and proliferation, followed by maturation of the cells to a functional state. Function is thus restored. Cell regeneration may occur if a portion of a cell is destroyed but the nucleus remains intact. Thus, the axon of a neuron may eventually regenerate if the cell body is undamaged.

Hyperplasia refers to an increase in the size of an organ or tissue in response to increased demands upon it. This is a normal adaptive phenomenon and it involves an increased number of normal cells. It tends to occur in organs of intermediate specialization such as the liver, pancreas, thyroid, adrenal cortex, and ovary. Hyperplasia of the liver may occur in response to an increased demand to detoxify a xenobiotic presented in high concentration or for a long period.

DNA Repair

DNA becomes damaged as a result of everyday exposure to toxic substances, the generation of free radicals, exposure to natural source radiation, etc. It is therefore not surprising that enzymatic DNA repair mechanisms exist. The characteristic of self-repair appears to be unique to the DNA molecule. Both constitutive (ongoing) and inducible (activated in response to an event) systems exist. The natural, endogenous repair system can be overwhelmed by massive or repeated exposure to genotoxic agents. In that case, primary DNA damage occurs (strand breaks, loss of bases, or adduct formation).

The ability to remove and replace damaged segments of DNA is central to the repair process. If repair can be accomplished prior to the fixation of the mutation, there may be no adverse affect from the DNA damage. DNA damage may inactivate both error-prone and error-free processes. Error-prone processes may actually lead to new mutations through nucleotide mismatches. Error-free processes will correct the damaged DNA site with the correct nucleotide sequence. Which process is dominant depends on a number of factors including species, (processes tend to be species-specific), cell type, the nature of the chemical mutagen, and the nature of the lesion. 3' and 5' endonucleases cleave the DNA on either side of the adduct and an exonuclease cuts out the damaged region, including nucleotides on either side of it. A DNA polymerase inserts the correct bases into the patch and a DNA ligase seals it.

Even in error-free repair, the wrong bases sometimes are incorporated into the DNA patch. In this case, the repair process may be repeated to effect the correct pairing. This process is called mismatch repair and it is triggered by nonhydrogen-bonding base pairs. Evidence suggests that the repair process can correct damage resulting from low levels of exposure to genotoxic agents with a high frequency of success. This raises questions about the degree of risk associated with such exposures. The processes of mutation and repair are illustrated in Figure 1.14.

Cell Repair and Regeneration in Toxic Reactions

The ability of a tissue to repair itself after chemical insult is dependent on the percentage of cells damaged, the nature of the damage, and the rate of cell division. Cells of the epithelium (skin, hair, nails) and mucous membranes (gastrointestinal tract, lung, and urogenital tract) are capable of rapid cell division and tissue regeneration. Sloughing of these tissues is protective because it carries away any toxicants accumulated in the cells from low-level exposures. Bone marrow also has a rapid rate of cell division, as do the germ cells of the developing fetus. Damaged bone marrow may regenerate if enough cells survive.

Conversely, such rapidly dividing tissues are vulnerable to destruction when exposed to higher levels of cytotoxic agents. It is for this reason that hair loss, anemia, diarrhea, and gastrointestinal hemorrhage may occur with the use of antineoplastic drugs and following exposure to radiation. These adverse reactions occur because the interval between cell damage and cell division is too short for repair to occur. Cell division cannot proceed or is so impaired that second-generation cells cannot themselves reproduce.

Cells of the liver, kidney, exocrine and endocrine glands, and connective tissue have less rapid rates of cell division but are capable of proliferation and repair.

Heart cells, peripheral nerve cells, and voluntary muscle cells regenerate poorly, but recent evidence suggests that these may be replaced to a greater extent than previously thought. Increased muscle mass from conditioning is due to increased actin and myosin content of the cells.

Fetal Toxicology

Teratogenesis

The existence of malformed infants and animals has been recorded for thousands of years. Early explanations tended to be supernatural. Deformed infants were often regarded as warnings from the gods. The word "monster" comes from the Latin "monstrum," meaning portent. Malformed animals were thought to be the result of copulation with humans. Those suspected of such an act were often put to death painfully.

Although it was recognized as early as 1932 that dietary deficiencies could result in birth defects, it was not until 1962 that the possibility of drug-induced teratogenesis was recognized as a result of the phocomelia observed in the thalidomide infants in Germany, Japan, the United Kingdom, and other countries including Canada. This tragedy resulted in the introduction of regulatory requirements for tests for teratogenic effects of new drugs in all industrialized countries.

The placenta is a semipermeable membrane that will pass many xenobiotics and their metabolites. Whereas most nutrients cross the placental membrane by energy-consuming, active transport systems, toxicants cross mainly by passive diffusion, so that lipid solubility is an important determinant of toxicity to the fetus. The exceptions are antimetabolites (anticancer drugs), which are analogs of natural substrates and which may utilize their transport pathways. Highly lipid-soluble agents will establish equilibrium between the maternal plasma and the fetus very quickly. More polar agents will take longer to accumulate, but there is no such thing as a "placental barrier." The placenta possesses biotransforming enzymes so that some detoxication may occur, but it is not enough to protect against any but very low exposures. The human placenta is capable of Phase I and II biotransformations but the enzymes are not inducible. Biological functions in the fetus, even near term, are poorly developed. The blood–brain barrier is imperfect; biotransforming enzymes are undeveloped as is the excretory function of the kidney. The earlier in gestation, the more this is so. The consequences of fetal exposure to toxicants are several. If the exposure is high enough, embryonic death will occur and, possibly, reabsorption of the embryo. This tends to occur early in gestation. If the exposure occurs during structural development, teratogenesis may occur and an anatomical defect may result, the nature of which will reflect the stage of development when the exposure occurred. There is thus a critical period during which a target site exists that does not exist before or afterward. This is what occurred when pregnant women took the drug thalidomide to control morning sickness.

If structural development is more or less complete, a functional defect may occur that may not become evident until later in life when the affected function would normally come into play. The newborn infant may thus be vulnerable to developmental toxicity for several weeks after birth until organ systems become fully matured.

Examples of known teratogenicity resulting from fetal exposure to toxicants include the following (others doubtless exist):

1. Folic acid antagonists (aminopterin) used as anticancer agents.
2. Androgenic hormones (natural and synthetic progesterones) used to treat breast cancers (and by athletes to increase muscle mass) will masculinize female offspring.

3. Thalidomide. A very potent teratogen, use for a single day during day 20–50 of pregnancy was associated with phocomelia (flipper-like limbs). Most animal species are much less sensitive than humans.

4. Alcohol. Fetal Alcohol Syndrome (FAS) involves impaired growth, impaired mentation, and distinct facial characteristics (small head, small eye openings, thin upper lip. Fetal Alcohol Effects (FAE) includes only one or two of these criteria.

5. Methyl mercury. Infants exposed *in utero* may develop severe neurological disturbances similar to cerebral palsy. This occurred in the Minamata exposure in Japan.

6. Actinomycin D or dactinomycin is an antibiotic that is also used as an antineoplastic agent in certain tumors including Wilms' tumor and Kaposi's sarcoma. It is well known to be teratogenic and possibly carcinogenic. It is a potent inhibitor of rapidly dividing cells, both normal and neoplastic. Its potent cytotoxic effect is due to its capacity to bind to DNA in its double helix state. The dactinomycin-DNA complex is very stable so that DNA transcription by RNA polymerase is blocked.

Developmental toxicity can occur in the absence of any maternal toxicity. Teratogenesis may occur from causes other than drugs and environmental chemicals. In fact, in 65%–70% of cases, the cause is never established. Of the remainder, the breakdown is shown in Table 1.4.

Transplacental Carcinogenesis

The development of cancer as a result of fetal exposure is a possibility if the reproductive system is involved. The best-known example is the occurrence of carcinoma of the vagina and cervix in young women exposed *in utero*

TABLE 1.4

Causes of Teratogenesis (Percent of Cases)

Known genetic transmission	11–18
Chromosomal aberration	3–5
Environmental causes	
Human-source radiation (therapeutic, nuclear)	<1
Infections (rubella, herpes, toxoplasma, cytomegalovirus, syphilis)	2–3
Maternal metabolic imbalance (endemic, cretinism, diabetes, phenylketonuria, virilizing tumors)	1–2
Drugs and environmental chemicals	2–3

to diethylstilbestrol (DES). DES was given to prevent impending abortion, mostly from 1950 to 1970. Carcinogens abound as natural substances in the environment. Table 1.4 compares some relative risks of these with synthetic ones. The problems associated with attempting to predict the carcinogenic potential of new chemicals from animal studies will be discussed in the next chapter. The problem is critical, due to the high cost and protracted period it takes to evaluate a new chemical.

There is now general awareness that the fetus' welfare can be threatened by smoking and alcohol consumption by the mother. Avoidance of ingestion, inhalation, or dermal contact of xenobiotics, including recreational drugs, therapeutic agents (where possible) solvents, and other volatile chemicals is to be avoided during pregnancy, especially during the first trimester.

Population and Pollution

It is generally accepted that the root of most anthropogenic environmental problems is overpopulation. This certainly includes chemical carcinogenesis that abounds in our environment (Table 1.4). The increase in the demand for consumer goods, food, and energy resulting from population growth leads to the creation of more waste and pollution. As of April 2012, the world population stood at 7.035 billion. The doubling time of the Earth's population is down to about 30 years, and this has created incredible pressures on our ability to provide food, living space, energy, and manufactured goods, and particularly on our ability to deal with the waste products of our society (Table 1.5). Our insatiable demand for petroleum products has resulted in drilling in more difficult and hazardous areas such as the arctic where the results of an oil spill would be catastrophic. This subject will be dealt with in more detail later in this text. There has been little official recognition of the population problem in the West, nor specific actions to control population explosion. The People's Republic of China has long had a restriction of two children per couple in an effort to limit population growth. Controls such as this are repugnant to Western democracies. Western attitudes regarding measures to control population growth are changing, but fear of reaction from religious groups and right-to-life activists, who tend to link population control to issues such as abortion and birth control, has made the issue unattractive to politicians. The lessons of nature should convince us of the need to address this issue. When a population exceeds the ability of its ecosystem to support it, it dies out.

Table 1.6 lists some carcinogens that may be encountered in the workplace, a striking illustration of how our insatiable demand for consumer goods and energy can put us at risk.

TABLE 1.5

Relative Carcinogenic Hazards in the Environment

Source	Cancers/100,000 Population
Average U.S. lifetime risk of cancer (all types)	25,000–30,000
Foods	
Four tablespoons (60 mL) of peanut butter/day (due to aflatoxin)	60
One pint (550 mL)/day (aflatoxin)	14
8 oz (226.8 g) broiled steak/week (nitrosamines, polycyclic aromatic hydrocarbons or PAHs)	3
One diet soda/day (saccharin in the United States, questionable)	70
Average U.S. fish consumption/day	33
Lake Michigan sport fish (based on median consumption levels and EPA potency values)	480–3,300
U.S. sport fish based on (Kim and Stone potency values)	77–340
Drinking water	
Average U.S. groundwater based on 2 L/day	1
Niagara River water based on 2 L/day (EPA potency value)	0.3
Air	
Various estimates (U.S. urban)	10–560

TABLE 1.6

Some Potential Carcinogens Encountered in Industry

1. PAHs (e.g., benzo[a]pyrene); produced by any combustion process, refining and distilling of petroleum, breakdown of lubricants, in welding and foundry processes
2. Benzene (32 million tons/year produced); chemical and petrochemical industries, solvents (paint industry), as an impurity, etc.
3. Carbon tetrachloride (CCl_4); solvent in industrial processes
4. Halogenated hydrocarbons (PCBs, PBBs, TCDD); in lubricants, transformer insulation, pesticides, etc.
5. Asbestos from mining, in insulation
6. Vinyl chloride in plastics industry
7. Formaldehyde (embalmers, pathologists)

Review Questions

1. Match the following terms with the appropriate definition:
   a. Pharmacodynamic
   b. Probit
   c. Economic toxicology

    d. Xenobiotic

    e. Anthropogenic

        i. A substance foreign to living systems

        ii. A chemical process in biological systems in which neither of two agents which interact are permanently altered

        iii. The application of toxicology to achieve an advantage for humankind

        iv. Resulting from human activity

        v. A probability unit used for making comparisons of potency or toxicity

2. Select the correct statement.

    a. A quantal response is one that is "all or nothing."

    b. The response to increasing doses of a drug or toxin continues to increase indefinitely.

    c. The minimum lethal dose value is applicable to all species.

    d. A quantal response can never be converted to a graded one.

    e. Semilogarithmic dose–response plots are used to clean up bad data.

For Questions 3–10 answer true or false:

3. If drug A has a Therapeutic Index (TI) of 500 and drug B has a TI of 10,000, drug A is safer than drug B.

4. An acute toxicity reaction is defined as one that occurs within 48 h of exposure.

5. Parallel dose–response curves suggest that the two agents in question probably work through the same mechanism.

6. If the $T_{1/2}$ of an agent exceeds the dosage/exposure interval, it will never accumulate in the body.

7. Insecticides that inhibit the enzyme acetylcholinesterase cause rapid loss of consciousness.

8. Nerve toxins may work by blocking axonal conduction, blocking the enzymatic destruction of a neurotransmitter, blocking the attachment of a neurotransmitter to its receptor, or blocking its release from the nerve terminal.

9. Methylmercuries are less toxic than elemental mercury.

10. Cadmium is a nontoxic heavy metal.

For Questions 11–16 use the following code:

    Answer A if statements a, b, and c are correct.

    Answer B if statements a and c are correct.

    Answer C if statements b and d are correct.

Answer D if only statement d is correct.

Answer E if all statements (a, b, c, d) are correct.

11. a. Most anticancer drugs are teratogenic.
    b. Genetic abnormalities may involve point mutations or chromosomal breaks.
    c. Cancer induction is likely a multistage process.
    d. Viruses have nothing to do with cancer.

12. a. Promoters increase tumorigenesis in response to other agents.
    b. Exposure to the promoter must occur before exposure to the initiator.
    c. Promoters do not bind to DNA.
    d. Promoters always cause cancer in their own right.

13. a. Selective toxicity is usually absolute, i.e., a substance is completely toxic for the target species and completely harmless for other ones.
    b. The portal of entry may significantly affect the toxicity and carcinogenicity of a xenobiotic.
    c. Cancer studies in animals are highly predictable for cancer risk in humans.
    d. Oncogenes may be activated to convert a cell to a malignant form.

14. Highly water-soluble chemicals are
    a. Excreted by the kidney without biotransformation
    b. Poorly absorbed from the gastrointestinal tract
    c. Very polar
    d. Cross the blood–brain barrier very poorly

15. Highly lipid-soluble chemicals
    a. Are well absorbed from the gastrointestinal tract
    b. Do not enter the cerebrospinal fluid
    c. Require biotransformation by the liver before being eliminated by the kidneys
    d. Generally have a very short $T_{1/2}$

16. Which of the following statements is/are true?
    a. Serum albumin has numerous binding sites for weakly acidic xenobiotics.
    b. Substances with epoxide bonds tend to bind irreversibly to macromolecules within cells.
    c. Mixed-function oxidase enzymes in the SER of the liver cells are responsible for much drug metabolism.
    d. Conjugation of a xenobiotic with glucuronide is defined as a Phase II reaction.

For Questions 17–21 answer true or false:

17. The kidney is the sole route of elimination of xenobiotics.
18. Water-soluble agents of small molecular size pass through the glomerulus.
19. Body composition has no influence on the fate of chemicals in the body.
20. Some agents may be secreted by an active process across the wall of the renal tubule.
21. The presence of one chemical in the body may influence the metabolism of another.

For Questions 22–24 use the following code:

> Answer A if statements a, b, and c are correct.
> Answer B if statements a and c are correct.
> Answer C if statements b and d are correct.
> Answer D if only statement d is correct.
> Answer E if all statements (a, b, c, d) are correct.

22. a. Proto-oncogenes are involved in regulating cell signaling cascades.
    b. Proto-oncogenes may be converted to oncogenes by retroviruses.
    c. Oncogenes generally cause uncontrolled cell proliferation.
    d. Retroviruses may become carriers of oncogenes.
23. a. Tumor suppresser genes display dominant characteristics.
    b. Oncogenes are recessive in nature.
    c. Tumor suppresser genes have no function in the normal cell.
    d. Tumor suppresser genes normally prevent uncontrolled cell growth.
24. a. "Sense" refers to short segments of DNA coding for some important aspect of cell function.
    b. "Antisense" is an altered sequence of oligonucleotides matched specifically to a "sense" sequence.
    c. Antisense sequences inhibit tumor growth *in vitro*.
    d. Antisense sequences generally contain 15–25 bases.
25. List five chemicals that can induce hepatic microsomal enzymes.
26. List five known industrial carcinogens.
27. Define the following:
    a. Teratogenesis
    b. Carcinogenesis
    c. Mutagenesis
28. List three biological variables that can affect the body's response to xenobiotics.

29. Which of the following statements is/are correct regarding gene therapy for cancer?

    a. Retroviruses are used as transporters for the therapeutic gene.
    b. The virus is encapsulated to protect it from the host immune system.
    c. Preventing expression of oncogenic proteins is a target for some gene therapies.
    d. None of a, b, or c is correct.
    e. All of a, b, and c are correct.

For Questions 30–36 answer true or false:

30. D-actinomycin is a teratogenic antibiotic.
31. Fetal alcohol effects are a manifestation of teratogenesis.
32. There are no natural teratogens.
33. All teratogenesis involves anatomical defects.
34. The human placenta can biotransform some xenobiotics.
35. Most toxicants cross the placenta by active, energy-consuming processes.
36. High levels of toxicants are generally more damaging to rapidly dividing tissues.

Answers

1. i = d, ii = a, iii = c, iv = e, v = b
2. a
3. False
4. True
5. True
6. False
7. False
8. True
9. False
10. False
11. A
12. B
13. C
14. E

15. B
16. E
17. False
18. True
19. False
20. True
21. True
22. E
23. D
24. E

Find the answers in the text for Questions 25–28.

29. E
30. True
31. True
32. False
33. False
34. True
35. False
36. True

Further Reading

Assennato, G., Cervino, D., Emmett, E.A., Longo, G., and Merlo, F., Follow-up of subjects who developed chloracne following TCDD exposure at Seveso, *Am. J. Ind. Med.*, 16, 119–125, 1989.

Bailey, D.G., Fruit juice inhibition of uptake transport: A new type of food-drug interaction, *Br. J. Clin. Pharmacol.*, 70, 645–655, 2010.

Bailey, D.G., Dresser, G.K., Leake, B.F., and Kim, R.B., Naringin is a major and selective inhibitor of organic anion-transporting polypeptide 1A2 (OATP1A2) in grapefruit juice, *Clin. Pharmacol. Ther.*, 81, 495–502, 2007.

Binetti, R., Costamagna, F.M., and Marcelo, J., Exponential growth of new chemicals and evolution of new information relevant to risk control, *Ann. 1st Super Sanita*, 44, 13–15, 2008.

BRCA1 and BRCA2: Cancer risk and genetic testing, National Cancer Institute Fact Sheet. http://www.cancer.gov/cancertopics/factsheet/Risk/BRCA (accessed September 22, 2011).

Carillo-Infante, C., Abbadessa, G., Bagella, L., and Giordano, L., Viral infections as a cause of cancer, *Int. J. Oncol.*, 30, 1521–1526, 2007.

Cattaneo, R., Miest, T., Shashkova, E.V., and Barry, M.A., Reprogrammed viruses as cancer therapeutics: Targeted, armed and shielded, *Nat. Rev. Microbiol.*, 6, 529–540, 2008.

Hardman, J. and Brunton, L.L. (eds.), Blumenthal, D.K., Murri, N., and Hilal-Dandan, R. (assoc. eds.), *Goodman and Gilman's The Pharmacological Basis of Therapeutics*, 12th Edn., McGraw-Hill Medical, New York, 2011.

Hortobagyi, G.N., Toward individualized breast cancer therapy: Translating biological concepts to the bedside, *Oncologist*, 17, 577–584, April 2, 2012 (Epub ahead of print.)

Klassen, C.D. (ed.), *Casarett and Doull's Toxicology: The Basic Science of Poisons*, 7th Edn., McGraw-Hill Medical, New York, 2008.

Klassen, C.D. and Watkins, J.B. III. (eds.), *Casarett and Doull's Essentials of Toxicology*, McGraw-Hill Medical, New York, 2010.

Klaunig, J.E., Kamendulis, L.M., and Yong, X., Epigenetic mechanisms of chemical carcinogenesis, *Belle News Lett.*, 9, 2–21, 2000.

Lichtenstein, P., Holm, N.V., Verkasalo, P.K., Iliadou, A., Kaprio, J., Koskenvuo, E., Pukkala, E., Skytthe, A., and Hemminki, K., Environmental and heritable factors in the causation of cancer-analyses of cohorts of twins from Sweden, Denmark and Finland, *N Eng. J. Med.*, 343, 78–85, 2000 (comments 135–136).

Marx, J., Research news: Learning how to suppress cancer, *Science*, 261, 1385–1387, 1993.

McClean, P., Eukaryotic cell cycle and the genetics of cancer. http://www.ndsu.edu./pubweb/~mcclean/plsc431/cellcycle/cellcycl4.htm (accessed on October 12, 2011).

McPhee, S.J. and Papadakis, M.A. (eds.), Rabow, M.W. (assoc. ed.), *Current Medical Diagnosis and Treatment*, 51st Edn., Lange Medical Books/McGraw-Hill, New York, 2010.

News and Comment, Animal carcinogen testing challenged, *Science*, 250, 743–745, 1990.

News and Comment, Experts clash over cancer data, *Science*, 250, 900–902, 1990.

Oberbauer, R., Not nonsense but antisense—Applications of antisense oligonucleotides in different fields of medicine, *Wein. Klin. Wochenschr.*, 109, 40–46, 1997.

Oncogene, The Broad Institute of MIT and Harvard. http://www.broadinstitute.org/node/1417 (accessed on January 17, 2012).

Putnam, D.A., Antisense strategies and therapeutic applications, *Am. J. Health Syst. Pharm.*, 53, 151–160, 1996.

Research News, Dioxin revisited, *Science*, 251, 624–626, 1990.

Stephens, P.J., Greenman, C.D., Fu, B., Yang, F., Bignell, G.R., Mudie, L.J., Pleasance, E.D. et al., Massive genome rearrangement acquired in a single catastrophic event during cancer, *Cell*, 144, 27040, 2011.

Timbrell, J.A., *Principles of Biochemical Toxicology*, 4th Edn., Informa Health Care, New York, 2009.

Van Loon, J. and Weinshilboum, R.M., Thiopurine methyltransferase isoenzymes in human renal tissue, *Drug Metab. Dispos.*, 18, 632–638, 1990.

Vineis, P., Schatzkin, A., and Potter, J.D., Models of carcinogenesis: An overview, *Carcinogenesis*, 31, 1703–1709, 2010.

Worldometers: Current world population. http://www.worldometers.info/world-population/ (accessed on November 23, 2011).

Yu, M.-H. (ed.), *Environmental Toxicology: Biological and Health Effects of Pollutants*, Taylor & Francis Group, Boca Raton, FL, 2011.

2

Risk Analysis and Public Perceptions of Risk

(Risky Business)

Introduction

As noted in the opening page of Chapter 1, the American Chemical Society adds about 4000 new chemicals every day to the roughly 60 million already on its registry. It is estimated that between 60,000 and 70,000 industrial and commercial chemicals are currently in use in North America; a number that doubtless is already out of date. The (U.S.) Occupational Safety and Health Administration (OSHA) lists over 10,000 chemicals it considers to be hazardous as of 2011. Under the Canadian Environmental Protection Act the Toxic Substance List contains only 120 chemicals, but this list is more akin to a controlled substance list. It should be obvious that only a very small portion of these have been studied sufficiently to conduct any sort of risk assessment regarding human health, and such studies use, characteristically, only one route of administration (portal of entry). Not only do the sheer numbers overwhelm any chance of conducting comprehensive toxicity testing, but the system itself is felt to be outdated with large gaps in the safety net to protect the consumer. The public seems unwilling to give up the advantages accruing from such chemicals (plastics, pesticides, petroleum fuels, etc.) but also it is increasingly vociferous in its demands to be protected from any adverse effects arising from their use. In 2011 several U.S. senators introduced a bill to Congress, the "Safe Chemicals Act of 2011," designed to overhaul the aging Toxic Substances Control Act. A key feature is that it would require "basic health and safety information for all chemicals as a condition for entering or remaining on the market." The reader can be forgiven for experiencing shock that this requirement is not already in place. As of April 2012, the Act had not been put to vote.

In 1937, the Massengil Company marketed a drug it labeled "Elixir of Sulfanilamide," which consisted of a solution of sulfanilamide in diethylene glycol. At that time, there were few laws protecting people against unsafe food or drugs. Seventy-three people died of its toxic effects before the danger was recognized. Because the drug was mislabeled (an elixir referred to a substance dissolved in alcohol) the Food and Drug Administration was able to have it removed from the marketplace. It was this crisis that led to the passing of the Food, Drug, and Cosmetic Act of 1938.

While human health and safety remain the paramount consideration regarding the safety of chemicals, the environmental damage caused by some of these agents is becoming more and more evident and indeed this is a major danger facing humankind. Nevertheless, legislators and regulators are faced with the task of making decisions regarding safe limits for thousands of chemicals, often on the basis of very limited data, and in the face of pressure from consumer groups, environmental activists, and industry lobbies.

Risk analysis has become a highly specialized area with certification of risk analysts in place in some jurisdictions. What follows is not to be viewed as a training manual for risk assessors but is meant rather to acquaint the student with the concepts and steps involved in the process.

Assessment of Toxicity versus Risk

Risk and its assessment is a concept that can be applied to many aspects of life including financial investment. Regarding xenobiotics, two parameters must be established for risk assessment to be performed. First, the *hazard* (similar to toxicity) associated with the chemical must be established. How toxic is it? Obviously a very toxic substance is more hazardous than one of low toxicity. This requires that a dose–response determination be performed. These data may be used to establish the maximum allowable concentration (MAC) in the environment (usually air or water). Then the level of *exposure* must be determined. For example, if a hazardous substance is found in drinking water, the exposure level would relate to the amount of water consumed in a given period of time. Hazard and exposure are used to determine *risk*, which is the likelihood of a deleterious effect occurring in the given set of circumstances. The well-established definition of risk is expressed in the equation

$$Risk = Hazard \times Exposure.$$

Thus, a substance like potassium cyanide (CKN) might be quite hazardous but if the exposure level is very low as it would usually be for the general public, the risk is slight. For workers in an industry employing CKN, however, the risk could be significant.

The difference between hazard and risk is not always appreciated by the public or by the news media. Thus, statements have appeared to the effect that dioxin is the "most potent poison known to man." In fact, *botulinum* toxin is 100× more potent in mice than dioxin and the toxicity of dioxins in man has not been fully established.

The question of risk must consider such factors as, using dioxin as an example,

1. The biological half-life of the substance (dioxins are very stable).
2. The partition coefficient (dioxins are very lipid soluble; therefore, they are sequestered in the body).
3. Does the toxin concentrate up the food chain? Yes, because of #2.
4. What are the long-term effects? Is the substance carcinogenic? Yes, in experimental animals. In humans, the evidence is much less conclusive.
5. What are the predicted risks to humans and the environment based on known levels of contamination? This is the area that causes most controversy because it is highly speculative.
6. What are the costs of avoiding these risks? This is very difficult to estimate and therefore controversial. While risk to the general public is difficult to assess and may be of a very minor nature, risks encountered by industrial workers may be much greater because of the higher exposures and because of the risk of accidental contamination. Populations in some regions, however, may be exposed to similar risks from industrial accidents or from uncontained dump sites.

Predicting Risk: Workplace versus the Environment

Acute Exposures

Information from industrial accidents and from pre-regulation exposures is very valuable because it eliminates the need to make extrapolations from test animals. Prediction of risk following defined exposures is, thus, fairly accurate as, for example, in the case of cholinesterase-inhibiting insecticides. Animal data are still useful, however, because they too deal with acute exposure.

Chronic Exposures

Predictions are less reliable due to biological variations in susceptibility to chronic, lower exposure levels. Individual susceptibility to lung damage from paraquat, for example, may vary considerably.

Very-Low-Level, Long-Term Exposures

It is more difficult to predict organ toxicity from animal studies with this type of exposure but they are still useful. Epidemiological data from human exposures are most useful if available. For example, extensive data have accumulated over many decades regarding pneumoconiosis (black lung disease in miners).

Carcinogenesis

At best, predictions from animal data can only be a rough approximation due to the need to extrapolate from very high, to extremely low exposures and the possibility of species differences. Differences in the nature of the exposure may further complicate extrapolations from animal data to the human situation. Moreover, predictions of risk due to low-level exposures are complicated by the presence of other risk factors, many of them from natural sources. For example, volcanic eruptions can pour huge volumes of gases and particulates into the atmosphere, equal to years of industrial pollution. The word "pneumoultramicroscopicsilicovolcanopneumoconiosis" was coined as the longest word in the English language after the Mt. St. Helen volcanic explosion. It refers to pneumoconiosis from inhaling volcanic ash. Tobacco smoking would be an example of an "anthropogenic" risk factor (i.e., of human origin).

Risk Assessment and Carcinogenesis

As already noted, this is the most complicated and least reliable area regarding the prediction of risk to human health in the general population from exposure to very low levels of environmental pollutants. There are several mathematical models for predicting carcinogenic risk, either by extrapolation from animal data or from human industrial exposures. Regarding animal studies, there is general agreement among these models for extrapolation to human exposures at high doses. At very low exposure levels, predictions of cancer risk can vary by several orders of magnitude, and this is the very type of exposure that creates the greatest concern in the public's mind. These differences arise because of the application of different theories of carcinogenesis to the development of models for calculating risk (see also Chapter 1). Here are some examples of these models.

1. Distribution models (log probit, logit) assume that every individual has a threshold below which no adverse effect will occur (a "No Observable Adverse Effect Level" or NOAEL).

2. Mechanistic models are based on presumed mechanisms of tumorigenesis and assume that a cancer can arise from a single mutated cell. The single-hit model assumes that the exposure of DNA to a single molecule of a carcinogen is sufficient to induce carcinogenesis. The gamma multihit model assumes that more than one "hit" is required. Multistage models assume that carcinogenesis is a process requiring several stages (a series of mutations, biotransformations) involving carcinogens, co-carcinogens, and promoters that can best be modeled by a series of multiplicative mathematical functions. Predicted dose responses are linear at very low exposure levels and assume that there is not a NOAEL.

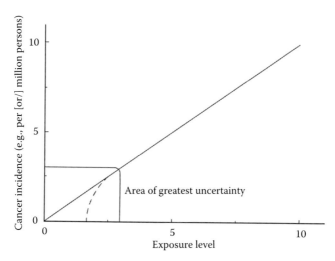

FIGURE 2.1

Exposure level and carcinogenic risk. The greatest inaccuracy in predicting cancer risks from animal data is in the area of low-level exposure. The question is whether there is a threshold for effect (dotted line) or no threshold (solid line).

All of these methods differ in the nature and shape of the dose–response curve at the low-exposure end. Figure 2.1 illustrates how these differences affect predictions.

The U.S. Environmental Protection Agency (EPA) uses the "Linearized, Multistage Assessment Technique," which assumes that there is no NOAEL and which involves the following steps (see Figure 2.2):

1. Evidence of carcinogenesis is obtained from animal studies in rabbits, rats, mice, with dose–response data for oral, inhalation, or dermal portals of entry (routes of administration).

2. From these dose–response data, the dose is calculated that would theoretically cause one cancer per million animals. The assumption is made that the dose–response curve is linear all the way to zero; i.e., that there is not a "no effect" level for the carcinogen.

3. An equivalent human dose is calculated that would cause the same incidence of cancer. This stage employs arbitrary factors to adjust for differences in absorption, metabolism, and excretion based on what data are available for humans, or simply uses a safety margin if no data are available. The 1/1,000,000 risk level is the "red line" that EPA has set for acceptable risk and it is used to determine safe limits in the environment.

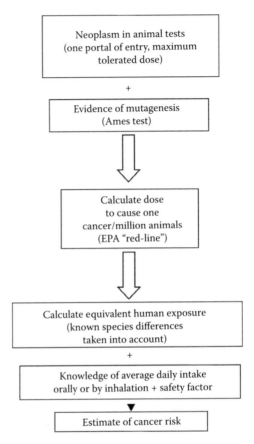

FIGURE 2.2
Stages in the process of cancer risk prediction involve several points of uncertainty.

4. Using knowledge of the average human intake of air and water, maximum allowable limits are set for the toxicant that would keep daily intake below the level that would induce one additional cancer per million people. An additional safety margin may be introduced, based on the lowest levels that can be achieved at an acceptable cost. In Canada, the Canada Environmental Protection Act (CEPA) defines the Tolerable Daily intake (TDI) as the maximum to be permitted. It uses a safety factor of 100× the threshold obtained from animal studies. It also uses the Exposure/Potency Index (EPI), a value that takes into account the level of environmental exposure as well as the known toxicity of a substance, to rank chemicals as to degree of risk. The linearized, multistage model assumes that there is no threshold for carcinogenesis, a reasonable assumption for electrophilic carcinogens affecting DNA, but this may not be true for epigenetic carcinogens such as dioxin. Canada and some

European countries set dioxin limits 170–1700 times higher than EPA limits because they do not apply the linear approach to dioxin risk analysis. The CEPA defines such "threshold" chemicals where possible and treats them separately from those where no threshold exists or where none has been demonstrated. These differences illustrate the fact that risk analysis for chemical carcinogens is a complex process and by no means is it a standardized one.

Sources of Error in Predicting Cancer Risks

Obviously there are several points in this method that require estimations and therefore there may be wide variations in resulting predictions. This is the greatest source of contention between governments and various special interest groups. Environmentalists generally press for reductions in allowable levels, whereas industry may lobby for higher levels if lower ones involve significant cost factors. Some specific sources of contention in risk analysis are discussed in the following.

Portal-of-Entry Effects

1. The method used may not be reliable when exposure of humans involves multiple portals of entry. Volatile chemicals, for example, may be inhaled, ingested, and absorbed through the skin.

2. Toxicity may be affected by differences in absorption or biotransformation occurring at the portal of entry, so that data obtained from one type of exposure may not be applicable to others. As an extreme example of portal-of-entry effects, the purest air can be fatal if injected intravenously, as can the purest water if inhaled. Ethyl acrylate produces a 77% incidence of tumors in rats at 200 mg/day orally. The same dose applied to the skin causes no tumors. Cadmium (Cd) is carcinogenic by inhalation but not orally or dermally. Conversely, epichlorohydrin will cause tumors at the point of contact with any epithelium. It has been stated that of the >500 risk assessments that have been completed, nearly all involve a single route. This applies both to carcinogenic and noncarcinogenic effects. Numerous examples of route-specific effects exist; for example, trichloroethylene causes central nervous system (CNS) depression at 7 ppm if inhaled, but the same concentration taken orally has no effect because of incomplete absorption.

3. The area of contact may affect uptake, even for the same portal of entry. Thus, if a large area of skin is exposed to a toxicant, more will be absorbed. Moreover, the skin of the forehead absorbs 20×, and that of the scrotum 40×, more effectively than the skin of the forearm. Transit time for ingested material in the intestinal tract may vary from 10 to 80 h, depending on age, diet, and other factors, so the

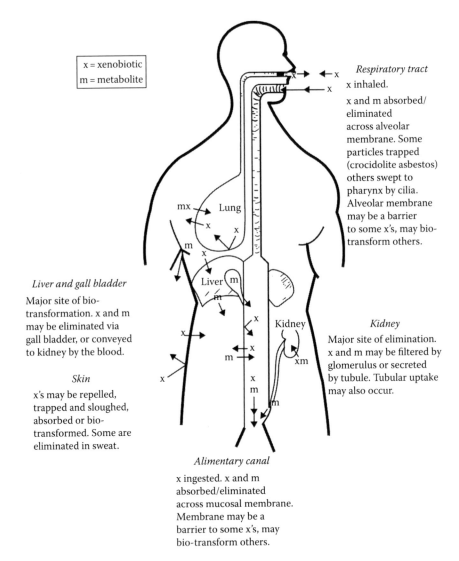

FIGURE 2.3
The possible fate of xenobiotics in the body.

time available for absorption will vary as well. The relatively rapid transit time through the small bowel may partly explain the rarity of cancer in this area.

Figure 2.3 summarizes the possible fate of xenobiotics (literally "foreign to life") that can occur at various portals of entry and thereafter. The mammalian body can be visualized as a thick-walled tube with the outer surface (skin) and inner surface (gastrointestinal tract) in contact with the environment.

Excretion of toxicants and waste products back to the environment takes place in sweat, expired air, feces, urine, and the sloughing of cells in contact with the environment.

Age Effects

The age of the population at risk may affect the degree of risk. Infants, especially premature ones, absorb chemicals through the skin much more efficiently than adults. Infants have died from absorbing pentachlorophenol used as an antibacterial agent in hospital bedding before the practice was abandoned. Even data from human industrial exposures usually deal with adult males and may not be applicable to the elderly or to females.

Exposure to Co-Carcinogens and Promoters

It is often difficult to control for the presence of co-carcinogens and promoters, even in animal studies. Regarding human data, such factors as smoking, alcohol consumption, and intake of nitrites and nitrates and saturated fats may differ considerably from an exposed, industrial population to the public at large.

Species Differences

These are the focus of considerable attention both from the scientific community and from animal rights activists who use them to trivialize the value of animal data. It must be stated at the outset that pharmacokinetic differences can be far greater among human beings than between them and experimental animals. Biological variation is a governing force in all living things. Nevertheless, there are important differences that are known and others that are only now being identified.

1. Human skin, for example, is much more impervious than that of laboratory animals, being more similar to that of the pig.
2. The rat forestomach is devoid of secretory cells and is a better model of squamous epithelium than of secretory tissue.
3. Moreover, it contains an active microflora that can alter chemicals, whereas the stomach and upper bowel of the human are virtually sterile because of the acidity.
4. This same acid medium can serve to denature and detoxify potentially harmful chemicals.
5. Anatomical differences in the branching patterns of bronchi exist in the lungs of rodents versus primates. This can result in vastly different deliveries of inhaled volatile toxins. The pattern in humans is described as "dichotomous-asymmetric," whereas that in the rat

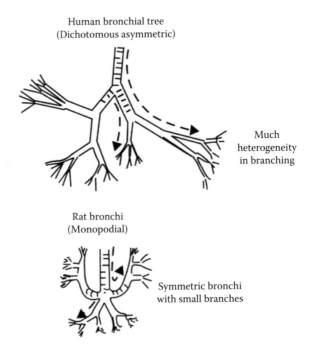

FIGURE 2.4
Comparative anatomy of human and rat bronchial trees. Dotted lines represent the differences in distances to the terminal bronchioles.

is "monopodial-symmetric." In the latter case, the primary bronchi penetrate deeply into the lungs and have secondary bronchi branching off their length. The distance to the terminal bronchiole may vary greatly and, hence too, will the target cell exposure (see Figure 2.4).

6. The rat has no gall bladder, so bile flow tends to be continuous and unaffected by food. Stasis of the bile, which can affect contact time, is rare.

7. There are numerous differences in the nature and location of biotransforming enzymes. Knowledge of these differences, for example, for cytochrome P450, can be exploited to select the most appropriate model for study. Chapter 10 further examines species differences and how they affect toxicity.

Despite the problems with extrapolation from animals to humans, it should be remembered that DNA varies from the human array by only 5% in mice, by less than 2% in most primates, and by less than 1% in chimpanzees. The similarities are far greater than the differences. Moreover, the extrapolation of risk to the general public from data acquired from industrial exposures, including accidents, has its own problems. Numerous differences usually exist between workers and the populace. The former tend to be predominantly males 18–65 years old, from the lower end of the socio-economic scale, and possibly with different habits regarding such health factors as smoking, alcohol consumption, and diet. There is also the need to extrapolate from moderate to high exposures (and possibly from very high single exposures as in an industrial accident) in the workplace to very low ones in the environment.

Extrapolation of Animal Data to Humans

One source of continuing dispute is the reliability of animal data in extrapolating cancer risk to humans. One critic of the current system is Bruce Ames (see under Carcinogenesis, Chapter 1), who feels that it is too sensitive, and thus it is predicting cancer risks that are artificial for many chemicals. Critics also claim that any substance, at high-enough doses, can be carcinogenic. The debate revolves around the use of the "estimated maximum tolerated dose" or EMTD, as the high dose level in cancer bioassays. This is defined as the highest dose in chronic studies that can be predicted not to alter the animals' longevity from effects other than cancer. According to the proliferation-mutagenicity theory, lower doses should not be carcinogenic if they do not induce cell proliferation. Defenders of the current (U.S.) National Toxicology Program, however, point out that about 90% of chemicals defined as carcinogens induced tumors at doses well below the EMTD. The World Health Organization's (WHO) International Agency for Research on Cancer (IARC) and the U.S. Annual Report on Carcinogens (National Toxicology Program) jointly publish a list of carcinogens divided by weight of evidence. Category 1 is for substances for which there is sufficient evidence for a causal relationship with cancer in humans. It lists 60 items not all of which are single chemicals. Some of the more familiar items include aflatoxins, alcoholic beverages, arsenic and its compounds, asbestos, benzene, cadmium and its compounds, nickel and its compounds, estrogens (diethylstilbestrol), tobacco smoke and products, and vinyl chloride.

Hormesis

The term hormesis refers to a U-shaped dose–response curve where the effects at the low end of the dosage/exposure scale are markedly different from those at the high end. The arms of the dose–response curve are separated by a flat area of no observable effect. Such low dose effects have often been shown to be beneficial. The phenomenon of hormesis has been demonstrated in both experimental animals and humans for a wide range of toxic substances. Lest it seem bizarre that toxic substances should have beneficial effects at low doses, it should be remembered that we exploit this fact every day, as in the use of chlorine in drinking water and fluoride in toothpaste. A publication put out by The University of Massachusetts

School of Public Health, *The BELLE Newsletter* (BELLE for Biological Effects of Low Dose Exposures) has devoted considerable attention to this subject. Exposure to low doses of ionizing radiation has long been held to impart some beneficial effects. Both experimental and epidemiological evidence suggest this. Experimentally, it has been shown that exposure to very low doses of uranium imparted resistance of cultured cells to higher, carcinogenic doses. Moreover, increased resistance to oxidizing agents such as hydrogen peroxide, and increased resistance to anticancer drugs also occurred. This adaptive response is felt to explain several instances of hormesis, but not all of them. The fluoride effect, for example, is due to hardening of dental enamel and increased resistance to cares. It is unlikely that a single, mechanistic explanation can be applied to all examples of hormesis.

The incorporation of hormetic effects into the risk assessment process has yet to occur and it is fraught with political pitfalls. The implications for the risk assessment process are enormous and likely to create considerable public controversy and resistance. Imagine the consequences of stating that low doses of dioxins might actually be beneficial to health! The concept that no dose of a carcinogen is safe would be radically altered. Little change is likely to occur until hard evidence of a hormetic effect is obtained for each specific agent.

Natural versus Anthropogenic Carcinogens

A more valid criticism perhaps, also raised by Bruce Ames, is the fact that there are hundreds of natural carcinogens in foods to which we are exposed daily, and of 77 that have been tested by the standard methods, about half (37) were carcinogenic. We are thus likely exposed to many more natural carcinogens than synthetic ones. Our natural defenses probably take care of many of these. We slough our epithelial layer regularly (skin, gastrointestinal tract, and lungs) and with it, its accumulation of toxins. We have killer lymphocytes that destroy abnormal cells, and we have detoxifying mechanisms that render many toxins harmless unless these defenses are overwhelmed by high doses. The natural decline in these defenses with age is a major factor in the increasing incidence of cancer in the elderly. In view of the abundance of natural carcinogens, elimination of all synthetic ones may not reduce cancer incidences as much as one might expect. The counterargument is that we have had several million years to evolve defenses against natural carcinogens and these may not work as efficiently against synthetic ones. One indisputable statistic, however, is that life expectancy has been steadily increasing for several decades. In fact, a great concern in all developed countries is that the ratio of working to retired persons has been declining steadily with the eventual consequence that there may not be enough revenues contributed to government pension and medicare plans to support all those who need them.

Reliability of Tests of Carcinogenesis

Another concern has been raised recently about the reliability of tests of carcinogenicity. A particular strain of mice, $B_6C_3F_1$, has been widely used in such tests because of its known tendency to readily develop tumors in response to a wide variety of chemicals. Information has been accumulating concerning significant differences in the metabolic processes of rodents and people, and it is likely that using this cancer-sensitive strain of mice may yield too many false positive findings of carcinogenicity. For example, the volatile solvent butadiene, used in the production of synthetic rubber, is exhaled unchanged in most animals and thus does not display carcinogenicity. In mice, however, much more is retained in the lungs and absorbed (33× as much as in monkeys). It is then oxidized to a mutagenic epoxide. The mouse has a much lower activity of a detoxifying enzyme "epoxide hydrolase" than do humans, so that the carcinogenic risk is many times greater in mice. This strain of mice also harbors a murine leukemia virus that has been shown to enhance carcinogenicity. The (U.S.) National Institute of Occupational Safety and Health (NIOSH) based its estimate of butadiene cancer risks entirely on studies of this strain of mice. The NIOSH model predicted that exposure to 2 ppm butadiene for 45 years would cause 597 excess cancers in 10,000 workers. In fact, 1066 workers who had been exposed to levels as high as 1000 ppm since the industry began in the 1940s, had only 75% of the cancer incidence in the overall population.

The cost of overestimating cancer risks can be horrendous as unnecessary and expensive protective measures are legislated. This can drive industry to seek homes in countries with less restrictive legislation, which may then lead to loss of control over other, truly hazardous, industrial chemicals. Such venues often have much lower labor costs, exacerbating the pressure to transfer operations.

Environmental Monitoring

Environmental monitoring occurs in two ways. Ambient monitoring refers to measurements in water or air downstream or downwind from the source and is primarily a measure of the state of the environment. So-called end-of-pipe or point-of-emission monitoring refers to the measure of effluent levels from drains and stacks and is used to ensure compliance with legislative regulations.

Bioassays are used to look for effects rather than to identify specific chemicals. For water, the water flea, *Salmonid* fingerlings and the opossum shrimp are used. Earthworms and germinating plants are used for testing soil. Sensitive bacteria are used to detect mutagens. The Ames test can detect a few mutations in several million cells and a newer test, the Microtox assay, measures reduced bacterial luminescence resulting from inhibited cell division. Genetically engineered species such as nematodes are being developed to detect contamination in indoor air—the modern version of the canary in the coal mine. EC_{50}, LD_{50}, and TLV values (see the following) can be calculated for these.

Recently, the question of monitoring was raised regarding the Alberta tar sands, widely held to be the greatest, single point source of air pollution in North America and potential major source of water pollution as well. In December of 2010, The Royal Society of Canada released its expert panel report entitled "Environmental and Health Impacts of Canada's Oil Sands Industry." One of its major criticisms was the lack of adequate monitoring of air and water quality, especially with reference to downstream native communities. This topic will be dealt with in more detail in Chapter 3.

Monitoring should not be restricted to air and freshwater. In our own research we have found that benthic marine species accumulate heavy metals, which may have biological consequences for these animals (see also Chapter 3).

Setting Safe Limits in the Workplace

As noted earlier, because of data collected after industrial accidents and from pre-regulation exposures, setting acceptable limits for toxic substances in the workplace can be done with somewhat greater confidence. Exposures are generally higher but of relatively short duration as compared to those in the environment. It should be noted that tolerance limits apply only to the particular type of exposure stated (inhalation, skin contact, etc.). Jurisdictions over worker safety vary widely from country to country. Most Western, industrialized countries have similar legislation. In Canada, provincial ministries are responsible for occupational health and safety. In Ontario, this comes under the Occupational Health and Safety Act, Revised Statutes of Ontario, 1980. Regulations made under this act deal specifically with biological, chemical, and physical agents in the industrial, construction, and mining settings and, most recently, regulations governing the Workplace Hazardous Information System (WHMIS). Under federal guidelines, each province has enacted comparable legislation.

Historically, the development of safety legislation in Ontario dates back to the Ontario Factories Act of 1884. The formation of the Royal Commission on the Health and Safety of Workers in Mines in 1976 led to significant additions to legislation dealing with workplace safety. In 1979, the Occupational Health and Safety Act was proclaimed together with regulations for the industrial, mining, and construction settings. In 1984, a Royal Commission on Asbestos resulted in a further regulation under the Act in 1985. In 1981 a list of "designated substances" was begun. These are chemicals that are considered to be especially hazardous and therefore to require special controls, restrictions, or even prohibition. The regulations apply to all workplaces and other projects (except construction sites) where designated substances are likely to be inhaled, ingested, or absorbed.

The 1990 list included acrylonitrile, arsenic, asbestos, benzene, coke oven emissions, ethylene oxide, isocyanates, lead, mercury, silica, and vinyl chloride.

In 1988, the WHMIS regulations were brought into effect. These define the information that must be on labels of chemical containers in the workplace and the information that must be readily available to the worker in the form of "Material Safety Data Sheets." Ontario Bill 208 expands on the Act. Some important features of Bill 208 are as follows:

1. It is designed to work on the "Internal Responsibility System," the key aspect of which is that workers have
   a. The right to be informed about the nature of the hazards they might be exposed to in the workplace
   b. The right to participate in the decision process concerning job safety
   c. The right to refuse to work in conditions they consider to be unsafe without fear of reprisal from the employer
2. Workplaces (as defined in the Act) must form a Joint Health and Safety Committee with representation from both the employer and the employees (union) who must have at least numerically equal membership. This committee may make recommendations concerning health and safety but the employer is not bound to accept them. This is a weakness in the current system.
3. Responsibilities for safety in the workplace must be shared by the employer and the employees.
4. The definition of a worker is "a person who performs work or provides services for monetary compensation" but does not include inmates of a correctional institution, owners or occupants of a private residence or their servants, farmers, or hospital patients. Persons performing work in their homes for monetary compensation are considered to be workers.
5. The Act contains a blanket clause to the effect that "the employer must take every reasonable precaution to ensure the health and safety of the worker" where specific regulations do not exist (they do for such things as protective clothing and equipment).

Some Important Definitions

TWAEV (time-weighted average exposure value): The average concentration in air of a biological or chemical agent to which a worker may be exposed in a workday (8 h) or workweek (40 h).

STEV (short-term exposure value): The maximum concentration in air of a biological or chemical agent to which a worker may be exposed in any 15 min period. If not specifically defined in the regulations, this is taken as 3× the TWAEV for up to 30 min.

CEV (ceiling exposure value): The maximum concentration in air of a biological or chemical agent to which a worker may be exposed at any time. If not specifically defined in the regulations, this is taken as 5× the TWAEV. Levels in air are expressed as ppm or mg/m³ of air (sometimes called an excursion limit).

The NOEL and the ADI are values that are employed in animal tests.

NOEL (no observable effect level): The highest level at which no effect is observed in experimental animals. The No Observable Adverse Effect Level (NOAEL) is the exposure level at which no toxic effect is observed. ADI (acceptable daily intake): This is the NOEL divided by an arbitrary number, at least 100. Reliability depends on the use of a large number of animals to determine the NOEL. The ADI is not a predictor for carcinogenic risk. ADIs are also calculated for human exposures to determine allowable levels in air, water, and food. The Act also provides regulations governing noise exposures. Standards are different for the mining and industrial sectors.

In the United States, somewhat different abbreviations are employed, but with similar definitions. For example,

TLV (threshold limit values): These apply mostly to vapors.

TLV-C (ceiling): That concentration that should never be exceeded (=CEV).

TLV-STEL (short-term exposure limit): The exposure that can be tolerated for up to 15 min without irritation, tissue damage, or sedation (=STEV).

TLV-TWA (time-weighted average): That level to which a worker may be exposed continuously for up to 40 h/week without adverse effects (=TWAEV).

Environmental Risks: Problems with Assessment and Public Perceptions

Establishing acceptable levels of toxic substances in the environment is difficult and inexact because

1. Exposure may be for a lifetime and no data usually exist regarding long-term, very low-level exposures.
2. Toxic effects may be difficult to identify unless they are unusual. Angiosarcoma of the liver was readily associated with vinyl chloride because it is otherwise very rare. For the same reason, Kaposi's sarcoma was associated with AIDS. If harelip were the teratogenic

effect of a chemical pollutant, however, the association would be difficult because it is a fairly common congenital defect.

3. Public perceptions of risk may be exaggerated, forcing the legislation of much lower maximum allowable levels than are necessary.

4. The concept of risk–benefit analysis may seem callous and calculating to segments of the populace. In their view, there may be no such thing as an acceptable level of risk, and they frequently have an imperfect understanding of the meaning of statistical probability. Moreover, the possibility that avoiding one risk may increase another is often overlooked. The debate over the expansion of nuclear versus fossil fuel power plants is a good example of this. The elimination of nuclear power plants would force greater reliance on coal-fired generators with increases in acid rain production, release of greenhouse gases, and a greater risk of fatal mining accidents (see Chapter 13).

5. Risk factors may be additive or even synergistic, making analysis even more difficult.

Psychological Impact of Real and Potential Environmental Risks

One of the greatest adverse effects of environmental pollutants is undoubtedly psychological in nature. Human beings have a natural fear of the unknown and incomprehensible, and the field of environmental toxicology brings both of these into play. The human mind is a highly impressionable instrument. Many years ago, when one city's drinking water was first fluoridated, it was announced that the fluoride would be added on a certain date. On that date, the switchboard of the municipal offices was flooded with calls complaining about the taste of the water. It was then announced that the fluoride had actually been added a week earlier. Only then did the phone calls subside. It is well established that after every major environmental accident there is a rash of vague medical problems such as headache, nausea, dizziness, etc. Cancerphobia (fear of cancer) leads to attributing every new case of cancer in the area of exposure to the accident. Such beliefs may persist even after extensive studies have failed to reveal any difference in incidence between exposed and nonexposed populations. A sheep farmer living near a nuclear generating station may attribute every abnormal birth and fetal deformity (not uncommon in lambs) to radiation from the power plant despite evidence from extensive testing indicating that there were no radiation abnormalities on the farm.

There is no question that there are real psychological consequences to people following an environmental disaster. A community survey conducted 1 year after the 1989 Exxon Valdez oil spill reported a two- to threefold increase in symptoms such as anxiety, posttraumatic stress disorder, and depression among responders who had had a high level of exposure to the spill or to cleanup efforts. The incidence of substance abuse (including alcoholism), domestic violence, and signs and symptoms associated with chronic illnesses also increased. Those who relied on the land for their living were especially vulnerable.

The Gulf coast area had not fully recovered from the devastation of Hurricane Katrina when the fire and explosion on the Deepwater Horizon occurred. The resulting oil spill continued for weeks. After the hurricane and the consequent destruction of health care facilities, there was a significant increase in the frequency of mental health symptoms. Not long after the spill, one study found that 20% of parents reported their children experienced mental health symptoms alone or in combination with physical symptoms that they attributed to the oil spill. Factors that are believed to increase susceptibility to mental health symptoms include extent of exposure, female gender, middle age, ethnicity or minority status, preexisting mental and physical health issues, and economic and psychosocial resources. Federal and state mental health assessments are ongoing. The mental health impact and the environmental impact will both be with us for a long time.

Voluntary Risk Acceptance versus Imposed Risks

Public pressure may force the expenditure of vast sums of money to avoid risks that are practically nonexistent. Conversely, a large segment of the public may steadfastly refuse to take steps that are inexpensive and proven to reduce premature deaths. The use of seat belts in cars is a prime example of this. It has been calculated that between 1975 and 2001, seat belt usage saved 135,000 lives in the United States. Neither Canada nor the United States has federal legislation mandating seat belt use. It is left to the provinces and states to do so. Motor vehicle accidents affect one of every 50–60 persons each year. The statistical probability of incurring such an injury during one's lifetime is thus quite high, but the risk of being injured on any given trip is very low, and this is what influences the public's perception of risk. Every state except New Hampshire has a seat belt law. Recent surveys indicate that, in the absence of state laws requiring their use, only about 15% of American drivers routinely use passenger restraints despite extensive efforts at educating the public. This pattern has not changed much in the last decade. Users differed from nonusers in a number of ways. Nonusers tended to be less well educated, were more often smokers, rated seat belts as uncomfortable or inconvenient more often, often considered seat belt legislation to be an infringement of personal liberty, and less frequently knew someone who had been injured in a car accident. Resistance to government control of their lives is a frequently stated reason for not wearing seat belts. The expression "nanny state" has become popular with those opposed to legislation enforcing safety on members of the public. This pattern appears to be being repeated when the mandatory use of helmets by bicyclists is newly introduced to a jurisdiction, with indignant letters to newspaper editors protesting this latest infringement on personal freedom of choice.

A recent report evaluated similar measures for horseback riders and found that the rate of serious injury per number of riding hours was greater than for either motorcyclists or automobile racers. In a recent 2 year period there were nearly 93,000 emergency room visits in the United States for riding-related injuries with over 17,000 head and neck injuries. Competition riders are required to wear safety equipment. In North America, the National Hockey League chose to make helmets for players optional, illustrating the degree of risk elements of the public will accept in some circumstances. A rash of concussions plus newer knowledge that there can be serious consequences later in life from these has fanned debate regarding violence in hockey. Motorcyclists are equally split on the issue of helmet usage. Some claiming it is an infringement on their personal freedom despite convincing statistics that they save lives and prevent brain injury.

People tend to accept significant risks when they think they are in control as when driving a motor vehicle, but become very apprehensive when they have to give up control to others as in a commercial aircraft.

Costs of Risk Avoidance

Because it is not possible to truly "save" a life, but only to postpone death, it is common in this field to refer to "premature deaths avoided." This requires an arbitrary decision regarding normal life expectancy and what constitutes a premature death and this definition will keep changing as life expectancy is extended. About one in three people will develop cancer by age 70 as a result of the gradual deterioration in immune and cellular defenses. If a carcinogen in the environment significantly increases the incidence of cancer but mostly in the >70 segment of the population, are these premature deaths?

The following are some calculated costs, in 1990 U.S. dollars, of avoiding a premature death by instituting some simple safety and health measures (your author was unable to find more recent figures):

1. Screening and education for cervical cancer—$25,000.
2. Installation of smoke alarms in homes—$40,000.
3. Installation of seat belts in autos at time of manufacture—$30,000.
4. Stopping smoking—Actually a saving of $1000/year plus medical costs avoided. Ironically, a Canadian group calling itself the "Smokers Freedom Society," using data from 1986, claimed that smokers actually save society money by dying prematurely and costing less in pension benefits and custodial care. They failed to take into account such factors as (1) the damage done to the health of others by side stream (second-hand) smoke, (2) the cost in workdays lost because of generally poorer health (a higher incidence of respiratory infections for example), and (3) the cost in lives and property from fires started accidentally by smokers.

5. Installing fire detection and control systems in commercial aircraft cabins—$200,000.

6. In contrast, banning of diethylstilbestrol, a synthetic estrogen used as a weight-gain promoter in cattle and suspected of being carcinogenic—$132 million.

A recent example of this cost-effectiveness problem comes from the experience of some American states that introduced compulsory AIDS testing for couples applying for a marriage license. While this approach may seem somewhat draconian in the current social climate, it is by no means a new concept. Not many decades ago, most jurisdictions required a Wasserman test for syphilis before a marriage license would be granted. As a result of the mandatory AIDS tests, 160,000 tests were performed at a cost of $5.5 million; 23 subjects were identified as HIV-positive, for a cost of $239,130 each.

Cost–benefit analyses frequently involve conclusions that may be repugnant to a segment of the public. The cost of detecting one breast cancer through annual mammography in 40- to 50-year-old women has been estimated at $144,000. In the 55- to 65-year-old group, it drops to $90,000. Legislators are required to make such difficult choices because of fiscal restraints, often in the face of severe criticism.

Some Examples of Major Industrial Accidents and Environmental Chemical Exposures with Human Health Implications

Nuclear Accidents

In 1979, a serious accident occurred at the Three Mile Island Nuclear Generating Station near Middletown, Pennsylvania. This was a disaster without noise, smoke, or visible evidence of damage. The information the public received came from the press, which, early on, labeled the event a manifestation of the "nuclear disease." Over 10 years later, the cleanup was still in progress. It produced 2.3 million U.S. gallons of weakly radioactive water, mostly from tritium. This could have been discharged into the river without exceeding federal standards, but public pressure forced the installation of a special evaporator at a cost of $5.5 million. The calculated cost of avoiding one premature death is as follows: The maximum exposure from the river discharge would have been 2 microrems (μrems). This is equivalent to a 4 min exposure to natural source radiation such as radon or cosmic radiation. The collective dose (mean exposure × number of persons exposed) would have been about 1 person-rem (prem). The calculated incidence of cancer for radiation exposure is 1/5000 prem. The computed cost of avoiding one premature death is about $25 billion.

The Chernobyl disaster will be discussed in Chapter 12. Suffice it to say that it remains the most catastrophic nuclear accident to date. In 1966, the Fermi 1 plant suffered a partial meltdown, although no release of nuclear radiation occurred. The plant shut down for extensive repairs and resumed operation but never reached its peak output again. It was decommissioned in 1975.

On March 11, 2011 a massive earthquake, 9.0 on the Richter scale, and the resulting tsunami struck Japan killing thousands and flooding a nuclear power plant, causing a partial meltdown and the threat of further nuclear disasters. To date the Japanese have been able to contain the threat and no radiation health issues occurred amongst the public. Large-scale evacuations were conducted around the plant.

Formaldehyde

In the late 1970s, a report was published indicating that formaldehyde fumes caused nasal cancer (squamous cell carcinoma of the mucosa) in rats and in one strain of mice. No evidence of carcinogenesis in hamsters could be shown and there were no human studies suggesting a carcinogenic potential for formaldehyde. This type of tumor is rare in people and no evidence of increased incidence in workers exposed to high levels of formaldehyde could be found. The EPA used the linear extrapolation multistage method to evaluate risk. The initial evaluation (1981) was overturned by a federal court on the grounds that there were not enough data to determine risk. Legal wrangles between the EPA and the chemical industry continued for the next few years. Urea formaldehyde foam insulation was banned in the United States and Canada and many homes were ripped apart and the insulation removed. Homeowners agitated for government subsidies to finance this. Over the next few years additional epidemiological data accumulated on populations deemed to be exposed to higher-than-average levels of formaldehyde. Whereas people exposed to 0.4 ppm (medical, dental, nursing, and science students) had cancer incidences little different from the general population, pathologists and funeral service workers exposed to 3.0 ppm had almost 2,000 additional cases of cancer/100,000 over the expected frequency. In 1987, the EPA defined formaldehyde as a probable human carcinogen. The EPA set the TWAEV at 1 ppm and the short-term exposure level at 2 ppm. An action level of 0.5 ppm was set. Above this, regulations come into force governing such things as medical surveillance, protective equipment, and training. Most people can detect formaldehyde by odor at 0.5 ppm. For example, the characteristic smell of new carpeting is due to formaldehyde used in the adhesive. Particleboard also contains it in the binding adhesive. The smell of formaldehyde is rarely if ever detectable in homes insulated with urea formaldehyde insulation, which releases it as the foam slowly breaks down. Levels are unlikely to reach the "action level" set by the EPA for industry, especially if ventilation is adequate, except in new homes where levels up to 1 ppm have been measured due to its release from building materials and carpeting. The Canadian government, however, subsidized removal of such insulation to the tune of $10,000 per home. An important source of toxic aldehydes as a potential health hazard is cigarette smoke. Acrolein is much more irritating than formaldehyde and the industrial TLV is set at 0.1 ppm. It is a major contributor to the irritant properties of cigarette smoke and photochemical smog.

Dioxin (TCDD)

On July 10, 1976, a serious explosion and fire occurred at a chemical plant near Milan, Italy. As a result, over 1 kg of TCDD (tetrachlorodibenzo-*p*-dioxin, the most toxic of the dioxins) was spread over the adjacent countryside. The chemical fallout was heaviest in the town of Seveso, where concentrations in some parts reached 20,000 μg/m^2 of surface area. By the end of July, 753 people were evacuated from the area; 3,300 animals died and 77,000 were killed, 500 people were treated for acute skin irritation, and 192 eventually developed chloracne. There also was evidence of liver damage as indicated by increases in serum enzyme levels. Several follow-up studies were reported, the most recent in 1996, 20 years later. The chloracne was completely reversible. All but one had recovered by 1983. Enzyme levels also returned to normal, and no other short-term abnormalities were reported. Because TCDD has been shown to be carcinogenic in animals, great concern centered on evidence of cancer in the victims. The 20-year study showed a slight but significant increase in total cancers in the most heavily exposed cohort with a noteworthy increase in lymphatic and hematopoietic cancers. However, a review of epidemiologic studies on TCDD and cancer published in 2011 did not confirm this in the most recent follow-up. These authors felt that there was no conclusive evidence linking TCDD exposure and cancer and that the use of total cancer statistics was not justified epidemiologically.

Agent orange, used to defoliate jungles in the Vietnam war, is a mixture of the herbicides 2,4-D and 2,4,5-T. The toxic ingredient is 2,3,7,8-tetrachlorodibenzo-*p*-dioxin (TCDD or dioxin). Questions about carcinogenicity emerged early on as veterans who had been doused by aerial spray returned home with numerous health problems. The American Cancer Society's website summarizes the existing evidence, both pro and con, and the best that can be said is that there is a high index of suspicion that dioxin is carcinogenic but that overwhelming evidence remains elusive. This topic will be dealt with in detail in Chapter 5.

But military use was not the only way in which TCDD was applied. In the 1950s, 1960s, and 1970s, an identical mixture was used to clear massive tracts of Crown (government) land in Ontario to get rid of unwanted broadleaf trees and allow the unimpeded growth of more economically desirable black spruce trees. The Toronto Star revealed recently that forestry workers, often Junior Rangers and high school and university students spent hours acting as human markers holding colored helium balloons to guide the spraying aircraft. Participants, who were drenched with spray, are now scattered across Canada and reports of cancer showing up are beginning to emerge. Non-Hodgkin's lymphoma has been found and this tumor has previously been associated with dioxin. Both the province and timber companies conducted aerial spraying operations. The Star's investigation included interviews with ailing former workers now scattered across Canada. The provincial government is now investigating and the Chief Medical Officer of Health of Ontario has been notified. The debate continues.

Some Legal Aspects of Risk

De Minimis Concept

Recently, courts in the United States and Great Britain have begun to apply an old legal tradition to the question of risk. This is the concept of "de minimis non curat lex," which means "the law does not concern itself with trifles." Applied to risk analysis, the "de minimis concept" means that in some cases the computed risk is so small that it does not justify regulation. For example, the U.S. Supreme court ruled in 1980 that the Occupational Safety and Health Administration (OSHA) could not further limit levels of benzene fumes in the workplace unless it could demonstrate significant risk to workers from existing exposure levels. Similarly, the appeals court of D.C. ruled that regulations governing plastic containers could not be introduced on the purely theoretical grounds that toxic substances might leach into the contents.

In 2007, the State of California introduced an emergency aerial spraying program to control the light-brown apple moth (LBAM) using a pheromone that interferes with the sexual cycle of the moth. Spraying occurred in Monterey and Santa Cruz counties, around Monterey Bay. It involved urbanized, populous areas, upsetting residents concerned about its safety. Attempts by citizen's groups in both counties to obtain court injunctions to halt the spraying were initially unsuccessful because according to California law at that time, definitive evidence of harm to people was required before spraying could be halted. This is another example of the de minimis concept at work. The question of aerial spraying over residential areas will be dealt with in more detail in Chapter 4. Paradoxically, this same state (California) has introduced some of the strictest laws regulating the presence of chemicals thought to be toxic or carcinogenic in products. These include

- A ban (2011) on the endocrine disrupter bisphenol-A (BPA) in plastic baby bottles and "sippy" cups (this is contrary to the aforementioned 1980 federal de minimis ruling)
- A ban (2008) on phthalates in flexible plastic toys intended for children under three

- A ban (2003) on two forms of flame-retardant chemicals, polybrominated diphenyl ethers
- A ban on lead (2006) and cadmium (2010) in children's jewelry

In 2010, Canada became the first country to ban BPA in baby bottles. It appears that the de minimis concept is giving way to the precautionary principle, i.e., if there is a reasonable suspicion that a substance or practice might be harmful to the public, regulatory action may be taken in the absence of definitive evidence of harm. The subject of endocrine disrupters will be dealt with in detail in Chapter 12.

Delaney Amendment

The Delaney Amendment to the U.S. Food and Drug Act has had a significant impact on worldwide public attitudes about cancer risks and on the responses of politicians to those attitudes.

Passed in its original form in 1958 it stated that, "No additive shall be deemed safe if it is found to induce cancer when ingested by man or laboratory animals or if it is found, after tests which are appropriate for the evaluation of the safety of food additives, to induce cancer in man or animals." The amendment was originally interpreted to include pesticide traces even though these are not truly food additives. The increasing sensitivity of test methods and the ubiquitous use of pesticides in agriculture led to attempts to have this interpretation changed, a move opposed by citizen and environmental groups. In 1996, however, the Food Quality Protection Act reversed this interpretation to exclude pesticides.

The U.S. Food and Drug Administration sought to apply the de minimis principle to two food colors that had been shown, in very high doses, to induce cancer in animal tests. A review court ruled against the FDA on the grounds that Congress intended the Delaney clause as an absolute prohibition against carcinogenic dyes in foods. Other authorities, however, are stating that it is time to reassess the Delaney clause in light of the extreme sensitivity of current methods of detecting impurities in food. In all areas of toxicology, the sensitivity of test methods has outstripped our knowledge of the significance of such low-exposure levels for human health.

Cancerphobia, an unreasonable fear of developing cancer, has spawned a new phenomenon; suing out of fear of developing the disease in the absence of any signs or symptoms. Some litigants are winning. In 1989 residents of Toone, Tennessee accepted an out-of-court settlement from the Velsicol Chemical Co. of $9.8 million because they had been exposed to contaminated drinking water as a result of corroding barrels in a toxic dump site owned by the company.

Early in 1989, the Natural Resources Defense Council, an American environmental activist group, published a report claiming that children who consume large amounts of apples and apple products could be at increased risk of cancer. Daminozide (Alar) was used to prevent premature windfalls of a variety of fruit. A breakdown product of daminozide, unsymmetrical dimethylhydrazine (UDMH), has been shown to be carcinogenic in animals. After some initial controversy, all Western countries have banned the use of daminozide for all food crops. The manufacturer had already withdrawn the product voluntarily. Canada has a maximum residue level (MRL) of 0.02 ppm. Alar is still allowed for ornamental plants and the MRL for foods recognizes that some cross-contamination could occur.

Statistical Problems with Risk Assessment

In toxicity studies, to reach the 95% confidence limit, one must have 12% responders in a group of 50 subjects, 30% in 20 subjects, and 50% in 10 subjects. To have a chance of seeing one single case in a population at risk, one must test a population three times as large. In other words, to detect an incidence of 1/100, one must test 300 subjects, or to detect 1/1000, one must test 3000, etc. This is an extremely important concept when dealing with toxic reactions that are not dose dependent. Aplastic anemia from bone marrow depression occurs very rarely in response to some chemicals. For example, the antibiotic chloramphenicol is estimated to cause this in one of 35,000–50,000 treated patients. Even assuming that the same genetic predisposition existed in a test animal as it appears to do in humans, one would have to test 150,000 animals in order to see one case. The effects of high doses, moreover, cannot always be extrapolated to low-dose situations, even if the effect is dose dependent. The time available for the repair of chemically damaged DNA may be adequate at very low doses, so that the defect is not expressed.

Risk assessment is thus a very inexact process when applied to low-risk situations relating to environmental pollutants. The public tends to overestimate risk and underestimate or ignore avoidance costs. Vested interests tend to do the reverse, and politicians are influenced sometimes by the most powerful lobby, which may be either an industry or an environmental organization. Before the Challenger disaster, NASA estimated the risk of a shuttle accident to be 1/100,000. Empirical data now suggest that the real risk was 1/25. Unconscious bias, small sample size, lack of human data, failure to consider interactive factors such as food chain biomagnification, public pressure resulting from unjustified fears, all may affect the decision process. It is a characteristic of human nature that people will accept significant levels of risk if they can exert some personal choice in the situation, but they will almost universally reject even the slightest degree of a risk if it appears to be imposed upon them by government or industry. Examples of voluntary risks discussed earlier include smoking, not wearing seat belts, not wearing protective helmets and, of course, there is a whole range of hazardous sports that involve considerable risk to life and limb. The question of relative risk assumes great significance when comparing various sources of energy. Is nuclear energy inherently more dangerous than that from coal-fired generators? This subject will be considered further in Chapters 4 and 13.

Risk Management

The Royal Society of Canada and the Canadian Academy of Engineering have formed a Joint Committee on Health and Safety. In 1993 it released a report entitled "Health and Safety Policies: Guiding Principles for Risk Management."

These principles can be summarized as follows:

1. Risk should be managed so as to maximize the net benefit to society as a whole.
2. The reference to net benefit recognizes the need to evaluate any negative aspects of an action, including cost, i.e., to conduct a cost–benefit analysis.
3. The desired benefit is quality-adjusted life expectancy (QALE). In other words, the determination of benefit must include an evaluation of how the quality of life has been improved, not just its length.
4. Health and safety decisions affecting the public must be open to public scrutiny and applied across the complete range of risks.

A number of problems are recognized in the report, some of which have been discussed earlier. Some individuals may be affected adversely by an action that benefits the majority of society. Jobs lost as a result of the closure of a polluting industrial site is an obvious example. The battle between environmental groups and the Canadian government over development of the Alberta tar sands has seen the government play the job creation card repeatedly. Measures taken to reduce one risk may increase another risk. (The debate over nuclear versus fossil fuel power generators is ongoing.)

A significant problem is how the public perceives a risk and how it may influence politicians to allocate huge sums to deal with minuscule risks. This has been touched on earlier but the Joint Committee report makes an important, additional point. Namely, that prevention of accidental death and injury is one of the most cost-effective risk management actions that can be taken, yet much regulatory effort has been directed at controlling occupational, environmental, and dietary cancer risks and such regulations are extremely costly. It is evident that, in keeping with the aforementioned third principle, much work needs to be done to educate the public about the differences between real and perceived risks.

Efforts have been made to devise systems that allow comparisons of risks. In 1993 the EPA introduced the IRIS method, Integrated Risk Information System. IRIS is an EPA database, updated monthly, that contains EPA's consensus positions on potential adverse effects of 500 substances. It contains information on hazard identification and dose–response evaluation, the first two steps in risk assessment (followed by exposure assessment and risk characterization). An IRIS chemical file may contain any or all of the following: an oral reference dose; an inhalation reference concentration; risk estimates for carcinogens; drinking water health advisories; U.S. EPA regulatory action summaries; and supplementary data.

$$\text{Oral reference dose (RFD)} = \frac{\text{NOAEL (mg/kg/day)}}{\text{UF} \times \text{MF}}$$

where the UF is an uncertainty factor used to allow for the variability in species and in the extrapolation to humans and the MF or modifying factor, which is essentially a fudge factor used to cover the possibility of unknown variables.

The Inhalation Reference Concentration is the same except that the NOAEL is multiplied by the Human Equivalent Concentration taken from industrial exposures where data exist. The Federal Register Notice 58 FR 11490 was first published on February 25, 1993.

Information regarding IRIS and its database can be found on the Internet at http://toxnet.nlm.nih.gov

---

Precautionary Principle

The precautionary principle advocates the "better safe than sorry" attitude toward risk management. The principle has been stated in many ways. Principle 15 of the 1992 Rio Declaration on Environment and Development, a heavily negotiated example of it, states "Where there are threats of serious or irreversible damage, lack of full scientific certainty shall not be used as a reason for postponing cost-effective measures to prevent environmental degradation." Although this expression relates to the environment, similar statements have been applied to potential threats to human health.

Implicit in all such statements, although not always expressed, is the concept that the cost of corrective or preventative measures must be commensurate with the estimated degree of risk associated with the potential hazard. Two uncertainties thus attend every attempt to apply the principle: (1) How great is the risk? (2) How much will it cost to reduce it?

There have been calls to reform or refine the existing expression of the Precautionary Principle (PP). In the past, legislative action has been taken based on the PP and subsequent research has shown the action to be unnecessary. Saccharin, for example, is no longer considered to be carcinogenic for humans because of newer and better scientific data. A particular risk also may have attending benefits. Moderate consumption of alcohol reduces the risk of heart attack but slightly increases the risk of hemorrhagic stroke. Consumer and environmental advocates may tend to take a one-sided view of such situations. It appears that the PP should be applied to the introduction of precautionary regulation. Vested financial interests may take the stance that if the PP is applied too rigorously, little progress would be possible. Risk taking, after all, is an integral part of development.

Case Study 1

A population of miners has been identified as having a high incidence of a rare form of cancer. It accounted for 18% of all deaths in workers who had been exposed 20–30 years previously. Those living close to the mine, and family members of miners, regardless of the location of their home, also had an elevated incidence of this same tumor (2%–3% of all deaths). The incidence of death from this cancer in the general population is only 0.0066% of all deaths. The incidence of lung cancer in miners who smoke is 60× the incidence in miners who do not smoke. In the general population, smoking increases the risk of lung cancer by about 20×.

Q. What factors, including biological and social, could account for this distribution of the rare cancer?

Q. Why is the incidence of lung cancer so much higher in the smoking miners?

Q. Why should family members of miners who live some distance from the mine have an elevated incidence of the rare cancer?

Case Study 2

Five men were employed in plugging leaks in, and waterproofing, an underground storage tank using epoxy resin paint. The tank contained several inches of water, so the ventilating fans that were available were not used for fear of electrocution. The tank measures about 20 yards/m long by 6 yards/m wide and was 3 yards/m deep. It was divided into three connecting compartments and a single ladder and hatch provided the only means of entry and egress. At 10:00 a.m. one worker left the tank because he was drowsy and nauseated. On reaching the surface, he vomited. At 11:30 another man left the tank to get a coffee. When both men returned, they found the remaining three men dead in the tank.

Q. What violations of safe working procedures occurred here?

Q. What safety measures could have been instituted to prevent these deaths?

Q. What can you say about the source and nature of the toxic substance involved in these deaths?

For both of the above case studies, apply common sense and some of the information provided in this chapter.

Review Questions

For Questions 1–6 use the following code:

Answer A if statements a, b, and c are correct.

Answer B if statements a and c are correct.

Answer C if statements b and d are correct.

Answer D if only statement d is correct.

Answer E if all statements (a, b, c, d) are correct.

1. Which of the following factors contribute to the degree of risk of a toxicant in the environment?
   a. The biological half-life of the substance
   b. The partition coefficient of the substance
   c. Its toxicity
   d. The level of exposure likely to occur
2. Which of the following statements is/are true?
   a. Individual susceptibility to a toxicant may vary considerably
   b. The prediction of risk associated with very low levels of exposure can be done with reasonable accuracy

   c. Mechanistic models assume that cancer can arise from a single, mutated cell

   d. Distribution models assume that there is no threshold below which a cancer-causing agent will induce tumor formation

3. Which of the following statements is/are true?

   a. The carcinogenicity of a substance is not affected by the portal of entry

   b. Toxicity studies are generally conducted using a single portal of entry

   c. The age of the population likely to be exposed to a cancer-causing agent does not affect the degree of risk

   d. Cancer incidence data acquired from accidental or industrial human exposures are of greater use for predicting risk in the population at large, if exposed to the same chemical, than animal data

4. Which of the following statements is/are true?

   a. A population of workers may not be representative of the population at large

   b. The DNA of other mammals differs by 50% compared to human beings

   c. Infants generally absorb chemicals through the skin much more efficiently than adults

   d. Smoking habits have no bearing on cancer incidence due to other carcinogens

5. Safety in the workplace requires

   a. A means of measuring levels of potentially hazardous substances in the work environment

   b. A knowledge of the toxicity of the substance

   c. A knowledge of the influence of level and duration of exposure on risk

   d. A knowledge of the effectiveness of appropriate safety measures such as respirators

6. Hormesis

   a. Involves a U-shaped dose–response curve

   b. Always refers to beneficial effects at the low dose

   c. Is not yet incorporated into risk assessment models

   d. Has been clearly proven in human exposures

7. Match the appropriate acronym to the definitions given below.

   a. STEV

   b. TWAEV

   c. NOAEL or NOEL

   d. CEV

    i.   The average concentration, in air, of a toxicant to which a worker may be exposed during an 8 h day or a 40 h week

    ii. The level of exposure at which no (adverse) effect is observed

    iii. The maximum concentration, in air, of a toxicant to which a worker may be exposed at any time

    iv. The maximum level of a toxicant in air to which a worker may be exposed in any 15 min period

For Questions 8–25 answer true or false

8. Industrial exposure to vinyl chloride has been associated with an increased incidence of angiosarcoma.

9. About one in three North Americans will develop some form of cancer by age 70.

10. The "de minimis" concept means that no level of risk from industrial pollutants is acceptable.

11. The cost of one "premature death avoided" from installing smoke alarms in homes is about $1000.00.

12. Most people can detect formaldehyde by odor at a concentration in air of 0.5 ppm.

13. The main toxic effect related to the dioxin accident at Seveso has been chloracne.

14. The rat stomach is a good model of the human one.

15. The skin of the forehead absorbs chemicals much more efficiently than the skin of the forearm.

16. The acceptable level of risk from environmental anthropogenic chemicals is defined by the EPA as one additional cancer per 1,000,000 population.

17. There are no natural carcinogens of great concern.

18. Patulin is a carcinogenic mycotoxin.

19. Highly lipid-soluble toxicants with long $T1/2$ values tend to concentrate up the food chain.

20. Skin does not have any biotransforming properties.

21. By sloughing our skin cells we may also eliminate some carcinogens that they have accumulated.

22. Daminozide is widely used now to prevent premature windfall of fruit.

23. Bisphenol-A (BPA) is banned for use on food crops in Canada.

24. The Delaney amendment now excludes pesticides from its included substances.

25. To detect a toxic reaction at an incidence of 1/3000 one would have to test at least 3000 test animals.

Answers

1. E
2. A
3. C
4. B
5. E
6. B
7. i = b, ii = c, iii = d, iv = a
8. True
9. True
10. False
11. False
12. True
13. True
14. False
15. True
16. True
17. False
18. True
19. True
20. False
21. True
22. False
23. True
24. True
25. True

Further Reading

Abelson, P.H., Testing for carcinogenicity with rodents (Editorial), *Science*, 249, 1357, 1990.

Abelson, P.H., Exaggerated carcinogenicity of chemicals (Editorial), *Science*, 256, 1609, 1992.

Agent Orange and Cancer, American Cancer Society, http://www.cancer.org/Cancer/CancerCauses/IntheWorkplace/agent-orange-and-cancer (accessed on April 20, 2012).

Aguilera, F., Mendez, J., Pasaro, E., and Laffon, B., Review on the effects of exposure to spilled oils on human health, *J. Appl. Toxicol.*, 30, 291–301, 2010.

Ames, B.N. and Gold, L.S., Pesticides, risk, and applesauce, *Science*, 244, 755–757, 1988.

Ames, B.N. and Gold, L.S., Too many rodent carcinogens: Mitogenesis increases mutagenesis, *Science*, 249, 970–971, 1990.

Assennato, G., Cervino, D., Emmett, E.A., Longo, G., and Merlo, F., Follow-up of subjects who developed chloracne following TCDD exposure at Seveso, *Am. J. Ind. Med.*, 16, 119–125, 1989.

Automobiles (Encyclopedia of everyday law), Seat belt usage, http://www.enotes.com/automobiles-reference/seat-belt-usage (accessed on April 21, 2012).

Boffetta, P., Mundt, K.A., Adami, H.O., Cole, P., and Mandel, J.S., TCDD and cancer: A critical review of epidemiologic studies, *Crit. Rev. Toxicol.*, 41, 622–636, 2011.

Carcinogen list, IARC and NTP carcinogen list, http://naturalrussia.com/pdfs/Carcinogen_list.pdf (accessed on April 14, 2012).

Cogliano, V.J., Farland, W.H., and Preuss, P.W., Carcinogens and human health: Part 3, *Science*, 251, 607–608, 1991.

Covello, T., Flamm, W.G., Rodericks, W.V., and Tardiff, R.G. (eds), *The Analysis of Actual vs. Perceived Risks*, Plenum Press, New York, 1981.

Environment California, Toxic-free future: Toxic chemicals in our bathrooms, kitchens, living rooms and bedrooms, http://www.environmentcalifornia.org/programs/green-chemistry (accessed on April 09, 2012).

Environmental and health impacts of Canada's oil sands industry, Report of the expert panel of The Royal Society of Canada, December 2010, http://www.rsc-src.ca

Fillion, M., Getting it right: BPA and the difficulty proving environmental cancer risks, *J. Nat. Cancer Inst.*, Epub ahead of publication, April 18, 2012, http://jnci.oxfordjournals.org

Flamm, W.G., Pros and cons of quantitative risk analysis. In *Food Toxicology: A Perspective on the Relative Risks*, Taylor, S.L. and Scanlon, R.A. (eds.), Marcel Dekker, Inc., New York, 1989, Chap. 15, pp. 429–446.

Formaldehyde-Council on Scientific Affairs (AMA) Report, *JAMA*, 261, 1183–1187, 1989.

Graham, J.D., Making sense of the precautionary principle, *Risk Perspect.*, 76, 1–6, 1999.

Goldfarb, B., Beyond reasonable fear, *Health Watch*, September/October, 14–18, 1991.

Ha, M., Lee, W.J., Lee, S., and Cheong, H.K., A literature review on health effects of exposure to oil pill, *J. Prev. Med. Public Health*, 41, 345–354, 2008.

Infante, P.F., Prevention versus chemophobia: A defence of rodent carcinogenicity tests, *Lancet*, 337, 538–540, 1991.

Joint Committee on Health and Safety (Royal Soc. Can. & Can. Acad. Eng.), *Health and Safety Policies: Guiding Principles for Risk Management*, Institute for Risk Research, Waterloo, Ontario, Canada, 1993.

Marshall, E., A is for apple, Alar, and … alarmist? News and comment, *Science*, 254, 20–22, 1991.

Marx, J., Animal carcinogen testing challenged, *Science*, 250, 743–745, 1990.

Osofsky, H.J., Palinkas, L.A., and Galloway, J.M., Mental health effects of the Gulf oil spill, *Disaster Med. Public Health Preparedness*, 4, 273–276, 2010.

Palinkas, L.A., Petterson, J.S., Russell, J., and Downs, M.A., Community patterns of psychiatric disorders after the Exxon Valdez oil spill, *Am. J. Psychiatry*, 150, 1517–1523, 1993.

Perera, F.P., Carcinogens and human health: Part 1 (letter), *Science*, 250, 1644, 1990.

Pesatori, A.C., Consonni, D., Rubagotti, M., Grillo, P., and Bertazzi, P.A., Cancer incidence in the population exposed to dioxin after the "Seveso accident": Twenty years of follow-up, *Environ. Health*, http://www.ehjournal.net/content/8/1/39 (accessed on April 16, 2012).

Rall, D.P., Carcinogens and human health: Part 2 (letter), *Science*, 251, 10–11, 1991.

Safe Chemicals Act of 2011, http://www.saferchemicals.org

Stone, R., New Seveso findings point to cancer, *Science*, 261, 1383, 1993.

Toxicity Testing in the 21st Century: A Vision and a Strategy, Committee on Toxicity Testing and Assessment of Environmental Agents, National Research Council, National Academic Press, Washington, DC, 2007.

Weinstein, I.B., Mitogenesis is only one factor in carcinogenesis, *Science*, 251, 387–388, 1991.

Weinstein, N.D., Optimistic biases about personal risks, *Science*, 246, 1232, 1989

Zeckhauser, R.J. and Viscusi, W.K., Risk without reason, *Science*, 248, 559–564, 1990.

Ziomislic, D., Star exclusive: Agent Orange "soaked" Ontario teens. www.thestar.com/news/canada/article/940243—star-exclusive—agent-orange-soaked-ontario-teens (accessed on April 26, 2012).

3

Water and Soil Pollution

Hang your clothes on a hickory limb but don't go near the water.

Introduction

Three components of the biosphere, soil, air, and water, can serve as toxicological sinks. These are often considered separately, but it should be obvious that they function as an integrated system. Thus, rain will transfer toxicants to soil and water, and evaporated surface water, and soil as airborne dust, can move them back into the air, where they may be transported over great distances by wind. Moreover, runoff from the soil, sewage, and industrial discharge are the main sources of water contamination. Seepage into deep aquifers from soil and surface water also may occur, and freshwater reservoirs are connected to the sea by rivers and estuaries. Thus, while this chapter tends to focus on water, both as an essential resource for human consumption and as marine and aquatic ecosystems, this should not detract from an understanding of the integrated nature of the biosphere. Soil often becomes the repository for our most toxic waste products and the consequences of this are touched upon later in this chapter. Chemicals may also enter foodstuffs grown in contaminated soil, and the spraying of crops with pesticides has been a matter of considerable public concern.

Water pollution is of considerable importance for several reasons. The most obvious is the possibility that xenobiotics may enter drinking water supplies and constitute a direct threat to human health. The contamination of fish and shellfish obtained both from the sea (marine organisms) and freshwater lakes and rivers (aquatic organisms) may further threaten human health when these foods are consumed. Larger (and older) fish often have higher levels of lipid-soluble toxicants, but younger ones have higher metabolic rates and may concentrate them more quickly.

Many toxicants are taken up initially by unicellular organisms that serve as a food source for larger (but still microscopic) ones, which in turn are food for bigger ones and so on. This process can lead to increasingly higher concentrations progressing up the food chain, and this is called *biomagnification*. Freshwater and marine organisms are themselves vulnerable to toxicants that may threaten their survival. Toxicants can shift the selection advantage for a species, so hardier ones may proliferate to the detriment of others. A classical example of this is the process known as *eutrophication*, which results when excessive phosphate and nitrate levels in water develop from fertilizer runoff from farmlands and from sewage effluent containing detergents. The high phosphate and nitrate levels favor the growth of certain algae and bacteria that bloom extensively and consume available oxygen until there is not enough to support other life-forms. Sunlight also will be blocked out, further altering the nature of the ecosystem.

Factors Affecting Toxicants in Water

All natural water contains soil and all soil contains water, but there is considerable variation in the mix. In fact, it is necessary to distinguish among various types because the behavior of pollutants differs in them. Moreover, the nature of the water itself may vary with regard to hardness, pH, temperature, and light penetration with consequences for the fate of pollutants. These modifying factors will be considered in more detail next.

Exchange of Toxicants in an Ecosystem

Figure 3.1 is a schematic representation of a body of water showing sources of contamination (rain, runoff, effluent discharge, percolation through soil) and some of the means of transferring toxicants to aquatic organisms. Of particular note is the layer of soil/water mix at the bottom interface. This is described as the "active sediment" and it contains, at the surface, a layer of colloidal particles suspended in pore water. The sediment contains organic carbon that tends to take up lipophilic substances. An equilibrium state is thus established with the pore water, which is in equilibrium with the body of water itself. The active sediment is a rich environment for many forms of aquatic life, and particle feeders may concentrate toxicants from the suspended particles, whereas filter feeders will do so from the pore water. Dilution of the toxicant in the principal body of water will shift the equilibrium and release more from the bound state. Thus, removing a source of contamination may not be reflected in improved water quality for some time. The active sediment can thus be both a sink and a source of toxicants (see next).

Limnologists accustomed to working in streams and small lakes encounter the floccular, low-density active sediment in a depth of only a few centimeters. Commercial divers on the Great Lakes, however, describe the phenomenon of sinking up to their helmet top in soft, bottom sediment; a disturbing sensation when first experienced.

Figure 3.1 shows that predatory fish (piscivores) may biomagnify a toxicant by consuming smaller, less contaminated species. Older and larger fish will

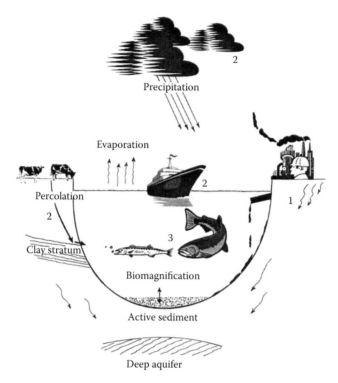

FIGURE 3.1
Dissemination of toxicants within an ecosystem.

generally have higher levels than smaller ones and fish that exist primarily on zooplankton will have the lowest levels of all. This is clearly illustrated by measurements of methylmercury in fish taken from the Wabigoon–Grassy Narrows river system in Northern Ontario (see Chapter 3, *The Grassy Narrows story*).

Factors (Modifiers) Affecting Uptake of Toxicants from the Environment

Modifiers are classified as *abiotic* (not related to the activity of life-forms) or *biotic* (related to the activity of life-forms).

Abiotic Modifiers

Abiotic modifiers include the following:

1. *pH.* As is the case in any solvent/solute interaction, the pH of the solvent will affect the degree of ionization (dissociation) of the solute. Since the nondissociated form is the more lipophilic one, this will influence uptake by organisms. The wood preservative

pentachlorophenol, for example, dissociates in an alkaline medium so that in theory at least, acid rain would increase its bioavailability by favoring a shift to the lipophilic form. Copper, which is very toxic to fish and other aquatic life forms, exists in the elemental cuprous (Cu^{2+}) form at more acidic pH, but as less toxic carbonates at about pH 7. Toxicity to rainbow trout decreases around neutral pH. An important aspect of pH concerns the methylation of mercury by sediment microorganisms. This occurs over a narrow pH range and is a detoxication mechanism that allows the microorganisms to eliminate the mercury as a small complexed molecule. About 1.5% per month is thought to be converted under optimal conditions (pH 7) (see also Chapter 6).

2. *Water hardness.* Carbonates can bind metals such as cadmium (Cd), zinc (Zn), and chromium (Cr), rendering them unavailable to aquatic organisms. Of course, equilibrium will be established between the bound and the free forms so that removal of the dissolved copper will cause the carbonate to give up some of its copper. There also is an intimate interaction between hardness and pH, so that the lethality curve for rainbow trout will be bimodal at a given degree of hardness, with dramatic increases in the LC_{50} (lethal concentration 50%) at pH 5 and 8. It should be noted that Canadian Shield lakes tend to be soft because they do not receive drainage from limestone. Figure 3.2 illustrates this relationship for copper toxicity at a single degree of hardness.

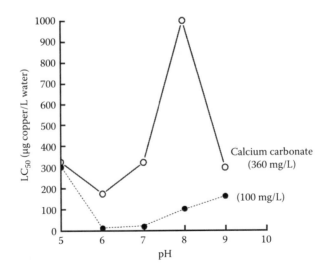

FIGURE 3.2
Effects of calcium carbonate and pH on copper lethality in trout. Bimodal lethality curve for copper in rainbow trout showing the influence of pH and water hardness.

3. *Temperature.* Apart from a few mammals, aquatic and marine species are poikilotherms, so that water temperature greatly affects their metabolic rate, which in turn will be reflected in the circulation time of blood through gills, the activity of transport processes and hence the rate of uptake of xenobiotics. Rates of biotransformation and excretion also may be affected. Temperature also will affect the rate of conversion of mercury to methylmercury.

4. *Dissolved organic carbon.* These will complex with a variety of lipophilic toxicants and serve as a sink for contaminants in sediment and suspended particles. Again, an equilibrium state will exist and if dissolved toxicants are removed or diluted, more will be released from the sink. Sediment typically consists of inorganic material (silt, sand, clay) coated and admixed with organic matter, both animal and vegetable, living and dead.

5. *Oxygen.* As noted earlier, oxygen depletion by algae blooms will compromise other life-forms that may be involved in processes of toxification or detoxification including the microbes that form methylmercury.

6. *Light stress (photochemical transformations).* Ultraviolet radiation may induce chemical changes in contaminants that may result in more toxic forms of a chemical. Thus, photo-oxidation can increase the toxicity of polycyclic aromatic hydrocarbons (PAHs) through the formation of highly reactive free radicals. In clear water, this effect can be significant at a depth of 6 m and it can have a marked impact on levels of toxicants.

Biotic Modifiers

These are similar to the factors that may affect a patient's response to a drug.

1. *Age.* Old trout are less sensitive than fry to some toxicants; larval forms of aquatic organisms usually differ metabolically from adult forms and may concentrate or metabolize toxicants differently.

2. *Species.* Many differences exist regarding species sensitivity to toxicants. Salmonid species are generally more vulnerable than carp, which can exist under a wider variety of environmental conditions. Disturbances of the natural ecosystem by the introduction of foreign species can have drastic consequences. The Great Lakes are especially vulnerable to the effects of introduced species since the introduction of the system of locks of the St. Lawrence Seaway connecting them with the sea. The Chicago Ship Canal also connects the Great Lakes with the Mississippi River. The question of invasive species in the Great Lakes is a very important one and will be considered separately later.

3. *Overcrowding.* This constitutes an additional stress factor that can influence responses to toxicants.

4. *Nutrition.* The level of nutrition will affect such factors as depot fat (an important storage site for lipophilic toxicants) and the efficiency of detoxifying mechanisms. The nutritional state in turn may be affected by abiotic factors.

5. *Genetic variables.* Unidentified genetic variables are undoubtedly at work, influencing the response of individuals to xenobiotics.

Invasive Species

Although not toxicants, invasive species can have a devastating effect on an ecosystem. They can crowd out native species, deplete food stocks in the food web and, if predatory, devour larger natives. They may also change the habitat by eroding river banks, changing the turbidity of water, and allowing the uncontrolled growth of some plants. The Florida Everglades have had a long-standing, often losing battle against them. Their subtropical climate and wetlands nature make them hospitable to a wide range of invaders, often released by exotic pet owners who no longer want them. The list included Burmese pythons, monitor lizards, nutria, about 100 plants, 15 insects, as well as snails and parasites. More information may be found at http://threatsummary.forestthreats.org.

The Great Lakes also have proven to be a suitable environment for numerous invasive species. Some made their way to the lakes by swimming up the canal system from the Atlantic and also via the Mississippi. The list includes the following:

1. The lamprey eel (*Petromyzon marinus*) arrived in the nineteenth century and is still a major problem for sport and commercial fisheries. It has a sucker-like mouth with a rasp surface and feeds on the blood of larger fish such as lake trout.

2. The alewife (*Alosa pseudoharengus*) is a small coarse fish that died by the thousands and fouled the beaches of Lake Huron in the 1960s until the introduction of the coho salmon controlled them.

3. *Coho salmon.* Also known as the silver salmon (*Oncorhynchus kisutch*), the coho is a Pacific native. It is not strictly speaking an "invader" as it was deliberately introduced in the 1960s to the Great Lakes where it has become an important asset for the sport and commercial fisheries. It is, however, a foreign species.

4. *Zebra mussels.* These natives (*Dreissena polymorpha*) of the Caspian Sea entered the lakes in 1986 in the ballast water of ocean-going vessels or "saltys" that pumped it (illegally) overboard. They clog water intakes and foul boat and ship's hulls. A single, 15 mm mussel can filter 1 L of water daily. It has been estimated that there are enough of them in Lake Erie to filter the entire volume of the lake every week. They are widely credited with clarifying its waters by consuming vast

quantities of algae but they are competing with native clams for this food source. They may also colonize the shells of native clams (they are hard-surface colonizers) so they are unable to open and the clam dies. Their larvae may be carried to other lakes on boat hulls or in bait pails.

5. *Quagga mussel.* This 20 mm, deep-dwelling mussel (*Dreissena bugensis*) is a native of the Ukraine. It arrived in 1989 also in ballast water. Unlike the zebra mussel, it is a bottom dweller able to colonize soft surfaces and survive in the face of a marginal food supply. It can rapidly deplete the food web. Natural transfer of mollusks from marine to aquatic environments is rare because the larvae are not strong enough to swim against river currents.

6. *Round goby.* The round goby (*Neogobius melanostomus*) is a native of central Eurasia including the Black Sea. It arrived in the usual hitchhiking fashion. This 10–22 mm fish provides a useful species to illustrate the complex interrelationship that can develop between an invasive species and the native inhabitants of an ecosystem.

    The round goby is aggressive, territorial, predatory, and voracious. It will consume any animal small enough to ingest. This includes everything from clam larvae to small fish. It is especially fond of the eggs of other fish. The bad news is that it can outcompete native species and deplete the food web. The good news is that it has been discovered as a food source by trout and salmon and they are thriving as a result. The goby also helps control zebra and quagga mussel populations by consuming their larvae. Recently, the round goby has established itself in tributaries including the Thames, Sydenham, Au Sable, and Grand Rivers. The round goby also threatens native mussels by reducing the numbers of small fish. Mussel larvae attach themselves to the gills of these as part of their natural life cycle.

7. *Asian carp.* A current impending invasion of great concern is that of the Asian carp. The term is a catchall that refers to the grass, black, silver, and bighead carp. These are all members of the *Cyprinidae* family and they have been cultivated for centuries as a food source in China. Escapees from fish farms have become well established in the Mississippi River basin. These carp can reach 45 kg (100 lb) in weight. Their natural defense, often in response to outboard motor noise, is to leap straight in the air. Fishers have been injured by fish landing on them. They are a potential threat to native species because they live on the plankton that is the basis of the food web and they require huge amounts of food to support their bulk. The main point of entry to the GL is the Chicago Ship and Sanitary Canal. It is the only aquatic connection to the Mississippi River basin. Electric fish barriers have been established in the canal in 2004 but opinion is divided as to their efficacy as Asian carp have been captured in all the Great Lakes except Superior, although there is as yet no evidence that a breeding population has been established.

8. *Other invaders.* The list of invasive species in the Great Lakes is long and includes plants and microorganisms as well as animals. Examples include the rusty crayfish, spiny water flea, ruffe (an Atlantic scale fish), white perch, the curly-leafed pondweed, Eurasian water milfoil, phragmites, purple loosestrife, and the virus of hemorrhagic septicemia. The extent of the damage that these invaders can inflict is not yet totally clear. The purple loosestrife was initially thought to be a threat to wetlands adjacent to the lakes but it appears that the threat is not as great as first anticipated.

Some Important Definitions

Acclimation: This refers to the process of adaptation to a single environmental factor under laboratory conditions. Acclimation to heavy metals such as cadmium occurs because of an increase in the levels of metallothionein, a metal-binding protein.

Acclimatization: This refers to the adaptation of an organism to multiple environmental factors under field conditions.

Anthropogenic: Any change resulting from human activity.

Bioaccumulation: This refers to the total uptake of the dissolved plus the ingested phases of a toxicant; for example, gill breathers absorb lipophilic substances through the gills and consume them in food.

Bioconcentration: This refers to the uptake of the dissolved phase of a toxicant to achieve total body concentrations that exceed that of the dissolved phase in the water, that is, against a concentration gradient.

Biomagnification: This refers to the concentration of a toxicant up the food chain so that the higher, predatory species, contain the highest levels, for example, PAHs such as benzo[a]pyrene (BaP) (complex ring structures, implicated as carcinogens, formed from incomplete combustion during forest fires or coming from oil spills) are concentrated but not metabolized by mollusks. These may bioaccumulate, and BaP has been detected in the brains of beluga whales taken from a polluted area of the St. Lawrence River.

Toxicity Testing in Marine and Aquatic Species

A wide variety of marine and aquatic organisms is employed for toxicity testing. This is important because of the biomagnification factor discussed earlier. Testing species only at the top of the food chain would not provide any information regarding the likelihood that those species might biocon-centrate and bioaccumulate the toxicant. Nor would it give any indication of how the toxicant might distribute in the aquatic or marine environ-ments. Species commonly employed include the organisms *Daphnia magna* (water flea, an aquatic crustacean 2–3 mm in length) and *Selenastrum* (duckweed), rainbow trout, and fathead minnows. Fish species are impor-tant because the gills are an important mechanism for uptake of toxicants. The gills will pass molecules smaller than 500 Da. Large molecules may clog the gills and suffocate the fish. A marine species gaining importance is the opossum shrimp *Mysidopsis bahia*. This is a tiny, live-bearing estua-rine species with a rapid life cycle and adaptability to laboratory culture conditions. It is being used as a bioassay for sewage effluent and petro-leum spill toxicity.

Water Quality

Liquid freshwater (as opposed to water vapor) exists on earth either as surface water (lakes, rivers, streams, ponds, etc.) or as groundwater. Groundwater may be in the form of a shallow water table that rather quickly reflects changing levels of xenobiotics at the surface, or as much deeper aquifers that acquire surface contaminants more slowly but just as surely nonetheless. An aquifer is a layer of rock or soil capable of holding great amounts of water. Subterranean streams and pools also exist. A significant difference between surface water and groundwater is the accumulation of sediments by the former. It is estimated that 50% of croplands in the United States lose 3–8 tons of topsoil/acre/year and another 20% lose over 8 tons/acre/year. Soil erosion contributes more than 700 times as much sedimentary material to surface water as does sewage discharge. Both surface water and groundwater can serve as a source for drinking and household and industrial use. Groundwater, however, provides a supply for 50% of all of North America, 97% of all rural populations, 35% of all municipalities, and 40% of all agricultural irrigation.

Numerous communities bordering the Great Lakes and even at some distance from them receive their municipal drinking water from intakes located offshore at some depth. Concern about the quality and safety of water from private wells spurred the move to establish reliable supplies of treated water. The Lake Huron pipeline, for example, draws water from the lake north of Grand bend and carries it to the city of London some 40 miles (64 km) away. It supplies many small towns and villages along the route.

Sources of Pollution

Sources of pollution include the following:

1. *Agricultural runoff.* Drainage systems conduct any soil contaminants to surface water and, by seepage, to groundwater. This includes agricultural chemicals (pesticides, chemical fertilizers), heavy metals leached from the soil by acid rain, atmospheric pollutants carried to the soil in rainfall, bacteria from organic fertilizers, seepage from farm dump sites (old batteries, used engine oil), etc.

2. *Rain.* Rain will transfer atmospheric pollutants directly to surface water. Gases may be dissolved directly in water.

3. *Drainage.* Drainage from municipal and industrial waste disposal sites and effluent from industrial discharge is an important potential source of contamination if not controlled.

4. *Surface runoff.* Runoff from mine tailings, which may be rich in heavy metals, can contaminate both surface and groundwater. In northern Ontario, a small town named Wawa recently launched a suit against a mining company that operated a mine, now defunct, in the area for many years. Arsenic contamination of soil from mine tailings has been detected to a depth of 10 cm. Heavy fall rains in 1999 contaminated the local water supply with arsenic to levels many times the maximum allowable level, forcing residents to use water trucked in tank trucks, or to purchase bottled water. This single incident clearly illustrates the close relationship between soil and water. In India, arsenic leached out of mountain soil and rock by rivers, a natural phenomenon, has made arsenic poisoning an epidemic problem (see also Chapter 6).

   The town of Picher, Oklahoma provided a striking example of the consequences of unbridled, long-term mining on a community. Picher was a twentieth-century boomtown peaking in 1926 with a population of 14,000. It was a major mining center for lead and zinc and provided most of the lead for American bullets in both world wars. In 1970, the mines went dry and the town has been in a downward spiral ever since, becoming virtually abandoned in 2010. Problems include huge mounds of "chat"—heavy metal-laden dust from the mining, which blows around coating everything. Health problems in residents and former residents are extensive and heavy metals are both inhaled and ingested. Because there are huge caverns under the town from the mines, subsidence is common and most buildings have been condemned for fear of collapse. The area has been declared a superfund site and the problem is further complicated by the fact that a native tribe, the Quapaw, still retain ownership of much of the area. It will be decades before the problems are all sorted out.

5. *Municipal sewage discharge.* Even if treated, this may carry phosphate detergents, chlorine, and other dissolved xenobiotics into water courses. *The Globe* and *Mail* (August 18, 1999) reported that major Canadian cities dump more than 1 trillion liters of poorly treated sewage into water courses annually. *The Globe* was quoting a study conducted by the Sierra Legal Defense Fund. Five cities actually dump raw sewage into rivers. This is illegal under the Federal Fisheries Act but some municipalities are chronic offenders. The average Canadian generates about 63,000 L of wastewater each year. A major problem is that many older drainage systems combine both sanitary and storm sewers. Thus, whenever there is a heavy rainfall, raw sewage is flushed into the water course. The Canadian federal government in 2010 passed legislation requiring that sanitary and storm sewers be separated but the timeline can be as long as 30 years. Over 20 municipalities will be involved and the expense will be huge. Environmentalists are concerned that not enough is being done to protect Canada's waters. In the Province of Ontario, the Walkerton town water crisis created impetus for improvement. This will be dealt with later in this chapter.

6. *Municipal storm drains.* These constitute another source of pollution through runoff. In the Great Lakes basin, salt is used extensively on roads to melt ice and improve traction for vehicles. The salinity of rivers and lakes is increasing as a result. Used engine oil from home oil changes in automobiles may be dumped down storm drains. In Canada, an estimated 30,000,000 L from such usage is not recycled annually. Calcium chloride also may be conducted to lakes, along with residues from vehicle exhaust.

7. *Natural sources.* Although the primary concern of many people is toxicants of anthropogenic origin, it must be remembered that natural toxicants such as methylmercury can form as a result of bacterial action on mercury leached from rock, and of special concern is the level of natural nitrates in drinking water. Nitrates form from nitrogenous organic materials derived from decaying vegetation. Natural levels are not usually a source of concern, but the addition of nitrates from agricultural activity (nitrate fertilizers, animal wastes) may increase the content to dangerous levels. Nitrates are converted by intestinal flora to nitrites that oxidize ferrous hemoglobin to ferric methemoglobin, which cannot transport oxygen. Infants are especially sensitive and cases of poisoning numbering in the thousands have been reported, with a significant mortality. Adults and older children possess an enzyme, methemoglobin reductase, that can reform normal ferrous hemoglobin. Normal nitrate levels in water are about 1.3 mg/L, contributing about 2 mg/day to the total intake of 75 mg/person/day. Levels as high as 160 mg/L have been reported

in some rural areas where wells serve as the source of water (see also section on "Food Additives" in Chapter 8). Both the U.S.-EPA and the Environmental Health Directorate of Health and Welfare Canada set maximum acceptable limits for toxicants in drinking water. For example, the EPA limit for nitrates is 10 mg/L measured as nitrogen. Water pollutants may be described as oxygen-depleting (contributing to eutrophication), synthetic organic chemicals (detergents, paints, plastics, petroleum products, solvents) that may be very persistent in the environment, inorganic chemicals (salts, heavy metals, acids), and radioactive wastes from nuclear generating plants. Low-level radioactive liquid wastes are produced in the primary coolant. Hospitals may also be a source of low-level radioactivity if medical isotopes are not properly disposed of.

Some Major Water Pollutants

Specific classes of xenobiotics will be dealt with in detail later in this text as they also may serve to contaminate soil, water, or air. Here we will briefly review the more important groups in water.

1. *Detergents.* A wide variety of substances are employed as wetting agents, solubilizers, emulsifiers, and antifoaming agents in industry and in the home. They have in common the ability to lower the surface tension of water (surfactant effect) and, as cleaning agents, this increases the interaction of water with soil, solubilizing the latter. Chemically, they possess discrete polar and nonpolar regions in the same molecule. The nonpolar region is usually a long aliphatic chain. Sodium dodecylbenzenesulfonate (an anionic detergent) and polyphosphates such as sodium tripolyphosphate are in this group. The latter, $Na_5O_{10}P_3$, is commonly known as STP, the engine oil additive. In sewage, it is readily hydrolyzed to form orthophosphate. Removal efficiencies for sewage treatment are typically 50%–60%, so that significant amounts can enter surface water to contribute to the process of eutrophication (discussed earlier). Despite a ban on phosphate detergents by most states and provinces bordering the Great Lakes, water phosphate levels have not dropped significantly. The ban has apparently been offset by the use of phosphate fertilizers. The average North American uses about 23 kg of soaps and detergents yearly. The biochemical, or biological, oxygen demand (BOD) is a measure of the organic material dissolved in the water column and hence of the oxygen requirement for its decomposition. It includes natural sources such as phytoplankton, zooplankton, and organic material from vegetation as well as nitrates. Pure water has a defined BOD of 1 ppm. BODs above 5 ppm suggest significant pollution. Pulp mill effluents may have levels over 200 and agricultural animal wastes may approach 2000.

2. *Pesticides.* This class of chemicals has generated great public concern, sometimes in the absence of any hard evidence of toxicity for humans at the level of exposure likely to be encountered. For example, the European Economic Community, in its "Drinking Water Directive" of 1980, set limits of 0.1 µg/L for any single pesticide and 0.5 µg/L for all pesticides combined without regard for their toxicity or their economic importance to agriculture. Included in this group are insecticides, herbicides, fungicides, rodenticides, and specific agents to kill snails (mollusckicides) and nematodes (nematocides). Nematodes (roundworms) from the Greek "nema" meaning thread, are a huge class of parasites that infect humans and animals as well as many plants. The galls that one sometimes sees on leaves of trees are usually due to nematode infestation. Although not strictly pesticides, the public tends to include other agricultural chemicals used to improve growth or ripening in this category.

Chemical Classification of Pesticides

Pesticides can be classified as follows:

1. Chlorinated hydrocarbons such as DDT, lindane, aldrin, dieldrin, heptachlor (also called organochlorine insecticides). Polychlorinated biphenyls (PCBs) are also chlorinated hydrocarbons but are not insecticides.
2. Chlorophenoxy acids include the herbicides 2,4-D and 2,4,5-T, which contain dioxins as impurities.
3. Organophosphorus insecticides such as parathion, malathion, DDVP, and TEPP.
4. Carbamate insecticides, which act like organophosphorus compounds (cholinesterase inhibitors) but which are derivatives of carbamic acid. There are also carbamate herbicides that lack significant anticholinesterase activity.
5. Bipyridyl herbicidal (paraquat) is an agent that is not considered to be an important environmental contaminant but which is extremely toxic to handlers if used incautiously.

Health Hazards of Pesticides and Related Chemicals

Chlorinated Hydrocarbons

Chlorinated hydrocarbons are very persistent in the environment. They are degraded slowly by bacteria and other microbes. In addition they are very lipid soluble and thus have very long biological half-lives. Although this group is considered to have low acute toxicity, the combination of lipophilicity and long T1/2 leads to biomagnification up the food chain and greater potential for chronic toxicity. This is not easily demonstrable in humans, but in nature, DDT (dichlorodiphenyltrichloroethane) and its metabolites have been shown to interfere with calcium metabolism, causing softening of the eggshell in many species of birds (gulls, peregrine falcon, bald eagle, brown pelican, etc.) with consequent loss of reproductive efficiency. Human fat may contain up to 10 ppm, with a clearance of about 1% of content/day. Acutely, DDT is a neuro-toxin, causing tremors and convulsions. The oral LD_{50} for humans is estimated at 400 mg/kg. PCBs have been used for many years for their insulating prop-erties and the fact that they can withstand temperatures up to 800°C. These properties make them ideal for use in electrical transformers, hydraulic fluids, brake fluids, etc. Their stability means that they are very persistent in the envi-ronment when contamination occurs through accident or improper disposal. In the United States, the Environmental Protection Agency (EPA) has set a maximum allowable level of 0.01 ppb in streams. In the Baltic Sea, PCBs have been incriminated in reproductive failure in seals (pinnipeds).

Considerable concern has been generated over seepage of PCBs into groundwater and streams from deteriorating containers in dump sites. Levels of 5–20 ppm have been detected in Lake Ontario fish. Toxic effects in the environment include reproductive defects in phytoplankton and, in mammals and birds, microsomal enzyme induction, tumor promotion, estrogenic effects, and immunosuppression. The potential for human toxic-ity is therefore high, but existing data are somewhat controversial (see later).

Chlorophenoxy Acid Herbicides

The chlorophenoxy acid herbicides 2,4-D and 2,4,5-T have been widely used on lawns and along road and railway rights-of-way. They mimic plant growth hormones so that accelerated growth exceeds the energy supply. 2,4,5-trichlorophenoxyacetic acid (2,4,5-T) is weakly teratogenic but the main concern is the presence of the contaminant 2,3,7,8-tetrachlorodibenzo-p-dioxin (TCDD, dioxin), a by-product of synthesis. It is teratogenic and very toxic to some animals. The LD_{50} for rats is 0.6–115 µg/kg. It causes degenerative changes in the liver and thymus, weight loss, changes in serum enzymes, porphyria, chloracne, and cancer. In humans, the only confirmed toxic effect is chloracne (see Chapter 2 on the Seveso accident). Although Vietnam veterans have been very concerned about the use of "Agent Orange" (which contains equal parts 2,4-D and 2,4,5-T) as a defoliant, several epidemiological studies have failed to conclusively con-firm long-term effects. Dioxin toxicity will be covered in detail in Chapter 5.

Organophosphates (Organophosphorus Insecticides)

Organophosphates irreversibly inhibit acetylcholinesterase and the symptoms of acute toxicity are those of massive cholinergic discharge and include miosis, vomiting, profuse sweating and diarrhea, abdominal pain, dizziness, headache, perfuse salivation, tremors, mental disturbances, convulsions, and death. Cardiac arrhythmias may occur, such as sinus tachycardia and more rarely, ventricular arrhythmias. Although parathion is fairly toxic for humans it does not persist in the environment and thus it is not a significant environmental hazard. These agents are water soluble but they are hydrolyzed to nontoxic by-products.

Organophosphorus compounds are either banned or strictly regulated in developed countries. The World Health Organization (WHO) maintains a classification system for poisons and most of these compounds belong to WHO Class Ia (extremely hazardous) or Class Ib (highly hazardous). But cholinesterase-inhibiting pesticides such as organohosphates and carbamates remain the most common cause of severe, acute pesticide poisonings worldwide, largely due to their unregulated use in developing countries coupled with the lack of knowledge regarding their safe use. Conditions contributing to this state include the use of leaky backpack sprayers that soak clothing and skin, lack of washing facilities, and long hours spent in conditions where exposure is bound to occur. Cholinesterase inhibiting insecticides known to be associated with incidents of poisoning include aldicarb, methyl parathion, parathion, terbufos, carbofuran, methamidophos, methomyl, monocrotophos, chlorypyrifos, and etoprophos.

Chronic toxicity also occurs. Signs and symptoms of impaired neurobehavioral performance are the most common manifestations. These include anxiety disorder, depression, psychotic symptoms, problems with memory, learning, attention, information processing, and also extrapyramidal signs and symptoms. These do not seem to correlate well with inhibition of acetylcholinesterase. Chronic exposure to organophosphorus compounds can also lead to DNA damage and genetic polymorphisms of xenobiotic-metabolizing enzymes.

Carbamates

Carbamates act generally like the organophosphates. The exception is aldicarb, which is not hydrolyzed and which has entered groundwater in several locations including Long Island, NY, where it is predicted to exceed maximum allowable levels of 7 ppb for up to 10 more years. It is highly toxic, but it is a reversible inhibitor and is rapidly degraded and excreted. Pesticides will be considered further in Chapter 9.

Bhopal Disaster

Carbaryl, also known as sevin, is a carbamate insecticide (1-naphthalenol methylcarbamate) still widely used throughout the world. Previously it was manufactured in two Union Carbide plants, one in West Virginia and the other in Bhopal, India. In Bhopal, at 1 a.m., on December 3, 1984, a safety valve on a large storage tank of methyl isocyanate blew out. This occurred because numerous safety features had been neutralized, allowing the tank to overheat and pressure to build up. Over 40 tons of methyl isocyanate (MIC) gas escaped, flooding low-lying areas (MIC is heavier than air). Nearly 4000 people died instantly. An estimated 500,000 received some level of exposure and health problems plagued the area for decades. It has been estimated that over 20,000 premature deaths occurred over the ensuing 25 years and thousands more suffered damaged health. The Bhopal disaster is widely held to be the worst industrial accident in history.

MIC has toxic effects on virtually every organ system in the body. Immediate deaths were probably a combination of asphyxiation as MIC displaced air, and severe irritation of the respiratory tract. Ocular, respiratory, gastrointestinal, psychological, and neurobehavioral effects persisted for months and years. Genetic damage has also been observed and there is evidence that cancer has been emerging as a late toxic effect.

Almost immediately Union Carbide Corporation (UCC) began to try and dissociate itself from legal responsibility, laying it off on its Indian subsidiary Union Carbide India Limited. The Bhopal plant had not been constructed, nor held, to the same safety standards as its North American counterpart. In this, UCC followed a practice that is still in place today; namely leaving it to local authorities to establish and enforce safety standards for industries including those owned by multinational corporations. In Canada, a private member's bill (C-300) that would have required Canadian companies to adhere to Canadian safety standards in their offshore operations resulted in a threat from the mining industry to move all of their headquarters out of Canada if the bill passed. It was defeated. In 2009, the Prospectors and Developers Association of Canada (PDAC), a powerful mining lobby group and a staunch opponent of Bill C-300, commissioned a report on mining behavior internationally. It was never released but was recently obtained by Mining Watch Canada, a watchdog organization. The PDAC report discusses 171 high-profile safety violations by several countries including Canada. Canadian mining companies were involved in more than four times as many violations as the next two offenders, Australia and India. The authors concluded that "… Canadian companies have been the most significant group involved in unfortunate incidents in the developing world. Canadian companies have played a much more major role than their peers from Australia, the United Kingdom and the United States. Canadian companies are more likely to be engaged in community conflict and unethical environmental behavior …"

Acidity and Toxic Metals

Acidity, largely from acid rain (the causes of which will be discussed in Chapter 5), can have several deleterious effects on water quality. This subject has already been introduced (see "Abiotic Factors" #1 above). Acidity can leach toxic metals such as lead (Pb), cadmium (Cd), and aluminum (Al) from soil into groundwater. It can contribute to the formation of more toxic methylmercury from mercury. It may also leach lead from solder in the plumbing of houses and cottages as has been shown in a study by Health and Welfare Canada. Water at pH 4.5–5.2 was allowed to stand in plumbing systems and reached maximum leaching rates after 2 h. After 10 days, the water showed levels of 4560 μg/L for copper (Cu), 3610 μg/L for zinc (Zn), 478 μg/L for Pb, and 1.2 μg/L for Cd. The upper limits recommended for Canadian drinking water are 100 μg/L for Cu and 50 μg/L for Pb. Arsenic (As) and selenium (Se) have also been detected. It is highly advisable to flush plumbing systems in houses and cottages that have been standing vacant for any length of time.

At pH 5 or less, aluminum can be leached from soil and transported as complexes with bicarbonate, organic materials, and in the ionic form. Acid surface water may contain 4–8 μmol/L, which can be toxic to fish. In humans, high concentrations of Al may be deposited in bones and brain tissue to cause osteomalacia and symptoms of dementia. Microcytic-hypochromic (i.e., small, pale cells) anemia can also occur. These problems have been encountered in hemodialysis patients due to the leaching of Al from the dialyzer into the blood of the patient and from oral aluminum hydroxide given them in antacids. Bauxite miners suffer respiratory problems from inhaling ore dust. Al also appears in drinking water because of the use of alum $[Al_2(SO_4)_3 \cdot 12H_2O]$ to precipitate suspended organic material in the third (tertiary) stage of water treatment. The first stage involves the removal of large solids by screens and the second stage removes most of the organic material with filters.

In 1980, the U.S. Congress commissioned the National Precipitation Assessment Panel (NAPAP), consisting of over 2000 scientists from virtually all of the major universities, to study the acid rain problem. In 1987 it issued a highly controversial interim report that concluded that the situation was not as bad as previously suspected. Of several thousand lakes studied in upper New York State, only 75 were seriously affected. Sulfur emissions were found to have declined by 25% in the last decade. The final 6000-page report was released in 1990 (total cost over $570 million) and concluded that 10% of eastern lakes and streams were adversely affected, that acid rain had contributed to the decline of red spruce at higher altitudes and to the corrosion of buildings and materials. A more controversial statement was that there was no evidence of widespread decline of forests in the United States or Canada. Acid precipitation, however, can be deposited hundreds of miles from its site of formation. Moreover, lakes that drain limestone bedrock areas are much more resistant to acidification because of their buffering capacity. Lakes that drain granite bedrock, however, are very susceptible because they have virtually no buffering capacity. This includes all of the lakes in the Canadian Shield area. Again, aluminum plays an important role. Scientists have discovered that fish in a laboratory setting can withstand a pH of 4.5 or less, while in the natural setting, such a low pH is frequently fatal to aquatic organisms. Aluminum silicates are a major soil component, and soft water lakes that drain soil areas acquire significant levels of these. A suspension of fine, aluminum precipitate forms in water. This blocks the sodium and oxygen exchange systems in the gills of fish, which then expire. Freshwater fish must take up sodium across the gills to compensate for that lost in urine. Freshwater fairy shrimp have "chloride cells" that also regulate sodium levels and that accumulate toxic levels of aluminum. Some success has been achieved in selectively breeding aluminum-resistant strains of aquatic plants such as duckweed, which may be used to revitalize dead lakes. Selective breeding of plant species was developed early in the century to combat the effects of acidic soils that poison plants by interfering, through metal solubilization, with calcium and phosphorus. Phosphorus, especially, binds to aluminum. Since sodium regulation in nearly all cells involves sodium/potassium ATPase (the sodium pump), the aluminum-bound phosphate can attach to, and disable, the sodium pump. This phosphorus link may be involved in the massive diebacks of forests exposed to high levels of acid rain, and selective breeding of resistant species may provide a partial solution.

Another hazardous chemical introduced as a result of water treatment is chloroform. It is introduced as a contaminant in the process of chlorination and it is a known carcinogen. Others, such as benzene and carbon tetrachloride, may enter groundwater from industrial sources.

Chemical Hazards from Waste Disposal

In addition to the types of hazardous contaminants discussed earlier, numerous substances may enter water from industrial, agricultural, institutional, and domestic sources. They may be solids, liquids, sludge, or gases, and may be corrosive, flammable, explosive, radioactive, or biologically toxic. Risks range from minimal to extreme and there may be short- or long-term effects on human health. Usual disposal methods for these substances include surface impoundments used in industry (48%), landfill sites (domestic and other, 30%), burning (10%), and other means, for example, disposal at sea (2%). An idea of the extent of the problem of buried toxic substances can be gleaned from the experiences of workers in tunnel construction projects. Traditionally, compressed air has been used in tunnel construction for the purpose of excluding water from the tunnel and also for stabilizing the surrounding ground. A new use is emerging, however. Increasingly, tunnel projects in urban areas (for sewer mains, rapid transit systems, auto tunnels under rivers, etc.) are encountering pockets of toxic materials such as gasoline, probably leaked from old service station storage tanks, chlorinated hydrocarbons, and other dump site toxins. The use of compressed air in the tunnel prevents the seepage of toxic fumes and liquids into the tunnel and provides a safer working environment. Rehabilitation of old service station sites has become an expensive proposition as soil is invariably contaminated with waste oil and leaked fuels. Such rehabilitation often must wait until the value of the land in question has appreciated enough to make it economically feasible.

Water from drinking wells continues to be a source of concern regarding chemical contamination. In 1987, a study of the Tutu well fields in St. Thomas, U.S. Virgin Islands, showed that 22 wells were contaminated with benzene, trans-1,2-dichloroethylene, trichloroethylene, and tetrachloroethylene originating from several sources. Although levels were low, an estimated 11,000 people were exposed for about 20 years. In Minnesota in 1981–1982, 41 of 137 private and commercial wells located downhill from an industrial complex were found to be contaminated with trichloroethylene and trichloroethane. Such wells generally should be sealed with concrete or clay and abandoned. Rural areas are frequently littered with abandoned wells or those kept for emergency livestock watering. Every hole dug or drilled into the ground becomes a potential conduit for contaminants to enter groundwater.

Love Canal Story

The Hooker Chemical Co., between 1942 and 1953, disposed of about 420,000 MT of around 300 organic chemicals by burying them in steel drums in the abandoned Love Canal near Niagara Falls, NY. The site covers about 16 acres and was about 10 ft deep. Subsequently, a subdivision was built over it. As the barrels rusted out, chemicals seeped into the groundwater. There were some early warning signs such as chemical odors in people's basements, sinking areas over collapsing barrels, some exposed pools of waste, and some minor health problems such as rashes and eye irritation after contact with exposed chemicals. Overall, the residents of Love Canal seemed unaware of any unusual health problems or circumstances. In 1976, however, the International Joint Commission on the Great Lakes undertook a study to find the source of the banned pesticide mirex in Great Lakes fish and the chemical contamination was discovered. The New York State Department of Environmental Conservation studied the situation over the next 2 years amid great controversy and in the face of emerging anecdotal claims of serious health problems. In August of 1978, a series of dramatic events occurred. The state health commissioner declared a health emergency and recommended pregnant women and children under 2 be evacuated. President Jimmy Carter declared the area a federal disaster zone, thereby creating a mechanism for federal assistance to be provided. The Governor of New York State announced that 239 families would be relocated at state expense. The immediate consequence was that Love Canal became a ghost town and considerable anxiety was created about long-term health effects, not only in the evacuees but also in those who lived near the edge of the arbitrary demarcation line.

In 1988, a federal district court found Hooker Chemical (by then Occidental Petroleum) to be liable for the cost of the cleanup estimated at $250 million. The state commissioner of health declared that some areas were safe to resettle, but new information challenged this decision (see later). Numerous health studies had been conducted in the intervening decade. They generally failed to reveal any significant evidence of an increased incidence of illnesses. Cancer registries were relatively new, and some states still did not have one. An analysis of the New York State cancer registry found that the incidence of lung cancers was generally higher throughout the Niagara Falls region, but no differences in the occurrence of liver cancer, lymphomas, or leukemias were noted in the Love Canal region. The increased incidence of lung cancer is intriguing in light of a study by The University of Toronto and Pollution Probe, which found that the mist from Niagara Falls contains PCBs, benzene, chloroform, methylene chloride, and toluene. Although the levels were not themselves high enough to pose a risk, they could be additive with other carcinogens in cigarette smoke, auto exhausts, etc. and the mist could settle out on crops to enter the food chain. Statistics compiled in the United States and by Health and Welfare Canada showed statistically significant differences in mortality from different types of cancer among and within counties bordering the Great Lakes but did not indicate that these populations suffered substantially higher cancer mortality than non-basin counties. Nor were there consistent differences among municipalities within Niagara County. One fairly convincing bit of data is that the families living closest to the canal had a higher occurrence of miscarriages, abortions, and low-birth-weight babies.

Several recent studies have attempted to identify differences in morbidity and mortality in former Love Canal area residents. One reported in 2011 found that there was a statistically significant elevated risk of preterm birth among children born on the Love Canal prior to the evacuation of residents from the Emergency Declaration Area when compared to the general population of upstate New York (standardized incidence ratio, or SIR, was 1.4). Significantly elevated SIRs were also noted for cancers of the bladder (1.44) and kidney (1.48). Although death from myocardial infarction and accident were also elevated, there was no difference in the overall mortality rate when compared to the mortality rates of New York State and Niagara County. Limitations of the studies were noted and included small cohorts for some events, a qualitative exposure assessment, and no data on deaths in the area prior to 1978. Since many of the recorded health problems have been previously linked to chemicals known to be in the contaminated zone, a direct cause and effect could not be ruled out. Psychological stress was also thought to be a contributing factor.

There is no question that a major component of the health impact of exposure to such toxic disasters is psychological in nature, which is not to say that it is not real. People tend to accept natural disasters with much greater equanimity. Flood victims do not appear to suffer the same prolonged mental stress as victims of a toxic disaster, nor do they attach blame as readily, even when there may be legitimate questions about the adequacy of flood control measures. Such things tend to be accepted as acts of God, and there may even be some essence of pioneer spirit to be taken from battling the forces of nature. In contrast, victims of toxic disasters tend to feel at the mercy of forces that are man-made and out of control. They are not just victims; they feel victimized by greed and callousness. Their sense of helplessness may be exacerbated by contradictory statements from government officials and by sensational reporting by the news media. An understanding of this phenomenon may be an important step toward minimizing the human price of such disasters. Following the Love Canal disaster, there was a high frequency of insomnia, severe depression, an increase in suicide attempts, lowered performance in school children, and other manifestations of emotional illness, including symptoms of malaise for which no organic cause could be found. This phenomenon will be discussed again in subsequent chapters. In April of 1989, a peer review panel consisting of 10 scientists familiar with the Love Canal situation met to examine charges that the area was not safe for resettlement. The charges claimed that the site had not been adequately cleaned up, and that an area in a zone designated as a control area for purposes of comparison had been shown to contain a "hot spot" with high levels of chlorinated organics under a church parking lot. The charges were brought by the Environmental Defense Fund, which also was responsible for the ALAR controversy (see Chapter 2). The EPA, however, concluded that the area comparison method was still valid for the Love Canal and that the habitability decision should not be reconsidered.

Problems with Love Canal Studies

1. Initial studies were incomplete and not well organized. Most of the chemicals involved were very volatile and would not show up in human blood or tissues except for a short time after exposure. Only lindane and dioxins were persistent enough to be detected later, and lindane was not found in the subjects who were monitored. Dioxin assays are expensive and time-consuming.

2. When the Centers for Disease Control was asked to conduct a study in 1980, those people with the highest risk of exposure had been evacuated. Little information was available on previous exposures.

3. The episode was highly emotionalized. These were the homes of working-class people who had invested their life savings in their properties.

4. The dispersion of the evacuated families made it hard to collect valuable data on health effects.

The story of Love Canal has been repeated in many other communities in smaller ways. In London, Ontario, a playground in St. Julien Park was built over a waste disposal site. There were anecdotal reports of local clusters of brain tumors and other problems, but a study conducted by local health officials failed to confirm increased health risks. Cleanup and preventive measures have been instituted. A very definite hazard of landfill sites is the production and accumulation of methane gas from decaying organic material. Seepage into basements has resulted in explosions with deaths and injuries. Landfill sites that are covered over and landscaped or built upon should always have vents installed to prevent the accumulation of methane.

Chemical spills in the St. Clair River have on several occasions threatened the water serving Wallaceburg, Ontario and the Wallpole Island First Nations Reserve.

Toxicants in the Great Lakes: Implications for Human Health and Wildlife

While the Great Lakes are not threatened by acid rain because of their large size and buffering capacity, they are definitely threatened by toxic chemicals. In 1991, the Joint Commission issued a report on toxicants present in Great Lakes water, which they felt posed a potential threat to the populations of provinces and states in the Great Lakes basin by virtue of their ability to interfere with reproduction. Table 3.1 is a partial list of these. An Environment Canada website states that an estimated 1000 chemicals are in the Great Lakes system. Over 360 chemicals have been quantified in the GL,

TABLE 3.1

Some Great Lakes Contaminants with Potential Reproductive Effects

PCBs: linked to embryolethality, deformities, development

DDT: now banned, it disrupts hormone balance

Dieldrin and aldrin (pesticides)-linked to the death of adult bald eagles

Chlorinated dibenzo-furans (carcinogenic)

Toxaphene: an insecticide used in the cotton belt; it has been found as far afield as the high arctic

Dioxin (TCDD): probably carcinogenic for humans exposed to high levels

Polycyclic aromatic hydrocarbons (PAHs)—carcinogens

Hexachlorobenzene: fungicide, causes organ toxicity, is carcinogenic, and may cause infertility

Mirex: carcinogenic insecticide, causes reproductive problems

Mercury: toxic to the CNS, liver, and kidney

Lead: toxic to the CNS

and reproductive abnormalities possibly related to their presence have been seen in 14 species including lake trout, fish-eating birds such as cormorants, terns and bald eagles, mink, and snapping turtles.

It is important to emphasize that these threats are largely potential, but there is an obvious need to improve the situation dramatically. Studies of white suckers from Lake Ontario have shown that, despite the fact that these bottom-feeding fish contain large amounts of PAHs, only about 5% showed cancer, and these all had parasitic liver disease. A study at the University of Guelph suggested that glutathione-S-transferase (GST) enzymes in the liver normally protect the fish by detoxifying the PAHs but the efficiency of these enzymes is destroyed by the liver parasites. This illustrates that there are many factors, including genetic ones, that likely must combine before a tumor will develop. Nevertheless, there is abundant evidence that numerous pollutants present in the Great Lakes are toxic to wildlife. These include lead, mercury, hexachlorobenzene, PCBs, PAHs, dioxin (TCDD), mirex, DDT, dieldrin, and toxaphene. Embryos and chicks of fish-eating herring gulls have been shown to suffer from a disease termed "expanded chick edema disease." This is very similar to a disease that occurs in domestic poultry exposed accidentally to high levels of polychlorinated dibenzodioxins (PCDDs) and PCBs. TCDD (2,3,7,8 tetrachlorodibenzodioxin) has been shown to have an LD_{50} in lake trout eggs of 80 parts per trillion (ppt), and in hamsters, 58% fetal mortality at the ninth gestation day has been shown at a dose of 18 µg/kg. Neurobehavioral and neurochemical abnormalities have been detected in fish-eating birds that consumed contaminated fish from Lake Michigan. Abnormalities have also been shown in laboratory rats fed a diet containing 30% salmon from Lake Ontario.

One of the difficulties in attempting to estimate the risk to humans of exposure to toxicants in the environment is the fact that little is known about the combined effects of several of these. An attempt has been made to deal with this problem as it relates to the polyhalogenated aromatic hydrocarbons that are structural analogs of TCDD. Recent evidence indicates that all of these are inducers of a hepatic microsomal monooxygenase enzyme, aryl hydrocarbon hydroxylase (AHH). The potency of these agents in inducing this enzyme in cultured hepatocytes correlates well with their experimental toxicity and this has led to the development of a set of "toxicity equivalency factors" (TEFs). The EPA has adopted the TEF method as a means of determining the toxicity of mixtures of these agents, as their effects on AHH seem to be additive. This approach, however, has limitations. The use of an in vitro culture system cannot take into account differences in pharmacokinetic parameters among different compounds, nor the extent to which they bioaccumulate. For example, one of the hexachlorobiphenyls has been shown to be a significant contaminant of mother's milk even though it is one of the less common environmental contaminants.

A combined report issued by the EPA and Environment Canada, "The State of the Great Lakes 2009," provides some encouraging data. Although there are areas of concern regarding chemical pollutants where there has been little or no improvement, the overall picture indicates that much progress has been made.

Evidence of Adverse Effects on Human Health

Effects on people are not well understood. Although residents adjacent to the Great Lakes have detectable levels of many toxicants, they are similar to those of residents of other industrial areas. It is axiomatic that toxicity to humans resulting from exposure to PAHs is related to total body burden. Portals of entry include the lungs (contaminated air) and the gastrointestinal tract (water and diet). PCDDs and polychlorinated dibenzofurans (PCDFs) are known products of municipal incinerators. The Ontario Ministry of the Environment has calculated a worst-case scenario of 126 pg TEQ/day for people living near a municipal incinerator with inadequate pollution controls. In contrast, because of their poor solubility in water, drinking water is estimated to contain less than 2 pg TEQ/L. Diet is probably the most significant source of exposure, and may account for 95% of all human intake. Levels of TCDD in human fat have been calculated to average 10 ppt worldwide, with little variation from place to place (evidence of the ubiquity of the substance). EPA has set an acceptable daily intake of TCDD of 1.0 pg/kg/day.

Accidental poisonings in Yusho, Japan and Yucheng, Taiwan resulted in total body burdens of PCDDs and PCDFs 200–300 times the North American average. These levels were associated with nausea and anorexia, increased frequency of premature births, low birth weights, impaired growth, impairment of neuromuscular and intellectual development, and a higher frequency of health problems. The Michigan State Department of Health has assembled three, large cohorts of individuals exposed to increased levels of organohalogens and it maintains a registry of these individuals.

Included in their records are farmers who were exposed accidentally to high levels of PBBs. This fire retardant was accidentally labeled as a mineral supplement for cattle. Farmers who were exposed to chlorinated biphenyls ("silo farmers") and a population of Lake Michigan anglers and their families, who consumed large amounts of fish from the lake and who were exposed to a mixture of contaminants are also included. These cohorts have been followed for over 20 years. Mothers in the "angler" families consumed an average of three meals of contaminated fish per month from the Great Lakes while they were pregnant. Levels of PCBs in the fish were low, and believed to be nontoxic, and the degree of impairment of the infants was much lower than in the Yusho and Yucheng incidents. The differences were nonetheless significant, even when a number of confounding variables such as smoking and alcohol consumption were taken into account. The effects were greatest in infants whose umbilical vein serum PCB levels were greater than 3.5 ng/mL. It must be emphasized that the effects seen in the Michigan infants were largely subclinical, and therefore it is difficult to determine the true degree of risk to the population at large.

In 1989, a workshop was held at which findings from these studies were reviewed and compared to a similar study from North Carolina in which 800 infants believed to have had moderate *in utero* exposure to PCBs (estimated from maternal breast milk levels) were followed for over 1 year. Most studies showed a delay in cognitive performance as measured by the Brazelton and Bayley scales. These decrements in performance were detectable in newborn infants and in children at age 4, as were deficits in body size.

The Michigan studies have been criticized for flaws that include

1. Using anecdotal reports (of the type and amount of fish consumption going back over 6 years)
2. Differences between control and test groups (e.g., in alcohol consumption, use of medications, caffeine consumption, and in the frequency of assisted births)
3. Limiting the study's statistical power by restricting participants to one-third of those exposed

If nothing else, this illustrates the difficulties in conducting epidemiological studies when looking for subtle, subclinical findings. Nevertheless, the potential for toxicity and the significance of these warning signs cannot be ignored. The large body of evidence from laboratory studies and from the examination of numerous wildlife species provides ample indication that many of the halogenated hydrocarbons (organohalogens) have serious reproductive consequences if sufficient amounts are consumed. The questions of extrapolation to humans and of the effects of much lower exposures are as troublesome here as they are for carcinogenesis (see Chapter 5 for a further discussion of these accidental exposures).

Evidence for Adverse Effects in Wildlife

There is certainly increasing concern about the consequences of environmental, estrogen-like chemicals for reproduction in aquatic species. Salmon with mixed male/female sex organs have been observed in the Great Lakes and abnormal sexual behavior has been seen in other fish. The subject of hormone modulators in the environment will be expanded upon in Chapter 12. Some of the invasive species in the Great Lakes were discussed under "Biotic modifiers" section.

Global Warming and Water Levels in the Great Lakes

In 1998, The International Joint Commission on The Great Lakes issued a dire warning. Unless the population of North America took water conservation more seriously, water loss from the system due to drought and evaporation could reduce water levels to the extent that the shoreline could move 50–100 m out from its present position within a decade. In the summer of 1999, levels in the lower lakes were down nearly a meter, causing financial hardship for many marina operators. Pressure to export water from the system continues as world shortages of freshwater worsen. Concerns were expressed in strong terms at an October 1999 conference in Atlanta. It was pointed out that population growth of 30% is being predicted for the Great Lakes basin and that, even if the 3-year drought does not continue, this growth may well outstrip the available water supplies.

Determining the degree of loss of water from the Great Lakes is not as easy as it might seem as they are subject to normal fluctuations related to the extent of precipitation, the depth of the snow pack in the north, the extent of evaporative loss, water diversion, wind strength and direction, and even barometric pressure. The Canadian Hydrographic Service monitors water levels at numerous stations throughout the system. It publishes a monthly report as well as an annual month-by-month graph. These are available at http://www.waterlevels.gc.ca/c&a/bulletin_e.html. The bulletin for March 2012 noted that water levels for Lake Superior were below their all-time average for the time of year and below the level of chart datum. (Chart datum refers to conditions noted at the time of preparation of the most recent marine chart.) Conversely, levels for Lake St. Clair were near their all-time average for the time of year and above the level of chart datum.

Ninety-nine percent of the water in the Great Lakes is melt-water from the last ice age. It is therefore a nonrenewable resource. Only 1% comes from runoff, and the combined effects of global warming, which increases loss from evaporation, and several years of drought, which reduces runoff, could produce a crisis within a few years.

North Americans are the world's most profligate users of water. According to 2011 Conference board of Canada statistics, Americans use 5150.7 L/person/day and Canadians, 4383.6. By contrast, Sweden uses less than 849.3 and the United Kingdom 493.3 L/person/day. Industry accounts for 68% and domestic use for 20%. Poor conservation efforts, lack of metering and flat rate charges are cited as factors in this excessive use. An ironic situation has developed in the city of London, Ontario. Citizens responded so well to a water conservation campaign that the Public Utilities Commission is now unable to sell enough water to recover its costs.

Water has also become a political and economic issue. Several companies, mostly American, have applied for permits to export water from Canada to the United States and elsewhere. So far, Canada has banned bulk water exports but this policy is being challenged under the North American Trade Agreement (NAFTA). The court's decision will have significant consequences for the Great Lakes and other waters.

Marine Environment

The marine environment is not exempt from the effects of pollutants. There is an area in the Gulf of Mexico called the "dead zone" that is virtually devoid of oxygen and hence of marine life, and it is increasing in size.

It appears each summer off the coast of Louisiana and Texas, and in 1998 it was estimated to be 18,000 km^2 in size. It is widely attributed to the flow of agricultural fertilizers, especially nitrates and phosphates, into the Gulf from the Mississippi River, which drains 41% of the continental United States and a small bit of western Canada. Algal blooms deplete the water of dissolved oxygen, making the environment uninhabitable for other organisms. There could be serious consequences for commercial fishing. Catches have already fallen dramatically from 1983 to 1993, and the Gulf accounts for about 40% of the annual American commercial catch.

1998 was declared "The Year of The Oceans," an event that went largely unremarked, at least as far as the popular press was concerned. An exception was a feature article in the October 5 edition of *Maclean's* magazine. Appropriately titled "The Dying Seas," it documented the declining fish stocks worldwide, largely the result of overfishing, and lists several species that are already on the endangered list or are threatened. Included are the bluefin tuna, barn-door skate, Atlantic haddock, Pacific salmon, anchovy, abalone, grouper, and orange roughy. The disastrous effects of bottom trawling, likened to strip mining, are also noted. The impact of overfishing on the Atlantic cod is well documented. Oil pollution, drift nets, and chemical pollutants are taking a heavy toll on cetaceans.

Sources of Marine Pollution

Nonpoint Sources of Pollution

As noted earlier in this chapter, water pollution can result either from a point source or a nonpoint source. As noted earlier, the Mississippi River serves as a conduit for land-sourced pollution to reach the sea. Agricultural runoff, urban and industrial effluents, leaching from mine tailings, all can contribute pollutants. The author has conducted research into the metal content of numerous benthic marine species (filter feeders), sand, organic sediment, and water along the north Florida Gulf coast from St. Marks to St. Joseph Bay. Samples were analyzed for a panel of metals that included Cadmium (Cd), Mercury (Hg), lead (Pb), copper (Cu), zinc (Zn), and arsenic (As). Specimens included several members of the *Porifera* (sponges), *Tunicata* (sea squirts, sea pork), *Bryozoa*, as well as sponge crabs, hairy brittle stars, and barnacles. All specimens, as well as sand and sediment, from all locations, contained all metals at concentrations well above the limits of detection. The highest level of Cd (6.5 µg/g dry weight) was found in a sponge (*Haliclona panacea*) collected from Apalachee Bay (the most open water area sampled). Hg was highest (0.77 µg/g) in a sea whip (*Leptogorgia virgulata*) from the same locale. Lead levels were highest in samples collected from dock pilings, possibly due to years of the use of leaded fuels or paints dockside. Cu and Zn levels also tended to be highest in these specimens. Organic sediment tended to have higher levels of most metals than sandy ones. In general, water samples taken from estuaries tended to have higher levels of most metals than those from open waters. Generalizations are difficult due to the variations among species and collection sites. For more details readers are referred to the original paper (Philp et al., 2003).

Laboratory studies showed that metals (Cd) could interfere with the normal biology of sponges. Sponge tissue, when placed in calcium-free artificial seawater, sheds its individual cells so that a cell suspension can be obtained. If a solution of Ca^{++} is added to the suspension at the appropriate concentration, these cells will form aggregates; a process thought to reflect the colony-forming activity of the sponge. $CdCl_2$ was as effective a stimulus of sponge cell aggregation as $CaCl_2$. The calcium channel blocker verapamil reduced calcium-induced aggregation but had no effect on cadmium-induced aggregation, indicating that the mechanism of the latter did not involve L-type channels. Low salinity increased Cd concentration by sponge cells and low pH seemed to potentiate this effect. The results suggest that environmental conditions like salinity and pH can alter the susceptibility of sponges to toxicants in the water. Figure 3.3 shows a specimen of the sponge *Microciona prolifera* and Figure 3.4 shows sponge cells in suspension and after aggregation by a solution of $CdCl_2$.

Heavy metals have been shown to have deleterious effects on other marine species. These include

1. Disturbed immune function in bivalves
2. Reduced magnesium-ATPase in the gills of eels
3. Inhibition of metallothionein mRNA and the production of reactive oxygen species in oysters
4. Abnormal pigmentation in fiddler crabs
5. Inhibition of the aggregation of marine sponge cells as discussed earlier

FIGURE 3.3
(See color insert.) Specimen of *Microciona prolifera*, the "red-bearded sponge of Moses," collected from St. Joseph Bay, Florida. (Photo by Richard B. Philp.)

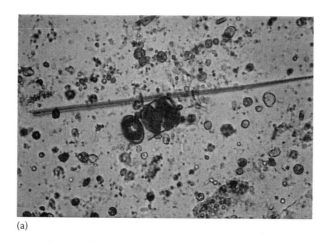

(a)

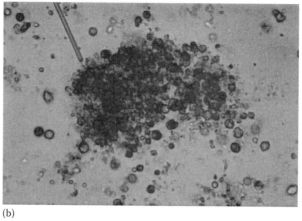

(b)

FIGURE 3.4
(See color insert.) (a) Sponge cells dissociated in a suspension of calcium/magnesium-free artificial seawater. (b) A large mass of aggregated sponge cells formed after the addition of $CdCl_2$ (final concentration 24 μM) (microscope power × 100). (Reprinted from *Comp. Biochem. Physiol.*, 118C, Philp, R.B., 347–351, Copyright 1997, with permission from Elsevier.)

Not all toxicants in water are anthropogenic. Many microorganisms such as algae, diatoms, etc. are capable of producing lethal toxins that can concentrate up the food chain. There is presently concern that algal blooms in the ocean are creating a hazard for marine life and people alike. This subject is discussed in Chapter 11.

Freshwater species also may be affected. Exposure of the freshwater sponge *Ephydatia fluviatilis* caused malformed gemmoscleres and prey attacks by juvenile bluegills were inhibited by exposure to as little as 30 μg/L.

Changing climatic conditions, whether of human origin or not, can have a significant impact on marine and aquatic environments. Salmon have already been detected in the Beaufort Sea and may be colonizing the Mackenzie River.

An increase in the seal population is contributing to shoreline erosion of this river, with associated permafrost melting. The 1992 *El Nino* caused a significant decline in the fish catch of the western coast of South America.

Acidification of the oceans has become a concern and not only because of acid precipitation. Some 35 billion MT of carbon dioxide are released into the atmosphere every year. A new, high resolution computer model suggests that this, plus deep water upwelling, could significantly increase the pH of seawater on the west coast of North America, having a negative impact on the habitat for shellfish.

We have noted in Chapter 1, the importance of the active sediment as a sink for toxicants and a site of exchange with organic carbon and water. In the sea, a microlayer at the surface, approximately 50 μm thick, concentrates certain toxicants. This sea-surface microlayer may have metal concentrations 10–1000 times that of subsurface water. Organisms that spend a few hours daily at the surface, often responding to sunlight, are at risk of severe, adverse effects including growth deformities, cancer, and death, often not manifested until later in their development. Phytoplankton and zooplankton including krill, larval stages of many organisms, and many other small crustaceans may be at risk. This has consequences for the entire food web as these are its foundation.

Plastics are the bane of the marine and aquatic environments. Environmental scientists now refer to the "Great Pacific Garbage Patch," a widely dispersed area rich in small, fingernail-sized or smaller bits of plastic. It is believed that anthropogenic sources of this plastic (there can be no other) have increased 100-fold in the past 30 years or so. It has already had an impact on the marine ecosystem. Nine percent of the fish caught in the patch had plastic in the digestive tracts. An ocean-going species of fly has adapted to the patch and is breeding successfully in it. Small fibers of wood pulp, believed to originate from the thousands of tons of toilet paper flushed into the oceans daily. All of this is largely invisible to the naked eye or even to satellite photography.

Point Sources of Pollution

Point sources of pollution also exist in the sea; in fact, there are likely thousands of them, shipwrecks from both world wars leaking toxic materials into the oceans of the world. The sunken German submarine U-864 was carrying a cargo of metallic mercury, which is now leaking into the sea and causing headaches for the Norwegian government. The u-boat was sunk in February of 1945 and became the first victim of a naval engagement fought entirely under water. She was sunk by the British sub *Venturer* and she carried many German and Japanese aeronautical engineers as well as plans for the German jet engine.

Another point source is French Frigate Shoals. Part of the Hawaiian Archipelago, this atoll was used as a supply depot and airbase during World War II. At the end of the war, much debris was dumped into the sea, including batteries, transformers, a fuel tank, and other potential sources of contamination. Elevated levels of metals and PCBs have been found in invertebrates and vertebrates in the area. Hawaiian monk seals in the area have elevated levels of organochlorines and PCBs. French Frigate Shoals is now a bird sanctuary.

Oil spills constitute a very large cause of point sources of pollution. The International Tanker Owners Pollution Federation (ITOPF) maintains statistics on oil spills from tankers. Between 1970 and 2008, over 1000 such spills occurred. The peak year was 1974 in which there were 89 spills of 700 MT or less and 28 that exceeded 700 tons. The frequency of such spills has been in steady decline. In 2008, only eight in total occurred and only one was in excess of 700 tons. The "poster child" for such spills is the *Exxon Valdez*.

On March 24, 1989, the *Exxon Valdez* ran aground on Bligh Reef in Alaska's Prince William Sound. It leaked 250,000 barrels of crude oil into the sea with devastating consequences for marine and bird life. Over 20 years later, scientists are reporting evidence of long-term damage. Although sea otters appear to have recovered without residual ill effects, the story regarding some diving ducks is quite different. The enzyme ethoxyresorufin-*O*-deethylase (EROD) is induced by PAHs such as are found in crude oil. EROD or cytochrome P450A1-dependent, has been shown to still be elevated in the livers of harlequin ducks (*Histrionicus histrionicus*) from an area that was contaminated by oil from the *Exxon Valdez* spill as compared to those from ducks not so exposed. The results suggest that the impact of the spill was still apparent some 20 years after the event. This same Simon Fraser University group conducted demographic studies of the harlequin duck and reported that in the immediate aftermath of the event their numbers in the oiled area fell by 25%. Survival rates remained depressed six to nine years later and did not recover fully for at least 11–16 years. Much of the recovery was due to ducks migrating from other areas into the affected area. The human health costs of the *Exxon Valdez* spill was discussed in Chapter 2 (see section on "The psychological impact of real and potential environmental risks" in Chapter 2).

The most recent and dramatic example of an ocean point source of pollution is the Gulf oil spill. On April 20, 2010, British Petroleum's Deepwater Horizon drilling rig, located about 50 miles (80 km) from the coast of Louisiana exploded and sank, killing 11 crew members. The pipe connecting the drilling rig to the wellhead some 5000 ft (1.5 km) below, bent and fractured, allowing crude oil to gush freely into the sea. During the ensuing 86 days until it was successfully capped it gushed 53,000 barrels of heavy crude oil every day. The impact on wildlife was immediate. Oil-covered seabirds, dying sea turtles, and dead fish washed up on the beaches of Louisiana. There were oil-coated beaches and salt marshes. Huge, toxic clouds of oil were detected moving beneath the surface of the water; one submerged oil plume was 10 miles (16.09 km) long. The tourist industry, commercial and sport fishing were all severely impacted. The financial hardship experienced by northern Gulf residents was incalculable. The National Oceanic and Atmospheric Administration (NOAA) closed fishing for 10 days from a point east of the Mississippi delta to the Florida Panhandle, a rich source of commercially fished species. By the end of May, the no-fishing zone had been expanded to 47,000 square miles (122,000 km^2). It will take years, possibly decades, before the full environmental impact of the Gulf oil spill will be known.

As was the case with the *Exxon Valdez* spill, stress-related and psychological effects have already been reported (see section on "The psychological impact of real and potential environmental risks" in Chapter 2).

Biological Hazards in Drinking Water

There is an oft-told story of a British physician who, in the nineteenth century, stopped a cholera epidemic in London by padlocking a communal pump. Cholera, caused by the bacterium *Vibrio cholerae*, and typhoid fever, caused by another bacterium *Salmonella typhi*, are two of many waterborne intestinal infections that spread when drinking water becomes contaminated by human feces, usually from untreated sewage. In the developed world these have largely become historical diseases, but they can resurface any time that water treatment facilities become overwhelmed and sewage contamination occurs. This is a serious problem during extensive flooding. Even in modern cities, common sanitary and storm sewer lines can result in flooded sewage treatment facilities with raw sewage being carried into water courses. Seafood contaminated with human sewage can also spread these infections. This is believed to have been responsible for an outbreak of cholera in Peru in 1991. There were 55,000 confirmed infections and 258 deaths. In January of 2010, an earthquake registering 7 on the Richter scale struck Haiti. In October, an outbreak of cholera struck and by 2011 nearly 6,000 had died of the disease and 216,000 had been infected.

Another group of bacteria, *Escherichia coli*, also can cause enteritis, the severity of which depends on the particular strain involved. Contamination of drinking water can result from sewage, runoff from manure piles (many strains infect both animals and humans), and even bird droppings. The frequent summer closures of beaches can be caused by agricultural runoff but also by concentrations of gulls fouling the bathing beaches. A period of high wave activity can stir up bottom sediment and temporarily increase bacteria counts in the water. *Aeromonas hydrophila* is a lesser-known but potentially serious bacterial contaminant of water supplies and a cause of enteritis. Infection can be spread by bird feces, and the population explosion of ring-billed gulls has become a significant source of water contamination. These birds are highly adaptable and can survive on virtually any source of protein, including garbage.

Intestinal, protozoan parasites can also be spread by water. One of these, *Giardia lamblia*, is the cause of the erroneously named "beaver fever." It can infest wildlife such as deer as well as domestic animals. In 1997, there was an outbreak in the Kitchener area of Ontario and in 1998 massive contamination of the drinking water of Sydney, Australia forced nearly four million inhabitants to boil their water for 2 weeks. Also found in their water was another parasite *Cryptosporidium parvum*. This parasite caused an outbreak in Milwaukee in 1992–1993 in which over 400,000 people were infected and there were several deaths. Small outbreaks have occurred in Ontario as well. It is especially troublesome because it survives standard chlorination and filtration procedures and requires the installation of special filters. It can be killed by ultraviolet light and UV units are now standard additions to water treatment facilities in Ontario.

Standard tests for water quality require that bacterial counts not exceed 100 cells/mL of water and there must be no *E. coli*. This organism is used as a marker for fecal contamination.

In 1991–1992, an extensive survey of groundwater quality was conducted in rural Ontario: 1300 farm wells were monitored for fecal coliform organisms, nitrate-N, several herbicides, and petroleum-based derivatives; 37% of all wells tested contained one or more of the target contaminants at levels above Provincial recommended limits; 31% had coliform levels above maximum acceptable limits; and 20% had fecal coliforms. The incidence was higher in wells located on farms with manure systems: 8% had detectable levels of herbicides and 13% had nitrate-N levels in excess of the maximum acceptable concentration. Nitrates have been a cause of poisoning and deaths in infants fed formula made with contaminated water. The source is usually chemical fertilizers (see also Chapter 8). It is obvious that special hazards attend the use of water from farm wells, especially shallow ones that collect surface water. Seafood, notably bivalves, may also be a source of infection with *Salmonella, E. coli*, and even hepatitis viruses. Contamination is almost always the result of the release of untreated sewage into the sea.

Walkerton Water Crisis

Walkerton is a town of 5000 located in southwestern Ontario about 40 km (24 miles) from Lake Huron. In early May of 2000, citizens, including children, began showing up at the local medical clinic with severe, sometimes bloody, diarrhea, vomiting, fever, sweating, weakness, and other signs and symptoms of a severe gastrointestinal infection. There was no common event such as a public dinner or a meal at any particular restaurant that pointed to food as the source of the infection. Samples were sent for culture and identification as the number of affected people continued to mount.

Infants were also being affected and soon the first death occurred. When the lab results came back, they showed that the offending organism was strain O157:H7 of *E. coli*. Nonvirulent strains of *E. coli* are common inhabitants of the gastrointestinal tracts of animals and humans, but this strain can be a killer and is the organism responsible for the so-called hamburger disease. The outbreak continued for several weeks, eventually affecting at least 2000 people and causing seven confirmed deaths, including the infant daughter of a local physician. Other deaths were suspected of being related to the outbreak. Testing soon revealed that the local water supply was the source of the infection. A "boil water" order was issued immediately and was predicted to be in effect until November. All water mains had to be scoured and disinfected right to the kitchen taps and thousands of liters of bottled water were shipped into the town over the summer. There followed charges and countercharges of improperly maintained and operated chlorination equipment, of inadequate testing, and of lack of communication by the provincial government. The source of contaminated water was traced to one of several deep-drilled wells (well #5) serving the community. This well was in fairly close proximity to land grazed by beef cattle that were shown to harbor the organism. Manure had been spread on adjacent fields and the problem was compounded by a period of very heavy rain, a so-called 100-year storm, which facilitated seepage of surface water into the contaminated well. Another pathogenic bacterium, *Campylobacter*, also was found in the water.

Two brothers who were responsible for monitoring water quality and maintaining the treatment system were charged and subsequently convicted (in 2003) of forgery (for falsifying water test results) and breach of public duty. Cost estimates, including medical costs and the cost of Justice O'Connor's judicial inquiry exceeded $64 million. Victims experienced health problems for years and hemolytic–uremic syndrome sometimes led to chronic renal failure requiring dialysis.

Over the course of the summer other rural communities experienced high coliform counts in their water supplies, and public health authorities issued several "boil water" orders. These episodes call into question the generally assumed safety of deep-drilled wells, and indicate that standard chlorination procedures may not protect against an overwhelming influx of contaminated surface water with a high sediment content and the presence of pathogenic organisms. The mounting reliance on intensive livestock farming, with its massive production of animal wastes, is an increasing cause for concern in rural communities. A hog operation of 7000 animals will produce as much sewage as a town with the same number of inhabitants, but the treatment of the sewage is much more rudimentary. The installation of drainage tiles in cultivated fields may also provide a conduit for manure runoff to enter ponds and waterways. Many jurisdictions are contemplating more rigorous regulations, with the predictable conflict between farmers and rural residents. Complicating this picture is the fact that rural areas of North America are littered with abandoned, shallow, dug wells that have not been filled in. These wells, often the first source of water for a farm before a deep-drilled well was installed, serve as conduits for surface water to enter the water table directly, instead of being filtered through layers of sand and gravel. Another potential source of contamination is private shallow wells that are cross-connected to municipal water systems. Farms and rural residents may draw from such wells for watering livestock, irrigation, or other outside uses. If common plumbing is employed, there is a chance that back flow may contaminate the general water supply.

E. coli is capable of producing a witch's brew of toxins. The relative amounts of each will vary with the strain. Strains are typed serologically and for the virulent strain O157:H7, the letters O and H represent specific antigens. In enteropathogenic strains, a plasmid-encoded (see Chapter 8) adherence factor, K88, promotes attachment of the bacterial cell to mammalian gastrointestinal epithelial cells, causing diarrhea. Enteroinvasive forms like O157:H7 cause a severe, hemorrhagic diarrhea. These strains produce plasmid-encoded outer membrane polypeptides similar to those produced by *Shigella*. A heat-labile exotoxin causes an increase in intracellular cyclic AMP levels with resulting increase in electrolyte and water secretion into the lumen of the gut. A heat-stable toxin, STI, acts similarly but on cyclic GMP. STII is thought to work via prostaglandin E2. Verotoxins 1 and 2, in O157:H7, inhibit protein synthesis after binding to specific glycoprotein receptors on the cell surface. Death from infection frequently involves kidney failure following massive fluid loss and hemorrhage.

As a result of the Walkerton tragedy, the Province of Ontario passed legislation that more tightly governed training of personnel, required more rigorous testing and more provincial oversight (Safe Drinking Water Act, 2002), and introduced a system to remove threats to municipal drinking water at their source (Clean Water Act, 2006).

Wildlife is not exempt from the effects of bacterial organisms. In November of 1999, thousands of Loons and Merganser ducks were found littering the beaches of Lake Huron and Lake Erie. Your author saw four lying in a perfectly straight line all facing inland, as if they had just fallen out of the sky, which indeed they had. The cause of this epizootic was the bacterium *Clostridium botulinum* type E. This a different strain of the organism that usually causes botulinum food poisoning (Type C) and the toxin has the same effect of paralyzing voluntary muscles by blocking acetylcholine receptors. It is significant that all of the affected birds were fish eaters (piscivores). Type E is prevalent in aquatic environments in cold and temperate regions of North America and Europe. Several outbreaks of Type E botulism have occurred in people as a result of eating warm-smoked, contaminated fish. In contrast, Type C causes western duck fever and more often affects dabblers that feed in warm, shallow ponds on vegetation. Whether the November 1999 outbreak was connected to human activity is uncertain because outbreaks occur naturally. The possibility cannot be ruled out.

Review Questions

For Questions 1–10, define each of the following terms (answers can be found in the text):

1. Abiotic
2. Biotic
3. Eutrophication
4. Bioconcentration
5. Biomagnification
6. Acclimatization
7. Acclimation
8. Bioaccumulation
9. Aquatic
10. Marine

For Questions 11–20, answer true or false.

11. Species is a biotic modifier.
12. Pore water is bottom water in which sedimentary particles are suspended.
13. pH has no effect on the methylation of mercury by microorganisms.
14. Metals bound to carbonates in water become more toxic.
15. Concentration equilibria are established between bound and unbound forms of toxicants in water.
16. Methylmercury never forms from natural sources.
17. DDT is an example of a chlorinated hydrocarbon.
18. Chlorinated hydrocarbons have a short biological half-life.
19. Human fat may contain up to 10 ppm of DDT.
20. Organophosphate insecticides are reversible inhibitors of acetylcholinesterase.
21. All infectious organisms can be removed from drinking water by standard chlorination and filtration techniques.
22. *Giardia* and *Cryptosporidium* are protozoan parasites that can cause intestinal infections when present in drinking water.

23. The microlayer at the surface of the ocean concentrates many toxicants.
24. Heavy metals may interfere with the function of many marine and fresh-water species.
25. Contamination of bathing beaches with *E. coli* and other coliform organisms does not pose a threat to human health.
26. Farm well water is always safer than municipal water supplies.

For Questions 27–31, use the following code:

Answer A if statements a, b, and c are correct.

Answer B if statements a and c are correct.

Answer C if statements b and d are correct.

Answer D if statement d only is correct.

Answer E if all statements are correct.

27. a. Dioxin (TCDD) is a proven carcinogen for humans at levels widely encountered in the environment.
   b. Chloracne (a skin rash) is the most common toxic manifestation of TCDD.
   c. TCDD is used as a herbicide.
   d. TCDD is a contaminant of the herbicide 2,4,5-trichlorophenoxyacetic acid.

28. Which of the following symptoms is/are not characteristic of organo-phosphorus poisoning?
   a. Profuse sweating
   b. Tremors
   c. Confusion and other mental disturbances
   d. Constipation

29. Acidity in water can contribute to
   a. Leaching of lead from solder joints in plumbing
   b. Acceleration of the transfer of metals from soil to groundwater
   c. A shift of copper to the elemental cuprous form (Cu^{2+}) from the carbonate form
   d. The existence of pentachlorophenol in the undissociated, more lipid-soluble state

30. Polychlorinated dibenzodioxins (PCDDs)
   a. Have been associated with accidental poisonings
   b. Are used industrially as pure chemicals
   c. Are products of municipal and industrial incinerators
   d. Are not themselves toxic

31. Which of the following statements is/are true regarding polycyclic aromatic hydrocarbons (PAHs)?

   a. Liver parasites may increase the toxicity of PAHs in fish by impairing their detoxication.

   b. Behavioral abnormalities have been observed in experimental animals exposed to PAHs.

   c. Herring gull embryos have shown signs of "expanded chick edema disease," which has been associated with exposure to PAHs.

   d. Acid rain is a threat to the Great Lakes by increasing the toxicity of PAHs.

32. Which of the following statements is/are true about the Great Lakes?

   a. They constitute the largest, single supply of freshwater in the world.

   b. Toxic effects of chemical pollutants have been detected in wildlife.

   c. The introduction of foreign aquatic species has been a major problem.

   d. Government regulations have effectively controlled the dumping of poorly treated sewage into the system.

33. Which of the following is/are true regarding soil contaminated with toxic chemicals?

   a. Those with long biological half-lives cause the greatest concern.

   b. Transfer of contaminants to groundwater seldom occur.

   c. Foodstuffs grown on contaminated soil may themselves become contaminated.

   d. Human activity such as mining is the only source of metal contamination of soil and water.

Answers

11. True
12. True
13. False
14. False
15. True
16. False
17. True
18. False
19. True

20. False
21. False
22. True
23. True
24. True
25. False
26. False
27. C
28. D
29. E
30. B
31. A
32. A
33. B

Further Reading

Austin, A.A., Fitzgerald, E.F., Pantea, C.I., Gensburg, L.J., Kim, N.K., Stark, A.D., and Hwang, S.-A., Reproductive outcomes among former Love Canal residents, Niagara Falls, New York, *Environ. Res.*, 111, 693–701, 2011.

Boehm, P.D., Page, D.S., Neff, J.M., and Brown, J.S., Are sea otters being exposed to subsurface intertidal oil residues from Exxon Valdez oil spill?, *Mar. Pollut. Bull.*, 62, 581–589, 2011.

Broughton, E., The Bhopal disaster and its aftermath: A review, *Environ. Health Global Access Sci. Source*, 4, 6, 2005. http://www.ehjournal.net/content/4/1/6

Chen, Y., Organophosphate-induced brain damage: Mechanisms, neuropsychiatric and neurological consequences, and potential therapeutic strategies, *Neurotoxicology*, 33, 391–400, 2012.

Doyle, M.P., Pathogenic *Escherichia coli, Yersinia enterocolitica* and *Vibrio parahaemolyticus*, *Lancet*, 336, 1111–1115, 1990.

Edelstein, M.R., *Contaminated Communities: The Social and Psychological Impacts of Residential Toxic Exposure*, Westview Press, Boulder, CO, 1988.

Effects of Great Lakes contaminants on human health, http://www.epa.gov/glnpo/health/atsdr-ref.htm (accessed April 26, 2012).

Essler, D., Trust, K.A., Ballachey, B.E., Iverson, S.A., Lewis, T.L., Rizzolo, D.J., Mulcahy, D.M., Miles, A.K., Woodin, B.R., Stegeman, J.J., Henderson, J.D., and Wilson, B.W., Cytochrome P450 1A biomarker indication of oil exposure in harlequin ducks up to 20 years after the Exxon Valdez oil spill, *Environ. Toxicol. Chem.*, 29, 1138–1145, 2010.

Exxon Valdez, http://www.exxonmobil.com/Corporate/about_issue (accessed April 26, 2012).

Flint, R.W. and Vena, J., (eds.) *Human Health Risks from Chemical Exposure: The Great Lakes Ecosystem*, Lewis Publ., Chelsea, MI, 1991.

Fujii, Y., Hayashi, M., Hitotsubashi, S., Fuke, Y., Yamanaka, H., and Okamoto, K., Purification and characterization of *Escherichia coli* heat-stable endotoxin II, *J. Bacteriol.*, 173, 5516–5522, 1991.

Gensburg, L.J., Pantea, C., Fitzgerald, E., Stark, A., Hwang, S.-A., and Kim, N., Mortality among former Love Canal residents, *Environ. Health Perspect.*, 117, 209–216, 2009.

Gensburg, L.J., Pantea, C., Kielb, C., Fitzgerald, E., Stark, A., and Kim, N., Cancer incidence among former Love Canal residents, *Environ. Health Perspect.*, 117, 1265–1271, 2009.

Global warming: How climate change will affect wildlife, agriculture and energy, *ON Nature*, 49, winter, 2009–2010, http://www.ontarionature.org

Gotfryd, A., Aluminum and acid: A sinister synergy, *Can. Res.*, June/July, 10–11, 1989.

Great Lakes (The), toxic substances and human health, http://www.on.ec.gc.ca/community/classroom (accessed April 28, 2012).

Gul, E.E., Can, I., and Kusumoto, F.M., Case report: An unusual heart rhythm associated with organophosphate poisoning, *Cardiovasc. Toxicol.*, 12, 263–265, 2012.

Henderson-Sellers, B. and Markland, H.R., *Decaying Lakes: The Origin and Control of Cultural Eutrophication*, John Wiley & Sons, Toronto, Ontario, Canada, 1987.

Humphrey, H., Environmental contaminants and reproductive outcomes, *Health Environ. Digest*, 5, 1–4, 1991.

Kimbrough, R.D., Health impact of toxic wastes: Estimation of risk. In *The Analysis of Actual Versus Perceived Risks*, Covello, V.T., Flammm W.G., Rodricks, J.V., and Tardiff, R.G. (eds.), Plenum Press, New York, 1981, pp. 259–265.

Kishi, M., Initial summary of the main factors contributing to incidents of acute pesticide poisoning: Overview of findings, http://www.who.int/heli/risks/toxics/bibliographykishi.pdf (accessed May 04, 2012).

Korkeala, H., Stengal, G., Hyytia, E., Vogelsang, B., Bohl, A., Wihlman, H., Pakkala, P., and Hielm, S., Type E botulism associated with vacuum-packed hot-smoked whitefish, *Int. J. Food Microbiol.*, 43, 1–5, 1998.

Malins, D.C. and Ostrander, G.K., Perspectives in aquatic toxicology, *Ann. Rev. Pharmacol. Toxicol.*, 31, 371–99, 1991.

Miao, X.-S., Woodward, L.A., Swenson, C., and Li, Q.X., Comparative concentrations of metals in marine species from French Frigate Shoals, North Pacific Ocean, *Mar. Pollut. Bull.*, 1049–1054, 2001.

Mishra, P.K., Samarth, R.M., Pathak, N., Jain, S.K., Banerjee, S., and Maudar, K.K., Bhopal gas tragedy: A review of clinical and experimental findings after 25 years, *Int. J. Occup. Med. Environ. Health*, 22, 193–202, 2009.

Neff, J.M., Page, D.S., and Boehm, P.D., Exposure of sea otters and harlequin ducks in Prince William Sound, Alaska, USA, to shoreline oil residues 20 years after the Exxon Valdez oil spill, *Environ. Toxicol. Chem.*, 30, 659–672, 2011.

Paneth, N., Human reproduction after eating PCB-contaminated fish, *Health Environ. Digest*, 5, 4–6, 1991.

Philp, R.B.,Effects of pH and oxidant stressors (hydrogen peroxide and bleach) on calcium-induced aggregation of cells of the marine sponge *Microciona prolifera*, *Comp. Biochem. Physiol.*, 118C, 347–351, 1997.

Philp, R.B., Cadmium content of the marine sponge *Microciona prolifera*, other sponges, water and sediment from the eastern Florida panhandle: Possible effects on *Microciona* cell aggregation and potential roles of low pH and low salinity, *Comp. Biochem. Physiol.*, 124C, 41–49, 1999.

Philp, R.B., Leung, F.Y., and Bradley, C. A Comparison of the metal content of some benthic species from coastal waters of the Florida panhandle using high-resolution inductively coupled plasma mass spectrometry (ICP–MS) analysis. *Arch. Environ. Contam. Toxicol*, 44, 218–223, 2003.

Rathmore, H.R. and Nollet, L.M.J. (eds.), *Pesticides: Evaluation of Environmental Pollution*, Taylor & Francis Group, Boca Raton, FL, 2012.

Roberts, L., News and Comment. Learning from an acid rain program, *Science*, 251, 1302–1305, 1991.

Scripps environmental accumulation of plastic expedition (Seaplex), http://seaplexscience.com (accessed April 26, 2012).

Service, R.F., Rising acidity brings an ocean of trouble, *Science*, 337, 146–148, 2012.

Singh, S., Kumar, V., Vashisht, K., Singh, P., Banerjee, B.D., Rautela, R.S., Grover, S.S., Rawat, D.S., Pasha, S.T., Jain, S.K., and Rai, A., Role of genetic polymorphisms of CYPA1, CYP3A5, CYP2C9, CYP2D6, and PON1in the modulation of DNA damage in workers occupationally exposed to organophosphate pesticides, *Toxicol. Appl. Pharmacol.*, 257, 84–92, 2011.

State of the Great Lakes 2009, prepared by the U.S. Environmental Protectation Agency and Environment Canada, http://binational.net/ (click on State of the Great lakes 2009 highlights).

Suppressed report confirms international violations by Canadian mining compa-nies, www.miningwatch.ca/en/suppressed-report-confirms-international-violationscanadian- mining-companies (accessed May 02, 2012).

Swain, W.R., Human health consequences of consumption of fish contaminated with organochlorine compounds, *Aquat. Toxicol.*, 11, 357–377, 1988.

Walker, C.H., Hopkin, S.P., Sibly, R.M., and Peakall, D.B. (eds.), *Principles of Ecotoxicology*, Taylor & Francis Ltd., London, U.K., 1996.

Walkerton water inquiry, http://www.walkertoninquiry.com (accessed April 20, 2012).

Water consumption, http://www.conferenceboard.ca/hcp/details/environment/water-consumption.aspx#water (accessed April 24, 2012).

Zebra mussels in the Great Lakes region, http://www.great-lakes.net/envt/flora-fauna/invasive/zebra.html (accessed May 03, 2012).

4

Airborne Hazards

The work is going well, but it looks like the end of the world.

S. Rowland, codiscoverer of the CFC effect, to his wife.

Introduction

When potentially noxious substances are discharged into the atmosphere at a rate that exceeds its capacity to disperse them by dilution and air currents, the resulting accumulation is *air pollution*. It may take the form of haze, dust, mist (which may be corrosive), or smoke and may contain oxides of sulfur and nitrogen and other gases that may irritate the eyes, respiratory tract, or skin, and other substances that may be harmful to the environment or to human health. Absorption may occur in amounts sufficient to cause acute or chronic systemic toxicity. Air pollution has been greatly underestimated as a cause of illness and death. In May 2000, acting Canadian Environment Commissioner Richard Smith quoted government statistics indicating that smog adversely affected the health of 20,000,000 Canadians and caused 5,000 premature deaths annually in 11 major population centers. This is in comparison to 4936 deaths from breast cancer, 3622 from prostate cancer, 3064 from motor vehicle accidents, and 665 from malignant melanoma. Air pollution obviously is an important health hazard.

Types of Air Pollution

Air pollutants may be gaseous or particulate in nature, and particulates may be either solid or liquid.

Gaseous Pollutants

These are derived from materials that have entered into chemical reactions or combustion processes. They include carbon-based compounds like hydrocarbons, oxides, and acids; sulfur compounds such as dioxide, trioxide, and sulfides; nitrogen compounds (ammonia, amines, oxides); and halogenated substances (organic and inorganic halides).

Particulates

Particle or droplet size may range from 0.01 to 100 μm in diameter. The smaller particles are referred to as aerosols and can remain suspended, scattering light and behaving much like a gas. Below 10 μm, particles are capable of penetrating to all sites in the respiratory tract. Industrial particulates are usually solid and are carbonaceous, metallic, oxides, salts, or acids, and their porosity is such that they will absorb other gases and liquids.

Smog

The word is a combination of smoke and fog and is a popular term for a fairly uniform mixture of gaseous and particulate pollutants that accumulate over urban centers and persist for a prolonged period. Smog is a brown or yellow haze and it usually occurs during the phenomenon of temperature inversion when a high-level mass of cold air traps warmer air beneath it to prevent mixing and dispersion. An especially bad "killer smog" occurred in London, England, in 1952. It lasted over a week and it was responsible for about 4000 deaths, mostly from respiratory diseases. As a result, the Clean Air Act was passed in 1956, banning the use of soft coal for home heating.

Sources of Air Pollution

Air pollution may arise both from natural sources and from human activities. Volcanic eruptions, forest fires, and dust storms are natural sources, the importance of which should not be underestimated. The 1980 Mt. St. Helen explosion in Washington State pulverized half of a mountain and released millions of tons of dust. It affected weather patterns as far east as the Great Lakes. In 1912, a similar explosion of a volcano in Alaska released about 30 times as much dust as did the Mt. St. Helen one. The more recent eruptions of Mount Pinatubo in the Philippines, together with smoke from the Gulf War oil fires, were blamed for unusually cool summers and excessive rainfall throughout most of North America in 1991–1992. Additional major eruptions in the "ring-of-fire" are predicted for the future.

Human sources include discharge from coal-fired electrical generating stations, nuclear generating stations (although very much less), industrial emissions, and domestic heating. Transportation sources include passenger autos, trucks, diesel locomotives, etc. Pollution may arise from all sources of combustion, industrial fuming and volatilizations, dust-making processes, and photochemical reactions. Biological sources include microorganisms such as viruses, bacteria and fungi, pollen, and chemicals from decaying organic matter. The breakdown of pollution sources in industrial countries is approximately as follows: transportation 60%, industry 18%, electric generating 13%, heating 16%, waste disposal 3%. Considerable concern is arising over the problem of indoor air pollution. The hazards of sidestream cigarette smoke seem firmly established and this has led to increased restrictions on smoking in the workplace and in public buildings. Recent studies have shown that 4-aminobiphenyl, a potent human bladder carcinogen present in both mainstream and sidestream cigarette smoke, has been found in fetal hemoglobin, indicating that it crosses the placenta.

The importance of smoking as a cause of cancer cannot be overstressed. Lung cancer is now the leading cause of cancer deaths among women in Canada. In 1994, deaths of women from lung cancer approached 5600, while those from breast cancer were about 5400. Between 1982 and 1989, the overall incidence of cancer increased by 0.3% for women and 0.5% for men. In contrast, the lung cancer incidence in women increased by about 43%, while in men, it increased by about 8%. Statistics released in 2012 indicated that lung cancer incidence in men had declined as older men increasingly gave up the habit. The same is not true of teenagers unfortunately. Deaths from lung (20,100) cancer continued to outnumber deaths from the other three leading types of cancer: breast (5,200), colorectal (9,200), and prostate (4,000), for a total of 18,400. Between 1988 and 2007, total mortality from all cancers in Canada dropped 21% in men and 9% in women. This has been attributed to earlier diagnosis, better treatment, and the reduced incidence of smoking in men. The problem of lung cancer will be discussed further later (see also Chapter 1).

Other indoor pollutants may include formaldehyde gas (see Chapter 2), other toxic chemicals, particulates such as asbestos fibers and fiberglass wool, and radon-source ionizing radiation (see Chapter 12). Airtight houses and buildings, constructed during the energy crisis of the 1970s, increase the risk of adverse health effects. Industrial indoor pollution is a special problem. In Ontario, the Ministry of Labor has jurisdiction over levels of air pollutants in the workplace and defines acceptable limits under various conditions (see Chapter 2).

Atmospheric Distribution of Pollutants

Air pollution generally starts out as a local problem, but it may become global if the pollutants enter the atmospheric circulating system. Pollutants may enter the atmosphere in the form of gases, vapors (from volatile liquids), aerosol droplets, or fine dust particles (see Chapter 3 for a discussion of distribution of pollutants in the biosphere).

Movement in the Troposphere

The troposphere is the air mass up to an altitude of about 10 miles. In the upper troposphere, the winds are predominantly westerly and average 35 m/s (mps) to disperse pollutants worldwide in about 12 days. Vertical movement circulates air north and south from the equator in systems called Hadley cells. In a band from 30°N Latitude to 30°S Latitude other cells called Ferrel cells circulate air toward the poles. Speeds may reach 30 mps. Microscopic particles are retained for 1 or 2 months in the upper and mid-troposphere, and about 1 week in the lower troposphere (<1 mile or 1.6 km).

Airborne dioxins and similar compounds, chiefly from municipal and industrial incinerators, can be distributed over a distance of 1500 km. Half of the dioxins reaching the Great Lakes came from as far away as Texas. One-twentieth of all sources of dioxins accounts for 85% of the dioxins deposited in the Great Lakes region. The fetus and the breast-fed infant experience the highest body burden of dioxins.

Mercury is widely distributed throughout the world, even reaching significant levels in the polar regions. During the winter, pollutants from Eastern Europe and Russia are transported to the Arctic. These include mercury and chlorinated hydrocarbons. They reenter the atmosphere through sublimation of snow and ice. The continued widespread use of chlorinated hydrocarbons such as DDT could increase the pollutant load in the Arctic. Polar regions are the first places where the effects of climate change can be observed. By 1996, polychlorinated biphenyls (PCBs) were being detected in the Arctic.

Movement in the Stratosphere

The stratosphere extends from 10 to 30 miles above the Earth. Movement occurs very slowly, at the rate of a few cm/s, but particles may stay for 2 or 3 years at an altitude of 20 miles, and about 1 year at 11 miles. Certain gaseous pollutants such as freon, chlorofluorocarbons (CFCs), and some rare radioactive isotopes (e.g., krypton-85 from nuclear reactors; the T1/2 is 10.5 year) are not readily removed by physicochemical means and may persist in the atmosphere for very long periods. Recent studies suggest that fluorinated gases will persist in the atmosphere for 300–2000 years or more, depending on the chemical.

Water and Soil Transport of Air Pollutants

The subject of the exchange of pollutants among various components of the biosphere was introduced in Chapter 3. Gaseous atmospheric pollutants may be dissolved in rainwater and solid particles carried in it mechanically. Precipitation thus carries them into the soil and groundwater, and they can reach oceans, lakes, and rivers by runoff and soil erosion and deep aquifers by seepage. The oceans are the ultimate repository for pollutants and surface evaporation may conduct them back into the atmosphere. Several studies have confirmed this biospheric circulation of toxicants. In the 1950s, atmospheric tests of nuclear bombs resulted in widespread dissemination of radioactive fallout. Of particular concern was the presence of strontium 90, which exhibits chemical characteristics similar to calcium, including deposition in bone. Strontium 90 reached significant levels in cow's milk, other dairy products, and in fruit and vegetables, and concern about its accumulation in the bones of children was a major factor in the discontinuation of atmospheric nuclear testing. The estimated North American exposure from all anthropogenic radionuclides is estimated now to be <1 mrem/year (1 mrem = 10 µSv). In 1969, contamination of Antarctic snow with DDT was identified. The only way it could have reached there was through precipitation. Presently, the most compelling concern is the problem of acid rain, the pH of which may be less than 4. Acid rain may be deposited far from its source.

Types of Pollutants

Gaseous Pollutants

These include the following:

1. Sulfur dioxide (SO_2), which forms acid rain as sulfurous acid
2. Sulfur trioxide (SO_3), which forms acid rain as sulfuric acid
3. Nitrogen monoxide (nitric oxide, NO), oxidized to nitrogen dioxide (NO_2), which is part of photochemical smog and acid rain
4. Carbon monoxide (CO), a product of incomplete combustion, which forms carboxyhemoglobin in red blood cells that is incapable of transporting oxygen to the tissues
5. Ozone (O_3), which contributes to photochemical smog
6. Hydrogen sulfide (H_2S), which is very toxic
7. Various hydrocarbons (C_xH_y), from automobile emissions
8. CFCs, freon, vinyl chloride, and radioactive isotopes
9. Methane from several sources (rice paddies, termite mounds, livestock, wildlife, wetlands)

In 1999, the Sierra Legal Defense Fund drew attention to the fact that gasoline marketed in Canada had one of the highest sulfur contents in the world. This tends to defeat the pollution control systems in automobiles and even destroy catalytic converters. The result is increased emissions of sulfur dioxide, sulfate particles, carbon monoxide, nitrogen oxides, and hydrocarbons.

SLDF lawyers have intervened on behalf of Friends of the Earth in a court case to be heard in the fall. "Friends" requested data regarding the sulfur content of various makes of gasoline from Environment Canada to give consumers a choice, but five major oil companies took Canada to court to prevent the release of this information, claiming that it would cause them financial harm and jeopardize their competitiveness. The federal government has since announced that it will reduce the sulfur content of gasoline to 30 ppm by the year 2005 as part of a plan to reduce all automotive emissions. There has been some success in reducing tailpipe emissions from vehicles. Cars and trucks produced since 2004 have significant lower emissions. As an example of how this is helping, the city of London, Ontario has seen a dramatic reduction in smog alert days, and air pollution levels since 2008 are 65%–75% lower than the early 1990s.

Particulate Pollutants

1. *Dusts*: Fine particle solids may arise from sawdust, cement, grains, metals, rock (in quarrying operations), incomplete combustion of fossil fuels (producing particles of <1.0 μ), that is, smoke, and any other substance including chemicals (pesticides, etc.) existing in powder form. Particulate emissions from internal combustion engines are thought to be a major contributing factor to poor air quality in urban centers.

2. *Liquids*: Any liquid that forms droplets 1.0–2.0 μ in diameter will remain in suspension in air as a "mist" (e.g., sulfuric acid). Droplets <1.0 μ are defined as an aerosol. The term is also applied to solid particles of this size. It is important to note that water vapor is by far the most significant greenhouse "gas," accounting for about 85% of infrared trapping, but its level fluctuates widely.

Health Effects of Air Pollution

Acute Effects

Short-term exposure to hazardous levels of air pollutants may result in irritation to the eyes and the respiratory tract. Populations at high risk include the very young and the elderly whose respiratory and cardiovascular systems are not fully functional, people with asthma, emphysema, heart disease, and heavy smokers. These groups had the highest mortality rates during the killer smog in London, England. The accidental release of toxic chemicals from industrial plants has caused serious health problems and death, the most tragic being the release of 40 tons of methyl isocyanate from the American Cyanamid plant in Bhopal, India (see Chapter 3).

Chronic Effects

Long-term exposure to lower levels of pollution may result in, or aggravate, chronic bronchitis, pulmonary emphysema, bronchial asthma, and lung cancer. Cigarette smoke will cause all of these problems. Excessive secretion of bronchial mucus and a chronic cough are the hallmarks of chronic air pollution effects. Dust and other allergens, including pollen, 1–90 μ in diameter, can induce or trigger allergic reactions in susceptible people.

The relationship between tobacco smoking and lung cancer is well established and has been discussed earlier. A new and disturbing trend is emerging, however: the increase in the incidence of lung cancer in "never smokers." This trend is more common in women than in men. A "never smoker" is defined as one who has smoked no more than 100 cigarettes in a lifetime. The age-adjusted incidence rates of lung cancer in these people aged 40–79 years were 4.8–13.7/100,000 person-years for men and 14.4–20.8/100,000 person-years in women. About 20% of women with lung cancer have never smoked. The disease appears to be different in this group and to respond differently to treatments. The cause of the disease is poorly understood but sidestream smoke, radon gas (see Chapter 13), and genetic factors have been considered contributing factors. The fumes from cooking oils at high temperatures have received special attention. This form of lung cancer is more common in Asian women in Hong Kong where this method of cooking is common. These "never smoker" patients seem more responsive to epidermal growth factor receptor inhibitors such as cetuximab. They also may have a mutation of the gene for anaplastic lymphoma kinase (ALK) and respond to a new drug, crizotnib, which stabilizes tumors with the mutation. There is also evidence that aspirin (acetylsalicylic acid [ASA]) has a protective effect. There is much yet to be learned about this disease but a possible relationship to air pollution should provide additional incentive to clean up the air.

There is a tendency to consider lung cancer as a single entity. It is, in fact, a complex, multifaceted disease. The following is a brief summary of the common lung cancer types:

Small cell lung carcinoma (SCLC) is of bronchial origin and accounts for 15%–20% of cases.

Nonsmall cell lung carcinomas (NSCLC) include the following:

1. Squamous cell carcinoma of the lung arises from the bronchial epithelium and is the most common form in smokers. It accounts for 25%–35% of lung cancers, and is more common in men than in women.

2. Adenocarcinoma of the lung usually arises from mucus glands and is more common in nonsmokers.

3. Large cell carcinomas (5%–10% of cases) are poorly differentiated and heterogeneous and do not fit other classes.

For a discussion on the genetics of carcinogenesis refer to Chapter 1.

Adverse Effects of Aerial Spraying

Spraying crops or forests from the air is a cheap, fast, and convenient method of applying pesticides, especially when large tracts of land are involved. The hazards of spraying dioxin (agent orange, TCDD), a contaminant of the herbicide 2,4,5-T were discussed in Chapter 2. DDT has also been sprayed aerially. The practice was banned in 1972 in North America due in no small part to Rachael Carson's book *Silent Spring*, which outlined the environmental damage it was causing, and to evidence that birth defects appeared to be more common in urban areas where spraying had occurred, mostly to control mosquitoes. At the time the defenders of DDT, mostly those involved in its development or employed in its manufacture, claimed that it was safe for humans. The same was commonly thought of the dioxin-contaminated herbicide sprayed over summer student employees in Ontario in the 1950s, 1960s, and 1970s. Cases of cancer are now showing up and investigations are under way as of 2011–2012. Surprisingly, the debate surfaced a few years ago over DDT's use for mosquito control in malarial areas of developing nations. The argument used then was that the benefits of malarial mosquito control far outweighed the damage to the environment. The fact that mosquitoes can become resistant to DDT did not seem to enter the discussion.

In more recent times biological pesticides have become available. They offer the advantages of no known toxicity to humans, other mammals, or birds, greater target specificity, and short biological half-lives. Two examples of these agents are the bacterium *Bacillus thuringiensis* (var. *kurstaki*), and lepidopteran pheromones, or more accurately their semisynthetic derivatives. *B. thuringiensis* attacks and destroys the digestive system of a wide variety of insects (it thus lacks much specificity within the class *Insecta*). Pheromones are volatile hormones released by a great variety of species that constitute a signal that the female is receptive to males. By flooding an area with the semisynthetic version, the males cannot home in on the females and the reproductive cycle is disrupted. While these agents are safe in themselves, aerial application requires that they be combined with several other chemicals to form a capsule-like substance that may also contain adhesives to facilitate deposition onto leaves and other surfaces. The formulation of these agents is usually protected by patent but what is known is that they may contain respiratory irritants and particle size may be small enough (less than 10 μm in diameter) to penetrate deeply into the lungs. The use of these agents over highly populated areas almost always (the author knows of no exceptions) results in an agitated populace with a small but significant component experiencing an array of symptoms that includes sore throat, bronchial congestion, stuffy nosed, headache, fatigue, nausea, abdominal pain, and skin rashes. The symptomology tends to be nonspecific and consists of rather common complaints, making it easy to dismiss as purely psychosomatic in nature. Government scientists all too often take this approach, perhaps partly because of a lack of controlled trials, most especially studying the complete formulation, demonstrating safety or lack thereof, partly because recognition of a health hazard might raise liability issues as most spraying is initiated by some level of government, and partly because these compounds appear to offer a safer alternative to chemical pesticides. Nonetheless, the frequency with which health issues follow aerial spraying should constitute a warning that although biological pesticides may be safer, they are by no means perfect. What follows are some illustrations of this (for a more detailed discussion of these events consult Philp, 2012).

Light-Brown Apple Moth

Light-Brown Apple Moth (LBAM) is an invasive pest with the capacity to damage an extremely wide range of important agricultural crops. In February of 2007 LBAM was detected along the coast of California, although some experts believed it had probably been in California for years or decades. That summer, the California Department of Food and Agriculture (CDFA) began preparations to aerial spray large areas around Monterey Bay in Monterey and Santa Cruz Counties with the leptidopteran pheromone active against LBAM. The spray program was begun early in the autumn (September 9–13). Local activists objected on the grounds that no environmental impact assessment had been conducted but the CFDA claimed that an emergency situation had been declared, exempting them from this requirement. By September 24, 81 reports of illness were recorded by The Concerned Citizens Against Aerial Spraying. The aerial spray used in these applications is a microcapsule manufactured by two companies, one being sold under the trade name "Checkmate," manufactured by Suterra Inc., and the other marketed by "Concep" but also a Checkmate product. The Material Data Safety Sheet (MDSS) for these products list several precautions and safety hazards for handlers and applicators including "potentially harmful if swallowed, absorbed through skin or inhaled. Causes moderate eye and skin irritation" and "Applicators should avoid vapor or spray mist." Significantly, the MDSS for Consep/Checkmate states: "The toxicity of this product is determined by the toxicity of the pheromone active ingredient." There is no mention of the other ingredients. On September 24, the Santa Cruz Sentinel published a partial list of the other ingredients of Checkmate and received a "cease and desist" letter from Suterra that threatened legal action. In October, HOPE launched a lawsuit against CDFA alleging that Checkmate had made over 100 people ill including an 11-month-old infant who was rushed to hospital with severe respiratory distress.

One of the ingredients listed in the Santa Cruz article that was subsequently removed was polymethylene polyphenyl isocyanate. This is a known respiratory and dermal irritant. Isocyanates are a known cause of occupational asthma. According to a literature review conducted in 2007, "… sufficient evidence already exists on the potential risks of isocyanate skin exposure and the importance of preventing such exposures at work and during consumer use of certain isocyanate products." Following the lifting of the temporary restraining order on spraying, Governor Schwarzenegger ordered CDFA to resume the LBAM eradication program and to release the list of ingredients "to the maximum extent possible under U.S. trademark law." This was done in a press release on October 20, 2007. Significantly, there was no isocyanate on the list. However, 1,2-benzisothiozolin-3-one (BIT) was on the list. BIT is known to have corrosive properties and to be a strong skin and eye irritant.

Painted Apple Moth and the Asian Gypsy Moth

In the early part of the twenty-first century, parts of New Zealand were experiencing moth infestations that were threatening crops. In Auckland it was the painted apple moth and in Hamilton it was the Asian gypsy moth. Agricultural officials elected to use an aerial spray preparation of *B. thuringiensis* (var. *kurstaki*) (Btk, F48B). Prior to the proposed spraying program in Hamilton, the Auckland Regional Public Health Service published an extensive report after surveying the literature. Following a practice that appears to be the norm for government reports on this issue, their report is a rather glowing endorsement of the safety of Btk aerial spraying. It does, however, concede that "Some people may experience minor eye, nose, throat and respiratory irritation." The (previous) HRAs (health risk assessments) raised the possibility of asthma aggravation, which was considered "biologically plausible" and "Some people find the odor of F48B unpleasant. Some people may experience nausea, headache and other symptoms if exposed to unpleasant smells." The report states that long-term effects have not been reported but concedes that there is little reliable information available on this. Gastroenteritis was discounted as a problem, but the experience in both Auckland and elsewhere belies this claim as stomach cramps, nausea, vomiting, and diarrhea were often reported. Spraying occurred from 2002 to 2004.

The Hamilton experience differs from most similar ones in that spraying took place several times yearly over several years (as was proposed for the Monterey Bay region). Public outrage reached such a peak over concerns about adverse health effects that two events resulted. First, the New Zealand Government Ombudsman was convinced to conduct an investigation into the spray program and its possible health effects, and second, The People's Inquiry into the Impacts and Effects of Aerial Spraying Pesticide over Auckland, New Zealand, was formed to conduct an independent investigation. Both reports became public in 2006–2007. The Ombudsman's report was very critical of the manner in which the Ministry of Agriculture and Forestry (MAF) had handled the situation. In essence he felt that the MAF trivialized the possible adverse health effects arising from exposure to the spray, did not adequately inform the public of these effects, and it and the Ministry of Health were dismissive of public concerns and complaints. He states in his report: "In particular, there needs to be a clear official acceptance that although the numbers of people may not be great as a proportion of the community in the spray zone, there will, in raw numbers, be a significant number who the evidence indicates will require medical attention, and in some cases removal from the area to be sprayed. It is no light thing to be sprayed, perhaps repeatedly, with some substance the ingredients of which are to some extent confidential, and to have one's life substantially disrupted for what may be quite a lengthy period of time."

The report of the People's Inquiry, unfettered as it was by political niceties was even blunter. The inquiry received hundreds of written and oral reports from individuals and families who documented long-term health effects after the spraying. In addition to the symptoms noted earlier, people reported excruciating itching of the skin and other skin problems, muscle spasms, persistent fatigue, and exhaustion. The MAF and its contracted physicians tended to dismiss these as "generic" or "work-related allergies." A persistent, racking cough was common enough to be dubbed "the moth cough." Debilitating asthma developed in both adults and children. Serious illnesses were reported to be exacerbated by the spray. Numerous hospitalizations were reported.

In the spring of 2009, forestry officials determined that an outbreak of gypsy moth larvae in wooded areas on the west side of the city of London, Ontario, Canada was severe enough to warrant a spraying program to control the moths. The areas of concern are in close proximity to residential suburbs. The gypsy moth *Lymantria dispar* (L.) is an invasive species introduced to North America in the 1860s from Europe. The city elected to institute a spraying program to control the moths, using Btk. A commercial preparation, Foray48B, was used.

Following the initial application a number of residents experienced adverse health effects that caused them sufficient concern for them to hold a meeting of local residents and subsequently arrange the meeting with the city councilors to discuss their concerns. The signs and symptoms were typical of those reported elsewhere and outlined at the beginning of this section. At least two pet dogs in the neighborhood also were sufficiently ill to require the services of a veterinarian.

Other Incidents

Reported adverse reactions to aerial Btk spraying seem to vary greatly from locale to locale. In the Vancouver spraying program, nearly 250 people reported adverse reactions, mostly allergy-like and flu-like symptoms, and ground spraying personnel were frequently affected. During a Washington State program, over 250 people reported similar symptoms and six required treatment in emergency departments. In neighboring Oregon State postspray cultures were taken from various body sites and fluids from people during routine clinical examination. Btk was cultured from 55 individuals, and in 3 with preexisting medical problems it was considered to be a possible infectious agent. The authors stated their belief that the role of Btk as a possible pathogen deserved further study.

Despite the claims of some government reports that Btk does not cause gastroenteritis, there is compelling evidence to the contrary, besides the fact that symptoms of gastroenteritis (GE) are commonly reported after exposure to Btk spray. *Bacillus cereus* is closely related to *B. thuringiensis* and a known pathogen. The two organisms share the enterotoxins that are responsible for GE. *B. thuringiensis* has emerged as a fairly common isolate from ocular infections. Another very recent concern has emerged. *B. thuringiensis* was isolated from 21 patients with nosocomial bacteremia in two hospitals. These patients all had serious, underlying disease. *B. thuringiensis* was isolated from catheter tips, gauze, and the hospital environment. The authors demonstrated the ability of this organism to form biofilms on hospital devices that constitute a threat to patient's health. In conclusion it seems that Btk is not the benign organism it was previously believed to be.

Spraying with Conventional Insecticides

Conventional insecticides are still being used for aerial spraying, even in populated areas. The organophosphorus insecticide Dibrom® also known as Naled® is being used for mosquito control in south Miami-Dade in Florida. It is being sprayed by the U.S. Air Force Reserve out of Homestead. The U.S.A.F. released a statement that "The amount of insecticide in the air should not affect people or animals." (See Chapter 9 for more on organophosphorus insecticides.)

Glyphosate is the active ingredient in the herbicide Roundup®, which, in agriculture, is normally used in combination with a genetically modified crop that is resistant to the herbicide. Glyphosate also is used alone and thus nonselectively, as an aerial spray in forested areas of northern Ontario. It has become the most recent source of controversy regarding aerial spraying. Although banned in Ontario for cosmetic use, forest spraying has resulted in adjacent populace being affected with resulting protests by angry residents and local politicians. This controversy over glyphosphate aerial spraying mimics those of the other pesticides discussed earlier. Because glyphosate use is increasing and will likely continue to do so, the topic deserves considerable coverage using a more conventional science-oriented format.

There is disagreement regarding the toxicity of spray products. Once again this is in part due to the fact that they contain several ingredients that differ from product to product and that toxicity studies generally do not accurately reflect aerial exposures. Williams et al. in 2000 reviewed existing studies by both government scientists and academics. These included glyphosphate and its major breakdown product aminomethylphosphonic acid (AMPA) as well as the frequently used surfactant polyethoxylated tallow amine (POEA). General findings were that oral and transdermal absorption was low, both parent and metabolite were excreted unchanged. Toxicity studies failed to reveal any long-term effects and transient ocular irritation was about the only side effect reported. Busse et al. (2001) studied the effects of glyphosate at 100 times the concentration employed in the field, on soil organisms taken from ponderosa pine plantations. Even repeated exposures appeared to have little detrimental effect on the soil organisms.

There is little doubt that glyphosate is toxic when ingested deliberately or accidentally. Talbot et al. reviewed 93 cases and reported that the ingestion of large amounts resulted in erosion of the gastrointestinal tract and gastrointestinal hemorrhage. Other organ systems were also affected but less often. Several studies have linked herbicides with toxicity including increased cancer risks in farm workers, but it is not possible to identify effects due to specific agents in these.

Once again suspicion has fallen on the "inactive" ingredients of glyphosate preparations. Animal and cell culture experiments with POEA have shown toxic effects. Cardiovascular toxicity and death occurred in piglets infused with it at concentrations similar to those used commercially. There is convincing evidence that aquatic organisms are vulnerable to glyphosate toxicity. Fish, crustaceans, and algae all have been shown to be sensitive to glyphosate alone and in combination with POEA.

There seems little doubt that there is a strong likelihood that widespread use of these products will result in environmental damage. It also is clear from the aforementioned incidents that whenever people are subjected to aerial spraying with pesticides, regardless of their chemical nature, anxiety, anger, psychological and possibly physical consequences will result.

Air Pollution in the Workplace

Systemic poisoning has occurred to workers inhaling toxic levels of metals such as lead, arsenic, mercury, manganese, zinc, cadmium, as well as pesticides and drugs. Oxides of all of these metals plus those of copper, tin and nickel, and brass dust can cause a febrile reaction (fever, joint and muscle aches) called "metal-fume fever." Cutting with an acetylene torch generates temperatures high enough to vaporize metals, including lead. Workers exposed to vinyl chloride gas have a high incidence of hepatic angiosarcoma, an otherwise rare tumor. Pneumoconiosis or coal miner's lung results from the inhalation of coal dust with the formation of localized lesions with silica crystals, emphysema, fibrosis, loss of vital capacity and, eventually, right heart failure due to increased cardiac output to compensate for inadequate oxygenation of the blood. Organic solvents may be hazardous because of their CNS-depressing action.

Some recent studies have suggested that the offspring of firefighters have a higher incidence of birth defects in locales where firefighters, or their spouses, are responsible for washing their work clothes. This presumably is the result of the absorption of toxic contaminants on the clothing through the skin, although absolute confirmation of this risk source has yet to be confirmed.

Asbestos

Asbestos workers are exposed to a variety of health hazards including "white lung syndrome" (asbestosis, a form of fibrotic pneumoconiosis), carcinoma of the lung, mesothelioma (cancer of the pleural and peritoneal membranes), and possibly gastrointestinal cancer, although animal studies have not been able to confirm this. Malignant mesothelioma is a rapidly fatal cancer occurring most often 30–50 years after the first exposure. While it occurs most commonly in workers who have been exposed for many years, there are cases on record where the duration of exposure has been quite brief. The linings of the chest (pleura), abdomen (peritoneum), heart (pericardium), and testes thicken, fluid accumulates, and widespread metastases occur. This cancer occurs rarely in people not exposed to asbestos. There are several forms of asbestos fiber, and not all of them may cause mesothelioma. There is no doubt that the form known as crocidolite is carcinogenic, but controversy has centered on whether the form known as chrysotile is also carcinogenic. In order to cause pleural mesothelioma, asbestos fibers must traverse the lung and appear in the pleura. Chrysotile fibers will do this and they have been shown to cause mesothelioma-like lesions in experimental animals. The risk associated with chrysotile fibers has not been firmly established in humans. According to some studies, a very large number of fibers must be inhaled for this to occur.

Both SCLC and NSCLC may occur in both smokers and nonsmokers but these carcinomas of the lung occurs 60 times more often in asbestos workers who smoke than in those who do not. Asbestos becomes a hazard for the general populace when building insulation begins to break down or is disturbed during construction. Wear of brake linings releases asbestos particles in the air. (Asbestos is no longer used in brake linings for this reason.) There is increasing concern that glass wool fibers can cause the same type of cancer as asbestos. An excess in cancer incidence has been shown in workers in the glass wool industry, but no direct evidence linking this to the inhalation of fibers has been uncovered. In the Fiberglass Canada plant in Sarnia, an increased incidence was shown in the 2500 workers, but it was not statistically significant. In the United States, NIOSH recommended that allowable air levels of glass fibers in plants be reduced.

Asbestos is used in many cases where it is contained within other material; so-called encapsulated asbestos. While asbestos use has been discontinued in most Western countries, because of the long latency after exposure new cases continue to appear. It is still used extensively in developing countries and Canada continues to export asbestos.

Silicosis

This results from the inhalation of silica particles or silicates or other mineral fibers. Histiocytes are transformed into fibrocytes, alveoli harden, resulting in loss of elasticity and lung function. Emphysema results, as it does from cigarette smoking.

Pyrolysis of Plastics

Prior to about 1980, firefighters did not routinely wear a breathing apparatus unless dealing with a fire involving known toxic fumes. There is some evidence (still largely anecdotal) that firefighters who attended fires involving plastics are beginning to show increased cancer rates. It is now known that when polyurethane smolders, fine particles of degraded polymers are produced that may have toxic chemicals adsorbed to them. These release lytic enzymes in the lung to cause massive tissue damage and edema.

Dust

Even barn dust can be an environmental hazard in the workplace. It may contain dried fecal material, animal dander, protein from feed grains and hay, skin parasites, and microorganisms. A Scandinavian study found a high incidence of respiratory and other health problems in farm workers who spent a lot of time in hog barns. Thirty percent of workers lost work time due to respiratory problems.

An additional hazard associated with fine particle suspensions in air is explosion. Coal dust, grain dust, and wood dust, all have caused explosions and fires in coal mines, grain elevators, and saw mills. The April 28, 2012, issue of the *Vancouver Sun* reported that since 2009, at least five mill explosions and fires have been linked to high levels of wood dust ignited by electrical sparks. Injuries and deaths have occurred from such dust explosions.

Methane

Since methane is commonly found in seams of coal, methane explosions are an ever-present danger in coal mines. It was for this reason that miners in earlier times carried canaries into the mines as these would die following exposure to fairly low concentrations of the gas, thus providing an early warning system. Hence we have the expression "the canary in the coal mine" to denote something that provides advanced warning of some danger. It is estimated that 6000 coal miners die annually in accidents around the world. The combination of methane and coal dust is absolutely lethal. On May 9, 1992 in the province of Nova Scotia, Canada, a methane explosion in the Westray mine trapped and killed 26 coal miners. Concerns about the safety of the mine had been expressed a year earlier by Mr. B. Boudreau, a member of the provincial legislature. Allegations of graft and corruption resulted in criminal charges.

On April 5, 2010, 18 years after the Westray accident, an explosion in the Upper Big Branch coal mine of West Virginia killed 29 miners. A depressingly similar sequence of events followed. Almost immediately methane and coal dust were suspected as stated in a preliminary report released by the Mine Safety and Health Administration (MSHA). This suspicion was supported when federal officials stated that a methane detector had been found indicating that the methane level was at 5% at the time of the blast. Methane becomes explosive when it reaches between 5% and 15% of the atmosphere. Allegations of corruption followed.

CO and NO_2

Chemicals that are involved in atmospheric pollution may sometimes become a problem indoors. There is increasing concern over indoor events that involve the use of internal combustion engines. These include tractor pulls, monster truck rallies, and mud races. CO levels have been shown to peak as high as 250 ppm during such events. Peak levels should not exceed 30 ppm. NO_2 levels may also be elevated because of incomplete combustion. CO is colorless, odorless, and nonirritating. It can produce headache, nausea, and mental impairment. NO_2 is irritating and may cause pulmonary edema. High concentrations may be fatal.

There is growing evidence that particle pollution at levels encountered in the environments of most large urban centers may be more hazardous than previously believed. There are elevated incidences of premature deaths, hospital admissions, and a variety of health problems. There is a statistically significant association between acute exposures to particles and increased mortality regardless of the source of the particles or the climatic conditions prevailing at the time of exposure. This seems to suggest that the particles are the primary cause, although the mechanisms involved are not yet known.

The Centers for Disease Control in Atlanta issued a report that 23 million Americans were at risk because of exposure to particles <10 μm in diameter and concentrations >155 $\mu g/m^3$ of air (the 24 h average acceptable level is 150 $\mu g/m^3$). The EPA is considering setting new levels at a much lower concentration.

Air pollution in the workplace can take some strange forms. A 1999 report dealt with two workers in a cattle breeding station who were found unconscious on the floor of the laboratory in which samples of bull semen were frozen for storage. A tank of liquid nitrogen had been leaking, and displacing the air in the room. If they had not been discovered in time, they could have asphyxiated.

Multiple Chemical Sensitivity

Multiple Chemical Sensitivity (MCS) is one of a group of conditions that includes fibromyalgia and chronic fatigue syndrome, and probably sick building syndrome, for which there are no definitive diagnostic tests. There has been considerable debate and skepticism in the medical community over whether they are purely psychosomatic in nature or have a pathophysiological basis. Sometimes they are grouped into the general category of "idiopathic environmental intolerance" (IEI). MCS could be viewed as a form of (mostly indoor) air pollution that affects a very sensitive subpopulation.

MCS, previously called the environmental disease, the twentieth-century disease, chemical AIDS, total allergy, is a term applied to a vague array of symptoms that can be somatic (headache, fatigue, dizziness, nausea, muscle, and joint pain), cognitive (impaired memory, mental confusion, and poor concentration), apparently neurological (clumsiness and numbness), or affective (irritability, anxiety, and depression). Hyperreactivity to sound, light, or touch also occurs. Chronic fatigue syndrome (CFS), fibromyalgia, and acute anxiety syndrome may share some of these symptoms. It is of interest that higher levels of chlorinated hydrocarbons have been reported in a study of patients with CFS when compared to non-CFS controls, strengthening the possibility that chemical exposure plays a role; moreover, 20%–37% of patients with CFS report a significant degree of chemical intolerance as do 23%–47% of fibromyalgia patients. Patients with acute anxiety syndrome also have an increased incidence of MCS and of fibromyalgia.

MCS is often associated with an aversive reaction to an initial "sensitizing" exposure involving a chemical such as volatile solvents, perfumes, glue, marker pens, mothballs, fuels, and newsprint. Since diagnosis cannot rely on laboratory tests, these following criteria have been proposed as a basis for a working definition of MCS:

1. The initial symptoms are associated with an identifiable environmental exposure.

2. The symptoms involve more than one organ system.

3. Symptoms occur and recede in response to the presentation and withdrawal of predictable stimuli.

4. The symptoms are elicited in response to low-level exposure to a wide variety of diverse chemicals. A hallmark of the condition is that the level of exposure that will elicit a response is far below that which will produce a detectable effect in the general population.

5. No standard test of organ system function can explain the symptoms.

6. It might be added that some of the symptomatology can resemble an allergic reaction, although no specific antigen can be identified.

Women are more frequently affected than men. It has been estimated that 12%–16% of the U.S. population is affected to some degree, more than 80% of them women.

There does not appear to be a consistent pattern of laboratory findings suggestive of a specific immune defect. Moreover, there have been no animal experiments that show an immune deficit developing as a result of exposure to low levels of multiple chemicals.

Another term sometimes used to identify the group of conditions listed above is "unexplained medical symptoms" (UMS). The term "subjective health complaints" has been proposed as a less judgmental one. There is frequently an association with stress and depression, especially in persons with poor coping skills and excessive feelings of helplessness and hopelessness. Several studies from Scandinavian countries noted that almost all (96%) of several thousand subjects reported experiencing one of musculoskeletal (80%), pseudoneurological (65%), or gastrointestinal (60%) symptoms in the preceding 30 days but only a small percentage (13, 5, and 4, respectively) defined the complaints as substantial. This cluster of complaints is the most common cause of inability to work, long-term disability claims, and repeated visits to family physicians, often with unsatisfactory results.

The conversion of anxiety into physical symptoms is called somatization and is felt to be related to MCS. Anxious persons may detect fear-related information earlier than normal subjects and their complaints may be driven by their suggestibility, possibly reinforced by the physician's preoccupation with physical symptoms. A Scottish study of over 5000 men found that the prevalence and incidence of angina increased with the frequency of perceived stress but that the angina correlated poorly with objective measures of coronary blood flow. The cognitive activation theory of stress (CATS) defines learning as either stimulus expectancy or response outcome. The brain stores relationships between stimuli and outcomes (classical conditioning) or between responses and outcomes (instrumental conditioning). Lack of a perceived relationship between the act and the result can cause feelings of helplessness. When the relationship between act and result is disastrous, hopelessness develops. These can generate persistent high stress levels. This "sustained arousal" can interfere with the activity of pain pathways leading to sensitization. CATS may be regarded as a "fight or flight" reaction in which neither response is possible, thus leading to somatization as an alternate escape mechanism.

To determine whether MCS, or idiopathic environmental intolerance (IEI) is indistinguishable from somatoform disorder (SFD), that is, the somatization response to stress, 54 subjects with IEI, 54 with SFD (but no IEI), and 44 with neither were compared with respect to symptom scales, psychological questionnaires, and structured interviews for IEI, SFD, anxiety, and depression. Over half of the IEI subjects met the criteria for SFD as defined by the Diagnostic and Statistical Manual of Mental Disorders. This subgroup shared both symptoms and psychological features of somatization with the SFD group. The other IEI subjects were less impaired by their chemical sensitivity but were still different from the control group. The authors conclude that IEI is a variant of SFD.

The Minnesota Multiphasic Personality Index (MMPI-2) has been used to compare subjects with MCS, epileptic seizures, and nonepileptic seizures (another medically unexplained condition). The results strongly suggested a psychological component for MCS.

The search for measurable physiological changes associated with MCS has been ongoing. Some early studies reported modest changes such as depression of leukocyte, lymphocyte, and T-cell counts, and changes in levels of complements. This led to theories that there was an immunological component to MCS, possibly resulting from combined low levels of numerous chemicals. Lack of consistency in these parameters did not support this theory however.

Pregnant women have been shown to be more sensitive to odors than nonpregnant ones and less tolerant to noxious odors encountered in their daily lives. This sensory hyperreactivity reaction (SHR) did not apply to noise. The authors hypothesize that this feature may have evolved as a mechanism to protect the fetus from potentially harmful toxic agents (such as smoke from fires, natural toxic gases, etc.). While not directly related to the question of MCS, these observations raise the possibility that hormonal changes might alter the sensitivity of olfactory receptors and that these, and other peripheral and central receptors, could be affected by other factors as well.

One receptor candidate for this phenomenon is the vanilloid receptor. This TRPV1 or VR1 receptor is widely distributed both centrally and peripherally. It is activated by a broad range of chemicals, including many of those implicated in MCS such as volatile solvents (but not apparently pesticides). Stimulation of this receptor can increase nitric oxide levels and stimulate N-methyl-D-aspartate (NMDA) receptors. Both of these responses have been proposed as important to the proposed central role in MCS. The vanilloid receptor can be activated by a host of stimuli including some mycotoxins. The authors propose that this could account for some cases of sick building syndrome, which frequently progresses to MCS.

Evidence for SHR in patients with chemical sensitivity comes from studies using the capsaicin-induced cough reflex. Patients with upper and lower airway symptoms induced by scents and chemicals were compared a group of normal controls. All subjects displayed a dose-dependent cough response to the capsaicin, which was more pronounced in the patients. After the challenge the patients had a significant increase in nerve growth factor (NGF) levels that correlated with their cough sensitivity. It was concluded that there is a real and demonstrable SHR pathophysiology in patients with respiratory symptoms related to MCS.

MCS sufferers appear to process odors differently from nonafflicted individuals; show greater physiological responses to low levels of common chemical odors and report subjective symptoms more often; required more time to develop stable baselines when exposed to the scents. One study compared MCS subjects to normal subjects exposed to capsaicin. The cough reflex was induced in all subjects but the MCS ones were sensitive to a lower dose than the controls. Nasal lavage revealed a dose-dependent concentration of NGF in the MCS subjects.

While there is as yet no definitive explanation for the cause or causes of MCS, it seems probable that both psychological and neuronal factors will be involved. Smell is the most evocative of the senses and the nature of the odor (pleasant or unpleasant) does not necessarily determine the nature of the memory. Odors can trigger aversive reactions in sensitized individuals. The aversive response is frequently associated first with the workplace probably because it is often a stressful place.

There is a strong psychological component to MCS and other IEIs. Two or more of MCS, acute anxiety disorder, fibromyalgia, and sick building syndrome often coexist in the same individual and there are common symptoms. Greater suggestibility and higher body awareness have been proposed as predisposing factors. Lack of coping skills, feelings of helplessness and hopelessness resulting from a perception of being trapped in a situation (a job, unhappy relationship) play a role. Somatization becomes an escape route. But it would be a mistake to consider IEIs as exclusively psychological in nature. There is mounting evidence of an underlying atypical neurophysiological component in patients with IEIs. Some of the evidence discussed earlier that point in this direction includes the following:

1. Increased odor sensitivity in pregnant women suggests that hormonal factors could play a role. The fact that MCS occurs four times more often in women than in men reinforces this possibility.
2. The sensitivity of the vanilloid receptor to a host of inhaled chemicals makes it an attractive candidate for a central target in MCS.
3. Increased production of NGF in chemically sensitive individuals exposed to a capsaicin challenge indicates a real neurophysiological response.

As a working hypothesis it would seem best to view individuals suffering from MCS and other IEIs as having an underlying atypical neurophysiological component that renders them more vulnerable to hypersensitivity reactions that become manifested in the presence of external psychological stressors.

For every theory proposed, there are counterarguments. No one theory is accepted universally, nor is there incontrovertible evidence for any one of them. There is no question that, for the sufferer, the condition is very real and debilitating. Despite a scientific literature now numbering in the hundreds of papers we seem no closer to an answer. There is now little doubt that the mind can exert a great influence on somatic function, with accompanying neurochemical and endocrine changes.

Some hospitals have been establishing environmental suites, where no synthetic materials are used and the air is filtered with HEPA filters.

Victoria Hospital in Halifax started a pilot project a number of years ago that led to the opening of a full-time clinic devoted to environmental illnesses. They reported in the news media that one diagnostic test, the SPECT brain scan, held promise. SPECT stands for single photon emission computed tomography. News reports indicated that when a patient with MCS is exposed to a trace chemical, the SPECT showed subtle changes in brain chemistry.

Chemical Impact of Pollutants on the Environment

Sulfur Dioxide and Acid Rain

Over 100 million tons of sulfur dioxide from fossil fuels are emitted annually into the atmosphere around the world. Sulfur dioxide plus water (in atmospheric water vapor) forms sulfurous acid and, eventually, sulfuric acid in a complex series of reactions that involve shifts between gaseous and aqueous phases. A simplified summary is as follows:

$$2SO_2 \text{ (gaseous)} + O_2 \text{ (gaseous)} \rightarrow 2SO_3 \text{ (gaseous)}$$

$$SO_3 \text{ (gaseous)} + H_2O \text{ (aqueous)} \rightarrow H_2SO_4 \text{ (aqueous sulfuric acid)}$$

$$SO_2 \text{ (gaseous)} + H_2O \text{ (aqueous)} \rightarrow H_2SO_3 \text{ (aqueous sulfurous acid)}$$

Nitric acid also contributes to the acid rain problem. It can be formed from nitrogen oxides (NO_x) including nitrous oxide (N_2O), nitric oxide (NO), and nitrogen dioxide (NO_2).

$$NO_x + H_2O \rightarrow HNO_3$$

Nitrous oxide is released from the oceans and during biological processes in soil (the nitrogen cycle). It is a greenhouse gas as well as a source of acid rain.

The average retention time for sulfur dioxide in the troposphere is very short (about 2–4 days). The sulfuric acid thus formed is carried to the soil in precipitation (rain and snow). A pH as low as 1.7 was recorded in West Virginia in 1979. Battery acid and gastric acid are both about pH 1. Core samples of snow in the Arctic regions revealed a pH of 6.8 about 190 years ago versus 3.8 recently.

The author studied pH and salinity values (standard salinity units) in the coastal waters of the Florida panhandle and in river estuaries. There was a direct correlation between salinity and pH. At river mouths pH was lowest (5.5) as was salinity (5 units) and in the open waters of Apalachee Bay both pH (7.5–8.0) and salinity (25–30 units) were higher. Thus, the higher the concentration of freshwater, the lower is the pH. Evidence was also obtained that shifts in these parameters outside of certain ranges could affect biology of a species of sponge.

The anions in acid rain are SO_4^- (70%), and NO^- (30%). Acidity may affect the respiratory tracts of both people and animals. Asthma and allergy sufferers are prone to loss of lung function and respiratory disease.

The absorption of acid into the soil solubilizes metals such as aluminum, cadmium, and lead and facilitates their movement into vegetation and water, including drinking water. The accumulation of these metals may contribute to human diseases. Aluminum has been implicated in dementias, lead may affect the development of the central nervous system in infants and children, and cadmium can cause kidney disease (see Chapter 6). Acidification of lakes leads to a complete loss of aquatic life. It is estimated that up to 4000 lakes in Ontario have been so affected.

Paradoxically, although the Mount Pinatubo volcanic eruptions were partly responsible (along with *El Nino*) for the extremely cool, wet summer of 1992 in North America, the long-term effects are more likely to contribute to acid rain and global warming. There is some debate, however, about the extent to which air pollution and clouds may negate the effect of increased ultraviolet radiation exposure due to ozone depletion. The 1991 eruption injected 15–30 Mt of SO_2 into the stratosphere, which, within 1 month, was converted to H_2SO_4. This formed an aerosol that is expected to remain in the atmosphere for up to 3 years. The total aerosol load is estimated to be 10–20 times that produced by anthropogenic and other biological sources in the same year. Some models predict that ozone will be rendered more susceptible to degradation by atmospheric chlorine, and reflection of long-wavelength infrared may increase global warming (see later). In fact, a marked decline in atmospheric ozone began in 1991 but recovery was noted in 1993 and by 1994, it had returned to essentially normal levels. It is not known whether this is a long-term trend, or if the effect was attributable to the Mt. Pinatubo eruption.

For a more complete discussion of the chemistry of acid precipitation consult Baird and Cann (2004).

Chemistry of Ozone

In the stratosphere, at an altitude of about 20 miles, shortwave ultraviolet radiation converts O_2 to O_3, which, by direct absorption, prevents UV radiation from penetrating the Earth's atmosphere. When longer UV wavelengths are absorbed (>242 nm), O_3 is split back into O and O_2 and thus is recycled.

Ozone depletion is of considerable concern because it contributes to climatic change by allowing shortwave UV radiation to penetrate to the Earth's surface. Because this band is the ionizing form of UV radiation, ozone depletion is also associated with an increased risk of skin cancer. The layer is thinnest at the equator, so that the incidence of skin cancer in tropical and subtropical climates is greater than in the temperate zones. Light-skinned people are at greatest risk. The incidence of skin cancer in Texas is 3.8:1000 compared to 1.2:1000 in Iowa. The incidence of skin cancer is also increasing in northern climates as well, and warnings against unprotected sunbathing will be routine as each summer approaches. Sunscreen factors of 20 or more are recommended, as is avoidance of exposure between the hours of 1100 and 1500. An issue gaining traction recently is the excessive use of tanning beds, especially by young people under the age of 18. The UV radiation emitted by these devices is no different from that of the sun and the same dangers pertain. People overly obsessed with personal appearance have used tanning beds to an extent that seriously increases their risk of skin cancer especially melanoma. There have been suggestions that tanning beds should be regulated like tobacco and alcohol with age limits for their use.

The ozone layer is normally maintained at about 1 ppm but it can be depleted by the action of certain pollutants. Nitric oxide (NO) is a major offender in this regard. The following reactions can occur:

$$NO + O_3 \rightarrow NO_2 + O_2$$

$$NO_2 + O \rightarrow NO + O_2$$

Thus, NO will recycle to break down thousands of ozone molecules unless it reacts with another free radical; for example,

$$OH + NO \rightarrow HNO_3 \text{ (nitric acid)}$$

Chlorine

Chlorine, its oxides, and chlorine compounds such as CFCs, previously widely used as aerosol propellants and refrigerants, also contribute to ozone depletion. For example,

$$Cl + O_3 \rightarrow ClO + O_2$$

$$ClO + O \rightarrow Cl + O_2$$

For a complete review of the chemistry of ozone depletion refer to solomon, 1999. The appeal of CFCs was that they are chemically inert under nearly all conditions, and it was not until the impact of supersonic jet transports was studied that the effect of UV light on them was realized. The calculated ozone loss is presently 1% but this would increase to 10% if the release of 800,000 tons of CFCs annually were to continue.

CFCs are a particular concern because they are heavier than air. They reach the stratosphere by slow percolation vertically, driven by the concentration gradient. This means that even if their use were to stop immediately, their effect will not begin to decline for many years to come and it has not yet peaked. CFCs are themselves greenhouse gases, further contributing to the problem of global warming.

In 1988, the Montreal Protocol on Substances that Deplete the Ozone Layer was agreed to by most Western nations. The U.S. Environmental Protection Agency implemented the recommendations that required a 50% reduction on the use of fully halogenated CFCs within 10 years and a freeze on consumption of halons within 4 years. CFCs have been temporarily replaced by hydrofluorocarbons (HFCs) and hydrochlorofluorocarbons (HCFCs) in some jurisdictions. Although these destroy just as much ozone as CFCs, they dissipate much more quickly in the atmosphere. In November of 1992, 86 countries attended the United Nations Ozone Layer Conference in Copenhagen. The conference agreed to an accelerated ban on CFCs by 1995 instead of 1999 as originally proposed. France blocked a similar ban on, or reduction of, HFCs and HCFCs. Ironically, two physicians in Dortmund, Germany, may have rendered all of the hydro-, chloro-, fluorocarbon technology obsolete. The substitute, labeled Greenfreeze, is a simple, nonpatentable mixture of butane and propane that can be used in refrigerator units and which is not harmful to the ozone layer. The technology is available to anyone. Greenfreeze refrigerators are being marketed already by DKK Schjarfenstein in Germany.

Recent studies indicate that the Montreal Protocol is working. The atmospheric levels of ozone-depleting chemicals are declining and the ozone column is no longer decreasing. Midlatitude ozone is expected to return to 1980 levels by the middle of the twenty-first century. Outside of the Antarctic ozone hole, increases in UV-B radiation have been small. Since ozone depletion also contributes to the production of greenhouse gases, beneficial effects from the reduction in CFC levels has also been observed (see in the following also).

Climate Change

Global Warming Debate

The debate regarding the reality of global warming raged for decades. Mounting evidence of it has stifled the debate: evidence that includes shrinking polar ice caps, collapsing Antarctic ice shelves, retreating glaciers, disappearing northern sea ice causing onshore walrus stampedes with numerous fatalities, and the migration of species to previously uninhabitable areas. Salmon are showing up in the Beaufort Sea and appear to be colonizing the Mackenzie River. More seals are also showing up, and the banks of the Mackenzie are eroding in places because the permafrost is thawing. For the first time in their oral history, the Inuit at Sachs Harbor are experiencing thunderstorms with lightening, and seeing birds such as robins and barn swallows. Melting permafrost has caused mud slides, allowing an entire lake to spill into the sea.

Jeffrey Severinghaus of the Scripps Institute of Oceanography has been analyzing arctic ice for trapped isotopes of argon and nitrogen. (The polar ice caps are also a huge sink for CO_2.) His data suggest that there was an abrupt warming, by 16°C, at the end of the last ice age 15,000 years ago. This suggests that climate change may occur much more rapidly than previously thought. His work is published in the October 29, 1999 issue of *Science*. For a more detailed discussion of these aspects readers are referred to Philp (2012). Global warming skeptics have shifted their focus to the question of the anthropogenic causes, or lack thereof, of global warming. Despite that the some 2000 members of the International Panel on Climate Change (IPCC) seem unanimous in claiming that it is due in large part to anthropogenic activity, there is also a body of scientists who claim otherwise. Criticisms include the following:

1. Current warming trends are part of a natural cycle. There is no doubt that such cycles occur. According to an EPA climate change website, core samples from the Antarctic ice cap indicate several peaks and valleys of atmospheric CO_2 with peaks of 250–300 ppm occurring between 600,000 BC and year 1. This would reflect increased temperatures releasing trapped gas from ice and water. These peaks occurred eons apart. Conversely, between 1956 and 2006 atmospheric CO_2 rose from 320 to 380 ppm and the major spike occurred around 2000. This increase is greater and more rapid than any seen since 6000 BC when the level was 270 ppm. Similar spikes were seen for methane (CH_4) and nitrous oxide (N_2O). The natural factors affecting climate are discussed next.

2. Current models are too unreliable to predict future temperature increases. Computer models are always vulnerable to this charge. However, retro-analysis of historic data yielded "predicted" changes that closely matched the actual observed changes. The global average surface temperature rose 0.6°C ± 0.2°C since the late nineteenth century and 0.17°C/decade in the last 30 years. Current predictions are that from 1990 to 2100 the temperature will increase from 1.4°C to 5.8°C. A 2012 paper by Dutton and Lambeck sheds some rather disturbing light on the possible consequences of this increase. The last interglacial period was about 125,000 years ago. During that period sea level peaked at between 5.5 and 9 m higher than the current level. Temperatures were estimated at 1°C–2°C warmer than the present day. If history should repeat itself, coastal areas around the world will be flooded to a considerable distance inland.

3. There are scientists who argue that the cause of global warming is unknown.

4. Some who argue that global warming will have few negative consequences. Since negative consequences have already been observed, this argument seems specious.

5. There are some scientists who feel that global warming is most likely to be the result of a combination of natural and anthropogenic influences. To this author this seems the most plausible position. It raises the possibility that humankind could be the straw that breaks the "climate camel's" back by accelerating warming so that the tipping point (the point of no return) is reached more quickly. This position removes the justification for doing nothing or very little, while recognizing that natural factors play a role.

Chemistry of Climate Change

Water

Water vapor is undoubtedly the greatest single contributor to the greenhouse effect, accounting for up to 75% of heat trapping. High-flying jet aircraft produce millions of liters of water as vapor annually.

Carbon Dioxide

Another gas of concern regarding global warming is carbon dioxide (CO_2). CO_2 is a product of combustion and decay of organic matter. Green plants normally consume CO_2; these take it up during the day, converting it to O_2 and carbohydrates by photosynthesis. It is a basic rule that the oxidation of one carbon atom yields one molecule of carbon dioxide:

$$\text{Combustion, decay: } C + O_2 \rightarrow CO_2$$

$$\text{Photosynthesis: } CO_2 \rightarrow O_2 + \text{carbohydrates}$$

$$\text{Reaction in atmosphere: } CO_2 + H_2O \rightarrow O_2 + CH_2O \text{ polymer}$$

This last reaction reverses in winter and returns the CO_2 to the atmosphere. Depletion of rain forests for timber and increased CO_2 production from industrial sources and internal combustion engines have led to a dramatic increase in atmospheric CO_2. Solar energy is either reflected from the surface of the Earth or absorbed by it, in which case it heats the Earth. Very little heating of the air occurs, which is why air is cooler at high altitude. Most of the reflected energy passes back out into space but some is at a very long wavelength (50,000 nm) because the Earth is much cooler than the sun, and it is reflected back to Earth by water vapor and CO_2, the "greenhouse effect." The result is additional global warming (see in the preceding text regarding controversy).

By 2030, sea levels are predicted to rise 8–29 cm and continental interiors (the breadbaskets of the world) could go dry in summer. The American southwest has already experienced drought conditions and massive dust storms. The water displaced by huge volumes of ice sliding into the ocean could have serious consequences for many, low-lying coastal areas. (For a further discussion of polar ice, see Philp, 2012.)

Some lessons can be learned from history. In the late autumn of 1815, a huge volcano called Tambora in Indonesia exploded with a force many times that of Mt. St. Helen. The resulting dust spread over the surface of the Earth and 1816 became known as "the year without a summer." Air temperature dropped an average of 1°C. Crops were severely affected in the United States and it snowed in June in Maine. It was so cold and wet in Switzerland that summer that Mary Shelly challenged her husband to a writing contest for something to do. The result was "Frankenstein." Fortunately this situation was temporary, but it illustrates how little a change in temperature is required to effect drastic climatic changes.

Methane

Methane is a greenhouse gas, reflecting radiant energy back to the Earth. As a fuel, it has a lot of advantages (see later) but uncombusted methane reaches the atmosphere from fermentation in the intestinal tracts of animals (both wild and domestic) and from the fermentation (rotting) of vegetation. Rice paddies are a major source of methane. It has been suggested that the North American penchant for burgers is a contributing factor to the greenhouse effect by encouraging extensive beef production (and rain forest destruction) in South America. Animals and rice paddies each produce about 100 MT of methane annually, and since rice is one of the world's staple grains, a switch to a vegetarian diet might not result in a net methane saving. The biggest source of methane in the world is its termites. They produce about 200 MT annually. Since they digest dead wood, their population is expanding because of deforestation, and this may be the greatest threat to the environment from methane. There has been some controversy over the contribution of wetlands to methane production. While some authorities claim that this has been overestimated because of failure to account for uptake of methane by boreal soil, others have pointed out that fens, bogs, and beaver ponds are significant contributors because of their large area. Estimates of net contributions are 1.5–3.5 $g/m^2/year$. Like ozone levels, levels of polluting gases in the atmosphere, notably methane and carbon monoxide, have been returning to normal in recent years, and this too has been suggested to be the result of the declining influence of the Mt. Pinatubo explosion.

The ocean, and suboceanic oil wells, may be important, and underestimated, sources of atmospheric methane. At the low temperature and high pressure encountered on the very deep sea floor, methane hydrates will form and cover the sea bottom, with methane gas trapped below the hydrates. Underwater landslides can scour the hydrates, releasing vast amounts of methane into the ocean and atmosphere. Oil well blowouts also may release huge volumes of methane. Since gasified water will not provide buoyancy, any object on the surface, such as a ship or oil rig, will sink instantly. Over 40 rigs have been lost in this manner, with few survivors. There are also huge reserves of methane trapped as hydrates in arctic ice. These will be released as ice melts in response to global warming.

Sulfur Dioxide

SO_2, a major contributor to acid rain, may also contribute to global cooling. Aqueous-phase oxidation of SO_2 to SO_3 in clouds (rich in water vapor) occurs, and evaporation of the water releases a particulate aerosol of SO that backscatters solar radiation.

Motor Vehicle Exhaust

As noted earlier, the internal combustion engine is a major source of CO_2 and other pollutants such as lead and CO. Polar ice in Greenland has shown a sharp increase in carbon and lead deposits since the 1950s. Lead rose from 0.03 to 0.20 µg/kg snow or ice. Levels corresponding to the era of 800 BC were <0.001 µg/kg ice. In urban areas the lead content of the air is 5–50 times higher than in the country (1.1 vs. 0.02 µg/m³). In 1968, the gasoline combustion engine accounted for 181 of the 183 ktons of lead released into the atmosphere annually in the United States. Other sources included coal and oil combustion, and manufacturing processes including lead smelting. Vehicles using tetraethyl lead gasoline emit, on average, 1 kg of lead/year. Distribution and eventual inhalation of this lead by humans constitutes a health hazard. At levels of about 10 µg/dL (ingested) it is a potent neurotoxin to infants and children, causing impaired hearing, slowed neuronal transmission, and a variety of behavioral problems including hyperactivity, learning disabilities, and reduced mental capacities (see also Chapter 6). Lead from solder joints in plumbing is also implicated as a source. In the United States full phaseout of lead (as tetraethyl lead) in gasoline was enacted by the EPA in 1986. In Canada, regulation of lead content came into effect in 1990. Presently, 99.8% of gasoline sold in Canada is lead-free. Piston aircraft and some competition vehicles may use leaded gasoline.

Carbon monoxide (CO) is a product of all forms of combustion and it has been responsible for many deaths due to the use of heating devices (camp stoves, space heaters, defective furnaces, fireplaces, etc.) in poorly ventilated sleeping areas, including tents. Since the internal combustion engine is the principle source of CO (it has been used as an instrument of suicide many times over), it is convenient to consider CO toxicity here. Its natural concentration in the atmosphere is extremely low (1–2 ppm) and about 100 Mt yearly arise from vehicle engines in the United States. Other sources include natural processes (70 MT) and other human combustion sources (250 MT). Urban concentrations may range from 20 to 100 ppm depending on traffic and weather conditions. Firemen are at risk from CO poisoning if a breathing apparatus is not worn. Combination with free radicals in the environment helps to buffer the atmospheric levels of CO:

$$CO + OH^- \rightarrow CO_2 + H^+$$

CO combines irreversibly with hemoglobin (Hb) to form carboxyhemoglobin, which cannot combine with O_2 so that asphyxiation results. Levels greater than 6.4×10^3 ppm cause dizziness and headaches within 2 min and unconsciousness and possibly death in 15 min. Treatment consists of intravenous nitrates that cause the formation of methemoglobin. This has a much greater affinity for CO and binds with it to free the Hb. If a hyperbaric facility is available, O_2 at high pressure (2 atm) will provide adequate oxygenation through dissolved O_2 until enough Hb is free to assume oxygen transport.

Subtle Greenhouse Effects

Significant impact can occur long before the aforementioned catastrophic events. As temperatures climb, diseases of humans, livestock, and plants, which are normally constrained by subfreezing winter temperatures, will advance northward. Those depending on insect vectors will gain a foothold more rapidly as insect populations soar with increases in temperature. The cold barrier to Africanized honeybees will move north and their spread will be accelerated. Crop failures are bound to occur when crops are assailed by both drought and disease. This will necessitate increasing dependency on pesticides and irrigation, further stressing the environment. There is already evidence that expanding human populations modify the environment, and that diseases adapt to the new conditions, in some cases evolving to become new threats to human health.

Global Cooling: New Ice Age?

Studies of ice core samples from the polar ice cap suggest that the last ice age, in which the world was covered with ice to a depth of nearly 1000 m as far south as New York, resulted from a mean temperature drop of no more than 1.5°C. Ice ages result from irregularities in the Earth's orbit around the sun, which take it further away about every 200,000 years. Interglacial periods last about 10,000 years, and there is some evidence that the present one is ending soon.

Complicated, isn't it? It is even further complicated by the fact that the oceans act as CO_2 sinks, dissolving vast amounts of the gas. As gas is dissolved, the greenhouse effect is reduced, the temperature drops, and more snow and rain fall, contributing to the polar ice caps. As they expand, they reflect more and more UV rays back into the stratosphere, further lowering the Earth's temperature. The reflective property of materials is known as the albedo factor, and the efficiency of ice is nearly 100%. See the following for more on natural factors affecting climate.

Natural Factors and Climate Change

Although much has been made of the impact of human activity on climate change, the influence of natural events is often ignored, as is the fact that nearly all computer simulation models are based on a doubling of CO_2 without reference to other factors, including water vapor, discussed earlier. Methane is another natural greenhouse gas (see in the preceding text for climate debate, and Chapter 14).

Contrary to statements in many texts, the Earth's orbit is not a fixed ellipse, but will vary considerably. Moreover, its tilt may vary by several degrees. Both effects are due to the gravitational pull of Jupiter and they can have a dramatic impact on climate. Some estimates claim that they are responsible for up to 85% of climate variation. Solar activity such as flares and sunspots also correlates well with climatic changes. The 1988–1989 season was very active, with over 400 sunspots recorded. The activity of *El Nino*, the Pacific upwelling of warm water from the depths, also correlates well with sunspot activity and it has a marked effect on climate.

From 1675 to about 1725 there was a "Little Ice Age," with record cold temperatures and snowfalls. There was ice skating on the Thames River in London, England and elsewhere in Europe. Since then the climate has been warming. Should this trend have reversed? Cooling trends tend to occur fairly slowly, whereas warming trends are much more abrupt. Ice cores taken from the Vostock 4 site in Russia, going back 800,000 years, measured CO_2 levels in bubbles in the ice and these indicated a natural cycle of about 100 years. This cycle appears to be shortening dramatically. The observatory at Muana Loa in Hawaii recorded atmospheric CO_2 levels of 315 ppm in 1958, rising to 355 ppm in 1989. Methane levels are rising even faster, from 400 to 1800 ppb in the same period. There is little doubt that human activity has contributed to these changes, but the seas are a vast sink for CO_2 and a natural warming trend would also release CO_2 from this sink. Increased water vapor from the oceans would increase cloud cover, which would probably have a buffering effect on the warming trend. Nevertheless, some models predict a 4 ft (1.1 m) rise in sea level over the next 50 years unless the trend is reversed, submerging Bangladesh and the Maldives. In northwestern Ontario is the Experimental Lakes Area, which has been the subject of climatic, hydrologic, and ecological study for over 30 years. Records show that both air and lake-water temperatures have increased by 2°C during this period and that the ice-free season has lengthened by 3 weeks. The thermocline has also deepened. Evaporation has been higher, and precipitation lower, than normal, with the result that lake levels have dropped and pollution has concentrated, all of which have ecological consequences.

Glaciers and small ice caps respond much more rapidly to climate changes than do the polar ice caps and thus they are useful for computer modeling of short-term effects of warming. In the last 100 years, mountain glaciers have retreated in most parts of the world and have likely contributed to the 10–20 cm rise in sea level during this period. A recent study of 12 such areas where data collection has been in progress for several years indicted that for a 1 K rise in temperature, glacial melting would account for a rise in sea level of 0.58 mm/year, which is significantly less than earlier estimates.

Remedies

How all these factors will balance out is anybody's guess, but it is clear that we do not know enough to take chances. As one wag put it, Chicken Little only has to be right once. It is abundantly clear, however, that a reduction in fossil fuel consumption is essential. This means more fuel-efficient cars, changes in transportation habits with more public transit usage, and the development of alternative fuels. One such is natural gas (methane) for heating and for automotive fuel. Although methane obeys the one carbon–one CO_2 rule, and although it is a greenhouse gas itself, it produces more energy per carbon with fewer polluting by-products than petroleum oil-based fuels, so that the negative impact on the environment is much less. Further development of nuclear energy may also be essential. Despite the emotional reactions that it can generate, a well-designed, well-regulated, and well-operated nuclear power generator may be the safest source of electrical energy that we have. It is worth remembering that not one proven death has resulted in North America from a nuclear generator (see also Chapter 2) but hundreds have died from failures of hydroelectric dams and thousands from coal mining accidents and black lung disease. In 1979, a power dam in India collapsed, killing thousands. Time carried a few lines, compared to the several pages devoted to Chernobyl. In 1989 in the (then) USSR, a gas pipeline explosion killed hundreds of people in two passing passenger trains. The Atomic Energy Commission of Canada did a study indicating that nuclear energy is even safer than solar energy since the latter requires materials that need to be mined and refined with energy derived from coal. The calculation determined that solar energy was inherently 500 times more dangerous than nuclear energy. Fossil fuels (coal and oil) were 2000 times more dangerous; natural gas was the safest of all.

Hydrogen may also become a source of pollution-free energy for transportation through the development of hydrogen fuel-cell-powered vehicles. Fuel-cell-powered buses have already been tested in California and Honda has developed a fuel-cell-powered car, the FCX Clarity, already being marketed to a limited extent. Honda presented the keys to one to Canadian 2010 Olympic hockey team captain Scott Niedermayer. The limiting factors for these cars are the relative expense and the lack of an infrastructure for hydrogen fuel. Fuel economy has already improved greatly for some, but not all, models comparing 2012 fuel consumption to that of 2000. The Canadian government maintains a website listing fuel ratings for most vehicles and most years (http://oee. nrcan.gc.ca/transportation/tools/fuelratin).

A theory that was popular a number of years ago is undergoing a renaissance. Known as the Gaia hypothesis and developed by James Lovelock, it considers the Earth and its atmosphere as a living organism, with all parts interconnected and in balance, like a mammal in a state of homeostasis. Interference with one component of the system may have far-reaching consequences that may not be foreseen with our incomplete understanding of how the system works. One essential change that must occur is a dramatic reduction in the use of fossil fuels by developed countries, especially in North America. Northern developed countries use 7.5 kW/person/year, versus 1.0 for southern, developing countries. Given the populations of 1.2 billion for the former and 4.1 billion for the latter, the north consumes over double the energy of the south. Add to this the fact that 6 million hectares (1 hectare = 2.471 acres) of arable land are lost annually from erosion, development, and salination, whereas 4 million are created by clearing, which entails the loss of irreplaceable rain forests, and it becomes evident that major changes are essential. One of these is a check of the population explosion. The doubling time of the Earth's population is 30 years and of its energy consumption, 20 years. It is evident that we cannot continue in this manner if our species, and the planet, are to survive.

Case Study 3

In August of 1989 a previously healthy 4-year-old boy developed signs and symptoms that included leg cramps, rash, itching, excessive perspiration, rapid heartbeat, intermittent low-grade fever, personality changes, and peripheral neurological disorders. The interior of the house had been painted 1 month earlier using over 60 L of a latex-based paint. The house was sealed and air-conditioned.

Q. Is there likely to be any connection between the boy's illness and the redecorating?

The signs and symptoms strongly resemble a condition known as acro-dynia, a rare form of childhood mercury poisoning. A 24 h urinary mercury determination revealed a level of 65 µg/L. The boy's mother and two siblings had similar urine mercury levels.

Q. Do these results point to an environmental source of the mercury?

Q. What additional analysis or analytical information would be useful?

Case Study 4

In the same year (1989), two farm workers in their 30s were working in an indoor manure pit 25′ × 25′ × 5′ deep, attempting to clear a blocked pump intake pipe. Several hours later, both men were found face down in several inches of liquefied manure.

In an unrelated incident, five men died after consecutively entering a similar pit, each attempting to rescue those who had entered before. The first person to enter was attempting to replace a shear-pin in a piece of equipment.

Q. What is the underlying cause of death in both of these incidents?

Q. What simple safety measure could have prevented these tragedies?

(See also Case Study 2)

Case Study 5

Two machine shop workers were cleaning and degreasing equipment in a confined, poorly ventilated space using a CFC known as CFC-113. After about 20 min, one worker clutched his chest and collapsed. He was cyanotic and not breathing. The second worker called for help and began to administer artificial respiration (he was not trained in CPR) but he himself collapsed soon afterward. Both victims were evacuated. The first was pronounced dead on arrival at hospital, the second recovered in the ambulance after oxygen was given.

Q. What was the portal of entry of the toxicant?

Q. What organ systems were involved in the intoxication?

Q. What was the likely offending agent?

Q. What steps should have been taken to prevent this accident?

Case Study 6

Soapstone carving is an economic mainstay for the Inuit of the Canadian north. Recently, the Inuit Art Foundation created a comic-book superhero called Sanannguagartiit ("your carving buddy" in the Inuktitut language). With his flying snowmobile and his loyal sled dog Quimmiq, he crisscrosses the Arctic demonstrating masks and respirators and advising carvers to leave their work clothes at the carving hall and not to take them home.

Q. Why would it have been necessary to institute this program?

Case Study 7

Workers in a particular industry have been having a very high incidence of respiratory problems including

1. A high frequency (15.7% of workers) of chronic bronchitis
2. A sevenfold increase in restrictive lung function and a threefold increase in obstructive lung function compared to the general population
3. A 20% occurrence of shortness of breath and flu-like symptoms such as fatigue, muscle and joint pain, and general malaise
4. A 30% annual occurrence of absenteeism due to respiratory illnesses
5. A 35% occurrence of wheezing and tightness in the chest
6. A 50% frequency of chronic, productive cough and frequent colds

Q. Is this most likely a primary infection or a pollutant?

Q. If the latter, what is the likely portal of entry?

These men are farm workers in the pork industry.

Q. What contaminants could be present in the air in pig barns?

Q. What corrective measures should or could be taken?

Case Study 8

In Wisconsin an incident was reported to public health authorities in which 11 high school students were treated in two different emergency departments for acute respiratory symptoms including labored breathing (dyspnea), spitting (coughing) blood (hemoptysis), cough, and chest pain. Two of them required hospitalization. All of them had participated in an ice hockey tournament the previous evening in an indoor arena. Interviews with other players and spectators revealed that many of them had suffered respiratory problems that became progressively worse as the evening wore on, and many had central nervous system symptoms including headache, dizziness, sleepiness, nausea, and vomiting. Of 131 students who were interviewed, 48% reported symptoms. The frequency among players was double that among spectators.

Q. What is the likely portal of entry in this toxic reaction?

Q. What offending agent(s) would you suspect?

Q. What laboratory test would be helpful?

Review Questions

1. Which of the following sources, in industrialized countries, accounts for the greatest amount of air pollution?
   a. Industry
   b. Generating electric power
   c. Transportation
   d. Heating
   e. Waste incineration

2. One reason for stopping atmospheric nuclear tests was concern over the deposition of a radioactive isotope in the developing bones of children. The isotope of concern was
   a. Cobalt 60
   b. Strontium 90
   c. Iodine 125
   d. Cesium 133
   e. Carbon 14

3. The diameter of droplets or particles defined as aerosols is
   a. $>10 \, \mu$
   b. From 5 to 1 μ
   c. Less than 1 μ
   d. From 1 to 2 μ
   e. From 2 to 5 μ

For Questions 4–7, use the following code:

Answer A if statements a, b, and c are correct.

Answer B if statements a and c are correct.

Answer C if statements b and d are correct.

Answer D if only statement d is correct.

Answer E if all statements (a, b, c, d) are correct.

4. Which of the following can be sources of indoor, household pollution?
   a. Radon gas
   b. Cigarette smoke
   c. Asbestos fibers
   d. Formaldehyde gas

5. Pollutants that can contribute to acid rain include
   a. Sulfur dioxide
   b. Sulfur trioxide
   c. Nitric oxide
   d. Chlorofluorocarbons

6. Acid rain
   a. May solubilize toxic metals
   b. Can never be neutralized naturally
   c. Can kill aquatic life
   d. Is never the result of a natural phenomenon

7. Which of the following statements is/are true?
   a. The troposphere is the air mass in contact with the Earth's surface and it extends vertically for about 10 miles.
   b. Hadley cells are vertical air movements that circulate air northward and southward from the Equator.
   c. The stratosphere extends from about 10 to 30 miles above the Earth.
   d. Pollutants are rapidly cleared from the stratosphere.

8. List four groups of individuals who are at greater than average risk of respiratory problems from air pollutants.

9. List four conditions known to be associated with air pollution in the workplace.

10. List five natural factors that likely contribute to global warming.

11. What do you think is the single-most important underlying cause of anthropogenic environmental problems?

12. List three environmental consequences of global warming.

13. Complete the following equations:

$$SO_2 + H_2O \rightarrow$$

$$NO + OH \rightarrow$$

$$NO_2 + O^{\bullet} \rightarrow$$

$$Cl + O_3 \rightarrow$$

$$ClO + O^{\bullet} \rightarrow$$

$$HSO_3 + O_2 - uv \rightarrow$$

$$SO_3 + H_2O \rightarrow$$

$$SO_3 + NO + H_2O \rightarrow$$

$$NO + O_2 \rightarrow$$

Answers

1. b
2. c
3. e
4. a
5. b
6. a

Refer to text for remainder.

Further Reading

Abelson, P.H., Global change. Editorial, *Science*, 249, 1085, 1990.
Air Force to combat mosquitoes in South Miami-Dade, http://www.homestead.afrc. af.mil/news/story.asp?id=123264743 (accessed on February 05, 2012).

Atmospheric concentrations of greenhouse gases in geological times and in recent years, http://www.epa.gov/climatechange/science/recentac_majorghg.html (accessed on May 07, 2012).

Bailler, J., Witthoft, M., Paul, C., Bayerl, C., and Rist, F., Evidence for overlap between idiopathic environmental intolerance and somatoform disorders, *Psychosom. Med.*, 67, 921–929, 2005.

Baird, C. and Cann, M.C., *Environmental Chemistry*, W.H. Freeman, New York, 2004.

Bates, D.V., Asbestos: The turbulent interface between science and policy, *Can. Med. Assoc. J.*, 144, 554–556, 1991.

Binder, L.M., Storzbach, D., and Salinsky, M.C., MMPI-2 profiles of persons with multiple chemical sensitivity, *Clin. Neuropsychol.*, 20, 848–857, 2006.

Bradbury, S.M., Proudfoot, A.T., and Vale, J.A., Glyphosate poisoning, *Toxicol. Rev.*, 23, 159–167, 2004.

Brasseur, G. and Granier, C., Mount Pinatubo aerosols, chloroflurocarbons, and ozone depletion, *Science*, 257, 1239–1242, 1992.

Busse, M.D., Ratcliffe, A.W., Shestak, C.J., and Powers, R.F., Glyphosate toxicity and the effects of long-term vegetation control on soil microbial communities, *Soil Biol. Biochem.*, 33, 1777–1789, 2001.

Cancer deaths in Canada, http://www.nationalpost.com/2012/5/9/cancer-deaths-on-the-decline-in-canada/ (accessed on May 09, 2012).

Caress, S.M. and Steinemann, A.C., A national population study of the prevalence of multiple chemical sensitivity, *Arch. Environ. Health*, 59, 300–305, 2004.

Dockery, D.W., Pope, C.A., Xu, X., Spengler, J.D., Ware, J.H., Fay, M.E., Ferris, B.G. Jr., and Speizer, F.E., An association between air pollution and mortality in six U.S. cities, *N. Engl. J. Med.*, 329, 1753–1759, 1993.

Dryzek, J.S., Norgaard, R.B., and Schlosberg, D., (eds.), *Oxford Handbook of Climate Change and Society (The)*, Oxford University Press, Oxford, U.K., 2011.

Dutton, A. and Lambeck, K., Ice volume and sea level during the last interglacial, *Science*, 337, 216–219, 2012.

Epidemiologic Notes and Reports, Nitrogen dioxide and carbon monoxide intoxication in an indoor ice arena-Wisconsin, 1992, *Morbid. Mortal. Weekly Rep.*, 41, 383–384, 1992.

Eriksen, H.R. and Ursin, H., Subjective health complaints, sensitization, and sustained cognitive activation (stress), *J. Psychosomat. Res.*, 56, 445–448, 2004.

Friedlander, S.K. and Lippman, M., Revising the particulate ambient air quality standard, *Environ. Sci. Technol.*, 28, 148A–150A, 1994.

Global warming: How climate change will affect wildlife, agriculture and energy, *ON Nature*, 49, Winter 2009/2010, http://www.ontarionature.org (accessed on December 12, 2009).

Guide on lung cancer in 'never smokers': A different disease and different treatments, http://www.sciencedaily.com/releases/2009/09/090916173328.htm (accessed on May 06, 2012).

Guilherme, S., Gaivao, I., Santos, M.A., and Pacheco, M., DNA damage in fish (*Anguilla anguilla*) exposed to a glyphosate-based herbicide—Elucidation of organ specificity and the role of oxidative stress, *Mutat. Res.*, 743, 1–9, 2012.

Hillert, L., Musabasic, V., Berglund, H., Ciumas, C., and Savic, I., Odor processing in multiple chemical sensitivity, *Hum. Brain Mapp.*, 28, 172–182, 2007.

Hoekstra, G., Wood dust linked to at least five mill explosions in B.C., http://www.vancouversun.com/news/thewest/Wood+dust+linked+least+five+explosions+mills/6533191/story.html

Holmberg, L., Yong, L.C., Kolonel, L.N., Gould, M.K., and West, D.W., Lung cancer incidence in never smokers, *J. Clin. Oncol.*, 25, 472–478, 2007.

Intergovernmental Panel on Climate Change (IPCC), Climate change: Mainstream assessment, http://www.ipcc.ch/publications_and_data/publications_and_data_reports.shtml

Joffres, M.R., Sampalli, T., and Fox. R.A., Physiologic and symptomatic responses to low-level substances in individuals with and without chemical sensitivities: A randomized controlled blinded pilot booth study, *Environ. Health Perspect.*, 113, 1178–1189, 2005.

Kerr, R.A., The greenhouse consensus, *Science*, 249, 481–482, 1990.

Kerr, R.A., Pinatubo fails to deepen the ozone hole, *Science*, 258, 395, 1992.

Lee, H.L., Kan, C.D., Tsai, C.L., Liou, M.J., and Guo, H.R., Comparative effects of the formulation of glyphosate-surfactant herbicides on hemodynamics in swine, *Clin. Toxicol.* (Phila.), 37, 651–658, 2009.

Lelieveld, J. and Heintzenberg, J., Sulfate cooling effect on climate through in-cloud oxidation of anthropogenic SO_2, *Science*, 258, 117–120, 1992.

Lim, W.Y., Chuah, K.L., Eng, P., Leong, S.S., Lim, E., Lim, T.K., Ng, A., Poh, W.T., Tee, A., The, M., Salim, A., and Seow, A., Aspirin and non-aspirin non-steroidal anti-inflammatory drug use and risk of lung cancer, *Lung Cancer*, 77, 246–251, 2012.

Lipson, J.G. and Doiron, N., Environmental issues and work: Women with multiple chemical sensitivities, *Health Care Women Int.*, 27, 571–584, 2006.

Lloyd, S., The calculus of intricacy, *Sciences*, September/October, 38–44, 1990.

McDonald, J.C. and McDonald, A.D., Asbestos and carcinogenicity, *Science*, 249, 844 (Let), 1990.

McKenzie, R.L., Aucamp, P.J., Bais, A.F., Bjorn, L.O., Ilyas, M., and Madronich, S., Ozone depletion and climate change: Impacts on UV radiation, *Photochem. Photobiol. Sci.*, 10, 182–198, 2011.

McPhee, S.J., Papadakis, M.A. (eds.), and Rabow, M.W. (assoc. ed.), *Current Medical Diagnosis and Treatment*, 51st edn., Lange Medical Books/McGraw-Hill, New York, 2012.

Milqvist, E., Ternesten-Hasseus, E., Stahl, A., and Bende, M., Changes in levels of nerve growth factor in nasal secretions after capsaicin inhalation in patients with airway symptoms from scents and chemicals, *Environ. Health Perspect.*, 113, 849–852, 2005.

Mossman, B.T., Bignon, J., Corn, M., Seaton, A., and Gee, J.B., Asbestos: Scientific developments and implications for public policy, *Science*, 247, 294–301, 1990.

Nongovernmental International Panel on Climate Change (NIPCC), Climate change: Non-consensus views, http://www.nippcreport.org/reports/2009/2009report. html

Nordin, S., Broman, D.A., and Wulff, M., Environmental odor intolerance in pregnant women, *Physiol. Behav.*, 84, 175–179, 2005.

Novelli, P.C., Masarie, K.A., Tans, P.P., and Lang, P.M., Recent changes in atmospheric carbon monoxide, *Science*, 263, 1587–1593, 1994.

Oelermans. J. and Fortuin. J.P.F., Sensitivity of glaciers and small ice caps to greenhouse warming, *Science*, 258, 115–117, 1992.

O'Malley, M., Westray—Here's what happened, *CBC Online News*, May 9, 2002, http://www.teamstersrail.ca/TCRC_Westray_Horror.htm

O'Reilly, K.M.A., McLaughlin, M.B., and Beckett, W.S., Asbestos-related lung disease, *Am. Fam. Physician*, 75, 683–688, 2007.

Ottar, B., The transfer of airborne pollutants to the Arctic region, *Atmos. Environ.*, 15, 1439–1445, 1981.

Pall, M.L. and Anderson, J.H., The vanilloid receptor as a putative target of diverse chemicals in multiple chemical sensitivity, *Arch. Environ. Health*, 59. 363–375, 2004.

People's Inquiry into the impacts and effects of aerial spraying pesticide over Auckland, New Zealand (Acquired April 15, 2011). www.peoplesinquiry.co.nz/

Philp, R.B., Effects of experimental manipulation of pH and salinity on Cd^{++} uptake by the sponge *Microciona prolifera* and on sponge cell aggregation induced by Ca^{++} and Cd^{++}, *Arch. Environ. Contam. Toxicol.*, 41, 282–288, 2001.

Philp, R.B., *Environmental Issues for the Twenty first Century and Their Impact on Human Health*, Bentham Science E-Publications, Oak Park, IA, 2012.

Pirker, R., Pereira, J.R., Szczesna, A., von Pawel, J., Krzakowski, M., Ramlau, R., Vynnychenko, I., Park, K., Eberhardt, W.E., de Marinis, F., Heeger, S.M., Goddemeier, T., O'Byrne, K.J., and Gatzemeier, U., Prognostic factors in patients with advanced non-small cell lung cancer: Data from the phase III FLEX study, *Lung Cancer*, 77, 376–382, 2012.

Population at risk from particulate air pollution-United States, 1992. *Morbid. Mortal. Weekly Rep.*, 43, 290–293, 1994.

Ravishankara, A.R., Solomonm, S., Turnipseed, A.A., and Warren, R.F., Atmospheric lifetimes of long-lived halogenated species, *Science*, 259, 194–199, 1993.

Report of the Opinion of Ombudsman Mel Smith on Complaints Arising from Aerial Spraying of the Biological insecticide Foray 48B etc. (Acquired April 11, 2011), www.ombudsman.parliament.nz/cms/imagelibrary/100260.pdf

Robert, O., The threat of barn dust, *Veterinarian Mag.*, 3, 29–30, 1991.

Rodhe, H., A comparison of the contribution of various gases to the greenhouse effect, *Science*, 248, 1217–1219, 1990.

Savitz, D.A., Arbuckle, T., Kaczor, D., and Curtis, K.M., Male pesticide exposure and pregnancy outcome, *Am. J. Epidemiol.*, 146, 1025–1036, 1997.

Schindler, D.W., Beaty, K.G., Fee, E.J., Cruikshank, D.R., Debruyn, E.R., Findlay, D.L., Linsey, G.A., Schearer, J.A., Stainton, M.P., and Turner, M.A., Effects of climatic warming on lakes of the central boreal forest, *Science*, 250, 967–970, 1990.

Solomon, S. Stratospheric ozone depletion: a review of concepts and history. *Rev. Geophys.* 37, 275–316, 1999.

Talbot, A.R., Shiaw, M.H., Huang, J.S., Yang, S.F., Goo, T.S., Wang, S.H., Chen, C.L., and Sanford, T.R., Acute poisoning with a glyphosate-surfactant herbicide ('Roundup'): A review of 93 cases, *Hum. Exp. Toxicol.*, 10, 1–8, 1991.

Tsui, M.T. and Chu, L.M., Aquatic toxicology of glyphosate-based formulations: comparison between different organisms and the effects of environmental factors, *Chemosphere*, 52, 1189–1197, 2003.

Walker, C.H., Hopkin, S.P., Sibly, R.M., and Peakall, D.B., *Principles of Ecotoxicology*, Taylor & Francis Group, London, U.K., 1996.

Weiss, B., Neurobehavioral properties of chemical sensitivity syndromes, *Neurotoxicology*, 19, 259–268, 1998.

Williams, G.M., Kroes, R., and Munro, I.C., Safety evaluation and risk assessment of the herbicide roundup and its active ingredient, glyphosate, for humans, *Regul. Toxicol. Pharm.*, 31, 117–165, 2000.

Yanko, D., Are animal disease patterns changing because of global warming?, *Veterinarian Mag.*, 2, 18–21, 1990.

Ziomislic, D., Star exclusive: Agent orange "soaked" Ontario teens, www.thestar. com/news/Canada/article/940243—star-exclusive—agent-orange-soaked-ontario-teens

5

Halogenated Hydrocarbons and Halogenated Aromatic Hydrocarbons

Introduction

Halogens are the related elements chlorine (Cl), bromine (Br), fluorine (F), and iodine (I). They may exist as gases (Cl_2, F_2), a liquid (Br_2), or a solid (I_2). Halogenated hydrocarbons, also known as organohalogens, are a group of organic compounds of diverse structure to which one of these halogens has been attached. The core structure may be either simple, consisting of one or two carbons, or it may be a more complex aromatic one (halogenated aromatic hydrocarbons or HAHs). Because they have been implicated almost universally in toxic reactions in mammals and lower species (including carcinogenesis in some), it is appropriate to consider them as a group despite their chemical diversity. Polycyclic aromatic hydrocarbons (PAHs) are multiringed, planar chemicals that share many of the same toxicological properties as HAHs.

Early Examples of Toxicity from Halogenated Hydrocarbons

One of the oldest and simplest of these compounds is carbon tetrachloride (CCl_4), which was used extensively as a solvent and as a dry-cleaning agent until its hepatotoxic nature was discovered. In fact, it was used originally as a treatment for hookworm in humans and domestic animals and as a component in fire extinguishers. These uses may still exist in some parts of the world. CCl_4 owes its toxicity to the fact that it is converted in the liver to carbon trichloride (CCl_3), which is a free radical capable of inducing peroxidation of lipid double bonds and of poisoning protein-synthesizing enzymes. Older, halogenated hydrocarbons include trichloroethylene, and the anesthetics halothane and chloroform (now abandoned because of its toxicity). Figure 5.1 shows the variety of structural formulae included in this class of compounds. The earliest form of poisoning associated with a halogen was probably "bromism,"

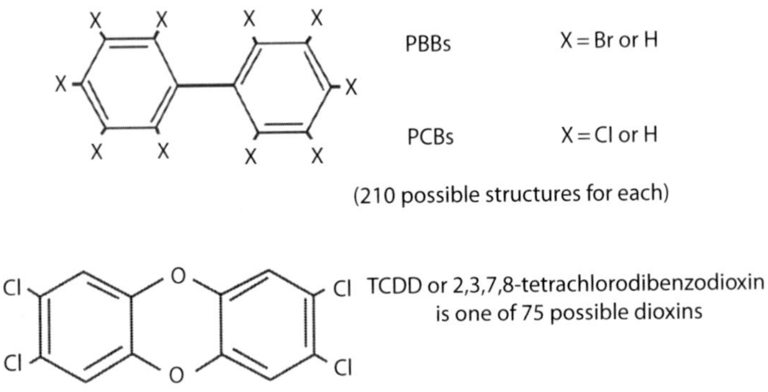

PBBs   X = Br or H

PCBs   X = Cl or H

(210 possible structures for each)

Cl TCDD or 2,3,7,8-tetrachlorodibenzodioxin
is one of 75 possible dioxins

FIGURE 5.1
Structural formulae of a number of halogenated hydrocarbons and PAHs.

a condition resulting from the use or abuse of sodium or potassium bromide as a sedative and sleeping potion early in this century. Symptoms included severe headache, stupor, delirium, cardiac problems, very bad breath (from the bromine), and an acneform skin rash of the type now called "chloracne." The expression "bromide" is used now to indicate a soothing but meaningless statement of the sort frequently uttered by certain politicians.

Physicochemical Characteristics and Classes of Halogenated Hydrocarbons

The characteristics of halogenated hydrocarbons that make them useful for a variety of applications are generally the same ones that make them hazardous to the environment and to humans. These include

1. High lipid solubility
2. Ability to survive heat >800°C
3. High resistance to chemical breakdown
4. Toxicity to microorganisms

These agents are used for a variety of purposes (see in the following).

Antibacterial Disinfectants

Hexachlorophene has been used for many years as a surgical scrub and, as a 3% solution, as a hospital disinfectant. It is also used as the active ingredient in deodorant soaps. In the late 1960s, a change was proposed in US-FDA regulations to permit the use of hexachlorophene as an antifungal wash for fruit and vegetables.

In light of the then-recent thalidomide tragedy, extensive testing was required for approval to be granted. Rats fed high levels of hexachlorophene developed weakness, ataxia, paralysis, and evidence of a type of brain pathology known as status spongiosus, indicative of axonal degeneration. In 1971, a study was done in which infant monkeys were washed daily for 90 days in 3% hexachlorophene. Neurological symptoms were observed and status spongiosus was seen in all specimens at post mortem. Significant blood levels have been detected in infants washed with 3% hexachlorophene, and those with severe diaper rash, burns, or congenital skin disorders are especially prone to absorb it, as the natural permeability barrier has been disrupted (see Chapter 1). Autopsies of infants dying from a variety of causes and who received high exposures showed evidence of *status spongiosus*. In 1972, hexachlorophene was accidentally added in high concentration to baby powder during its manufacture in France. Forty-one deaths of infants and young children were attributed to this error. A few years later, Dr. Hildegard Halling, a Swedish physician, published a report indicating that nurses who washed frequently in hexachlorophene (10–60 times daily) had a higher incidence of birth defects in their offspring (25/460 births) than those who did not (nil/233 births). Although this clinical study was criticized for design flaws, others with rats have revealed teratogenic effects. This product is no longer used in nurseries in North America and pregnant women are advised to avoid it.

Herbicides

This group includes 2,4-dichlorophenoxyacetic acid (2,4-D), 2,4,5-trichlorophenoxyacetic acid (2,4,5-T), and dioxins such as 2,3,7,8-tetrachlorodibenzo-*p*-dioxin (TCDD=dioxin). Agent Orange, used as a defoliant in Vietnam, was equal parts of 2,4,-D and 2,4,5,-T and contained TCDD as a contaminant. One of the best-documented human exposures to dioxin was the explosion at Seveso, Italy (see Chapter 2). Hundreds of individuals suffered from chloracne, which is the hallmark of toxicity of halogenated hydrocarbons.

The herbicides 2,4,-D and 2,4,5-T are used to control broad-leafed plants along highways, railways, and utility rights-of-way. They are hormonal growth promoters and force plants to consume energy at a greater rate than it can be replaced. The humans at greatest risk of toxic exposure are the workers who apply the sprays, and poisoning from dermal and respiratory absorption has occurred as well as from accidental ingestion. Signs and symptoms include peripheral neuritis, muscular weakness, and chloracne. Although 2,4,5-T is a weak teratogen in some animals, it is the presence of TCDD, or dioxin, that is the greatest source of public concern.

Dioxin (TCDD) Toxicity

Dioxins are a family of compounds of which TCDD (2,3,7,8-tetrachlorodibenzo-*p*-dioxin) has received the most public attention. There is significant species variation in TCDD toxicity with the guinea pig being most sensitive (LD_{50} 1 µg/kg) and the rat quite insensitive (LD_{50} 22 µg/kg). There are other manifestations of dioxin toxicity.

Hepatotoxicity

All species show enlarged livers and microsomal monooxygenase enzyme induction occurs in most. Rats develop fatty livers with triglyceride deposition. Hepatic fibrosis has been reported in humans. People exposed at Seveso had elevated serum enzyme levels (serum glutamic-oxaloacetic transaminase, SGOT, and serum glutamic-pyruvic transaminase, SGPT), indicating liver damage, for several weeks after the accident.

Porphyria

Porphyrins are pigments widely distributed in nature and they are present in the body as by-products of heme synthesis that is required for the formation of hemoglobin, myoglobin, and cytochromes. Heme is ferrous protoporphyrin IX. Hematin, the iron-containing molecule in catalase and peroxidases, is ferric protoporphyrin IX. The rate-limiting step in the synthesis of heme is ALA synthetase (ALA = gamma-aminolevulinic acid). Dioxin significantly increases the levels of ALA synthetase and hence ALA levels and the synthesis of porphyrins. This is not enzyme induction, but probably is due to interference with a feedback control system. The excess porphyrins are excreted in the urine, giving it a port wine color, and they are deposited in the skin, producing pigmentation. Because porphyrins are photoreactive, a condition known as *porphyria cutanea tarda* develops, which is characterized by photosensitivity, blistering, fragility of the skin, pigmentation, and hirsutism (hairiness). Congenital defects in porphyrin metabolism cause the same syndrome and this has been suggested as the explanation for the vampire and werewolf myths of Europe (characterized by hirsutism and the avoidance of sunlight). There is even an explanation of why drinking blood might have a therapeutic effect. Heme is the feedback substance that turns off ALA synthetase. Absorption of sufficient heme might thus inhibit porphyrin synthesis. In the late 1950s an extensive outbreak of *porphyria cutanea tarda* occurred in Turkey during a famine as the result of consuming seed grain treated with hexachlorobenzene as an antifungal agent. A simplified scheme of the steps involved in porphyrin synthesis is shown in Figure 5.2.

Chloracne

This skin disorder is typified by rash, cysts, and hyperpigmentation and this is the hallmark of poisoning with all halogenated hydrocarbons. It was the predominant toxic manifestation at Seveso.

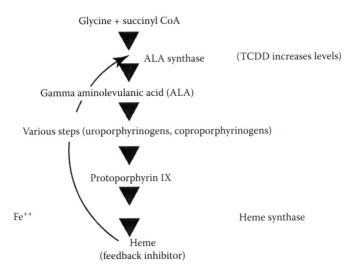

Glycine + succinyl CoA

ALA synthase   (TCDD increases levels)

Gamma aminolevulanic acid (ALA)

Various steps (uroporphyrinogens, coproporphyrinogens)

Protoporphyrin IX

Fe⁺⁺   Heme synthase

Heme
(feedback inhibitor)

FIGURE 5.2
Effect of TCCD on the (simplified) heme synthetic pathway.

Cardiovascular effects—A 10 year mortality study of the population exposed to TCDD after the Seveso explosion of 1976 revealed a significantly increased mortality from cardiovascular events.

Carcinogenicity

Dioxin is a potent hepatocarcinogen in mice and to a lesser extent in rats. There is a latency period before the emergence of the liver tumors. Evidence of cancer in several studies of Vietnam veterans, for whom claims of increased incidence of cancer have been made, has been inconclusive. A retrospective cohort study was conducted by scientists at the (U.S.) National Institute for Occupational Safety and Health (NIOSH) on 5172 workers at 12 U.S. plants in which TCDD was a chemical contaminant of the manufacturing process. Exposure was well documented and serum TCDD levels were obtained from 253 workers. The mortalities from several cancers previously associated with TCDD (stomach, liver, and nasal cancers, Hodgkin's disease and non-Hodgkin's lymphoma) were not significantly different from the overall population but the incidence of all cancers taken together was slightly but significantly increased. In a subcohort of 1520 workers with more than 1 year of exposure and more than 20 years of latency, mortality from soft tissue sarcoma was significantly higher than for the general population. The authors concluded that the results were not suggestive of the high relative risks of cancer reported for TCDD in previous studies. The slight risk of increased soft tissue sarcoma is weakened by the small numbers involved (only three cases) and confounding factors such as smoking and exposure to other chemicals.

In another epidemiologic study of 754 Monsanto employees exposed to high levels of TCDD in a 1949 accident, 122 of whom developed chloracne, there was no increased incidence of cancer in those who developed chlor-acne, although they were presumably the group with the highest exposure. Conversely, workers who were also potentially exposed to 4-aminobiphenyl, a potent bladder carcinogen, had increased mortality from bladder cancer, lung cancer, and soft tissue sarcoma. This suggests that TCDD might act as a co-carcinogen or promoter. Again, the effects of confounders such as smoking and exposure to other chemicals could not be ruled out, but recent experimental evidence supports the suspicion that TCDD could act in this way.

Walsh et al. studied the cell toxicity of aflatoxins in cultured human epidermal cells. AFB_1 was markedly toxic at 1 μg/mL. Neither AFB_2 nor AFB_1 dihydrodiol was toxic. TCDD alone was not toxic to the cells but at 5 nM, it dramatically stimulated AFB_1 toxicity at levels as low as 0.1 μg/mL. It also increased the formation of AFB_1 epoxides and a 20-fold increase in DNA adduct formation was observed. AFB_1 is the most carcinogenic of the aflatoxins (see Chapter 10).

The most recent report from Seveso (see Chapter 2 regarding this industrial accident) is suggestive of a carcinogenic effect in humans. It re-examined the data in 1996 and confirmed an excess risk of lymphatic and hematopoietic tissue neoplasms in the most exposed zones (A and B). The exposed population was divided into three groups according to their likely level of exposure, zone A being nearest the explosion and zone C furthest away. Unlike earlier studies, the 1996 one also found an increased risk of breast cancer in zone A and noted a need for further monitoring of this effect.

Weaknesses of this study include the inability to control for confounding factors such as smoking, and lack of hard data regarding real exposure levels. Despite the conflicting results of several epidemiological studies, most authorities now agree that there is a high index of suspicion for TCDD carcinogenicity in humans, especially for non-Hodgkin's lymphoma. A very recent (2011) critical review of epidemiologic cancer studies of TCDD determined that evidence for carcinogenicity was inconclusive and that the practice of comparing all-cancer incidences was not justified epidemiologically. The definitive word, however, remains to be heard, and the existing evidence comes from high industrial and occupational exposures that may not be relevant to environmental exposures encountered by the general population. If a threshold truly exists because of the TCDD/Ah receptor story, which follows, environmental exposures may never reach that threshold.

The mechanism of carcinogenesis in animals appears to be epigenetic. Recent evidence indicates that TCDD binds to a specific receptor, the Ah (for aryl or aromatic hydrocarbon) receptor and that a minimum number of Ah receptors must be occupied for TCDD to exert its effect (see the following). The implication of this is that there is, thus, a threshold dose and that the linear multistage carcinogenesis model is inappropriate for TCDD and for any other agent that works by this mechanism, such as PCDDs and PCDFs (see Chapter 3 and the following). The EPA is now reconsidering the use of the linear multistage model at least for TCDD and perhaps for a few other agents.

Neurotoxicity

Many toxic effects have been observed including impaired vision, hearing, and smell; depression; sleep disturbances; and others. These were observed in workers exposed to TCDD in the 1949 Monsanto accident.

Reproductive Toxicity

Testicular atrophy, necrosis, and decreased spermatogenesis have been seen in laboratory animals.

Metabolic Disturbances

Weight loss and depletion of adipose tissue occur in laboratory animals.

Role of the Aryl Hydrocarbon Receptor (AhR) and Enzyme Induction

TCDD is one of the most potent inducers of aryl hydrocarbon hydroxylase (AHH) so far discovered. In some species it is effective at 1 µg/kg. It induces synthesis of hepatic microsomal cytochrome P450 (CYP1A1 and possibly 1B1). TCDD uptake into the cell is passive (i.e., concentration gradient-dependent). Intracellularly, it binds to the aromatic hydrocarbon cytosolic receptor (AhR), which is the product of a regulatory gene. The unliganded, aryl hydrocarbon receptor complex (AhRC) contains the aryl hydrocarbon receptor (AhR). Binding of an Ah ligand causes release of the AhR that is translocated to the nucleus by a translocator protein. The unliganded AhRC is heteromeric, but after binding to a HAH or a PAH the AhR is released and associates with the AhR nuclear translocator protein to form heterodimeric, transformed AhRC. Ligands for the AhR are typically hydrophobic aromatic compounds including the HAHs dibenzo-*p*-dioxins, dibenzofurans, biphenyls, and PAHs such as benzo[a]pyrene and many other benzo derivatives (products of combustion), naphthalene, naphthacene, and many others (see http://chrom.tutms.tut.ac.jp//JINNO/DATABASE/00alphabet.html).

In the nucleus, AhRC activates numerous, xenobiotic-responsive, structural genes. The information is transcribed to m-RNA and translated to protein synthesis and the production of cytochrome P-450 ($1A_1$, $1A_2$, $1B_1$). This is known as a pleiotropic response, i.e., it results in more than one phenotypic effect. TCDD is the most potent known inducer of AHH. The consequence of this induction is that several drug (xenobiotic)-metabolizing enzymes are induced, and reactive metabolites may be formed, which react with proteins and nucleic acids to cause mutations, teratogenesis, and carcinogenesis, as well as altered drug metabolism. Conversely, the reactive metabolites may be excreted or detoxified, as by conjugation with glucuronide. A schematic representation of this pathway is shown in Figure 5.3. This system has been most studied in the mouse, where it is inherited as an autosomal dominant pattern. Similar systems have

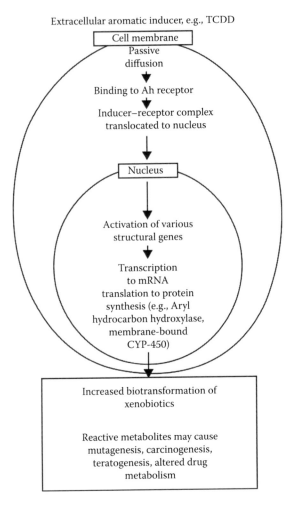

FIGURE 5.3
Activation of enzyme induction following binding of TCDD to the Ah receptor.

been identified in the rat, rabbit, and some fish. It is most heavily concentrated in the liver. In humans, there is considerable variation in the Ah locus. To date, no endogenous substrate has been identified for the Ah receptor. AhR "knock-out" mice, however, have small, fibrosed livers. In the absence of an exogenous ligand, AhR has been found in the nucleus of cultured Hela cells, behaving as if they were ligand-bound. It seems apparent that the AhR has an important physiological role. It is known that primitive species (bacteria, yeasts) utilize polycyclic hydrocarbons as an energy source and possess P450 metabolizing enzymes for them (camphor in *Pseudomonas*, benzo-[a]-pyrene in yeast), so these may have evolved as a detoxication system. The mechanism of TCDD toxicity is not known, but if there is a natural substrate for Ah receptors, its displacement by TCDD could be involved in the latter's toxicity. It is not clear whether TCDD itself or its metabolite(s) are responsible.

TCDD also has nonreceptor-mediated effects, including interference with calcium homeostasis and a variety of membrane-related changes. It is interesting that the chloracne associated with TCDD also occurs with bromides, which could not act as Ah ligands. For more on TCDD toxicity see Chapter 12.

Paraquat

This PAH herbicide (1,1'-dimethyl-4,4'-bipyridinium ion) is highly water soluble, therefore poorly absorbed across the skin or gastrointestinal mucosa. It is extremely toxic to humans when inhaled, however, and 5 g may be fatal. Pulmonary congestion, edema, and hemorrhage may result in almost complete functional destruction of the lung. In severe cases, lung transplantation has been tried as a last resort, with disappointing results. Although paraquat is poorly absorbed from the oral route, its highly toxic nature can result in lung toxicity days or weeks later. It is, thus, one of the "hit-and-run" class of toxicants. Liver and kidney damage and neurological damage also occur (see Chapter 9 on pesticides).

Insecticides

Chemical insecticides are used to increase food crop production, to protect livestock and household pets against insect pests (warble fly, botfly, screwworm, fleas), to control disease-carrying insects (anopheles mosquitoes that carry malaria), and to control destructive insects like termites. It was a chlorinated hydrocarbon insecticide DDT (dichlorodiphenyltrichloroethane) that first raised concerns over the impact of pesticides on the environment. In 1961, the author Rachel Carson brought out her book *Silent Spring* in which she documented the devastating effect that this chemical had on bird life because it weakened eggshells so that the eggs collapsed in the nest or the chicks were abnormal at hatching. The persistence of the chemical (it and its metabolite DDE have a biological T1/2 of 50 years) and its high lipid solubility result in its biomagnification up the food chain.

Predatory and fish-eating birds are especially vulnerable to DDT. The product was banned in the United States and Canada in 1972, but it is still used elsewhere in the world, including Mexico, and trace levels may be present in imported products. Levels of from 1 part/billion (ppb) to 1 part/ million (ppm) may be present in fish, oysters, and other seafoods and can contribute to human tissue levels of up to 10 ppm.

Halogenated hydrocarbon insecticides (chlorinated hydrocarbons) are principally neurotoxic, interfering with axonal transmission by altering sodium and potassium transport across the axonal membrane to prevent normal depolarization. Evidence of carcinogenicity also has been obtained in animal experiments (see Chapter 9 on pesticides for more details).

Industrial and Commercial Chemicals

Biphenyls

Polybrominated biphenyls (PBBs) are used as fire retardants in thermoplastics for TV and office machine casings. The more familiar polychlorinated biphenyls (PCBs) are highly stable and resistant to degradation in acids, bases, by oxidation and by heat (to 800°C).

These characteristics make them ideal insulators in transformers in the electric power industry and as hydraulic fluid and in brake linings. They are also used as plasticizers in polymer films. The same characteristics, however, make these agents very persistent in the environment. Exposure to these compounds is largely an occupational hazard, but exposure in the environment can occur as a result of contamination of groundwater from spills, improper storage of waste PCBs from old transformers and capacitors, or from fires in storage sites. Although the manufacture of PCBs was banned in the United States in 1977, they remain a problem because of their persistence and resistance to destruction. Forty percent of North Americans have body fat levels of 1 ppm or higher (these agents have high lipid solubility). Prior to 1970, 500,000 ton of PCBs were produced in North America.

Toxicity

1. *Animal.* The LD_{50} in rats may be 1–10 g/kg. Chronic toxicity involves skin lesions, hepatotoxicity, immunosuppression, and reproductive dysfunctions. Carcinogenicity has also been reported. Recently, genetic damage in cetaceans (whales, dolphins) and seals in the Baltic Sea has been ascribed to high levels of PCBs.

2. *Human.* Characteristic chloracne, impaired immune response, liver damage, gastrointestinal disturbances (nausea, vomiting, loss of appetite), CNS disturbances (weakness, ataxia), as well as reproductive problems and cancer have been associated with exposures or induced in experimental animals.

Pharmacokinetics and Metabolism

Because of the high lipid solubility, these compounds are well absorbed from the gastrointestinal tract (>90%) and stored in body fat. They are secreted in milk (toxicity has been shown in nursing mice and rats) and they constitute a hazard for the nursing infant (see study of Michigan mothers in Chapter 3). They have been shown to induce cytochrome P448 through the Ah receptor pathway and to form reactive intermediates and to deplete glutathione. Conjugation with glucuronide and renal excretion are the final detoxification mechanisms.

Biodegradation

Recent studies indicate that solar photolysis (near-UV light) can reduce the T1/2 in surface water to 1–2 years by dechlorination of PCBs but this has no effect on bottom sediments that may contain high levels. Fortunately, chemical dechlorination can occur here, and, recently, subsequent oxidation by anaerobic bacteria has been discovered in sediments of the Hudson River. These discoveries offer some hope that biodegradation of PCBs may occur at a faster rate than anticipated.

Accidental Human Exposures

In 1968, nearly 1700 people in the Fukuoka region of Japan developed chloracne as a result of using rice cooking oil contaminated with PCBs. The PCBs in turn were contaminated with tetrachlorodibenzofuran that is structurally and toxicologically similar to TCDD. These "Yusho" patients (Yusho means "oil disease") constitute the largest human population known to be exposed to toxic levels of PCBs and their health continues to be monitored closely to identify delayed effects. Five years later, 22 deaths were reported in 1200 of these patients, 9 from cancer, 2 involving the liver. Calculated body burdens were 5.9 µg/kg for chloracne and 4.4 µg/kg for nausea and anorexia. These are 200 times higher than average current levels found in North American populations. Similar mortality statistics occurred after an outbreak in Taiwan. In neither case were the cancer deaths considered excessive, and they were not age adjusted (see also Chapter 3). Several cases of contamination of animal feeds have occurred. In North Carolina, a leaky heat exchanger contaminated 16,000 ton of chicken feed, only 10% of which was recovered. No human health problems were directly attributed to the accident, but there was some evidence that children might have been affected *in utero* (see also Chapter 3).

A major human exposure to PBBs occurred in Michigan in 1973 (the year of Seveso). A PBB product intended for use as a fire retardant and marketed as "Firemaster" was accidentally bagged as "Nutrimaster," a magnesium supplement for dairy cattle. Whole herds of dairy cattle were afflicted with loss of appetite, open sores, weight loss, lack of milk production, sterility, and stillbirths. It took several months of detective work to trace the source of the problem; meanwhile, human food supplies were contaminated by milk containing the PBBs. Meat and eggs were also contaminated. By 1976 nearly 30,000 of Michigan's best dairy cattle had died or been destroyed, along with thousands of sheep and hogs and millions of chickens. Thousands of Michiganders had consumed unknown quantities of contaminated food, and hundreds began to complain of headache, fatigue, joint pain and numbness in fingers and toes. Farm families were the most severely afflicted. Little was then known about the human toxicity of PBBs. Long-term effects were as yet unidentified.

In recent years it has become clear that they exist. Studies published in 2011 found that among women who were exposed to PBB *in utero* because their mothers consumed foods contaminated in the 1973 accidental exposure had a significantly greater risk of spontaneous abortion than women not so exposed. In the midrange exposure group (1–3.16 ppb) and high exposure group (≥3.17) the odds ratio was 2.75. Exposure to PPB contaminated breast milk increased this risk.

There was also some evidence that those women who actually consumed contaminated foods at the time of the accident (or in the early 1970s) had a slightly increased incidence of a positive Pap smear.

Problem of Disposal

Recent evidence from a disposal site in Great Britain suggested that current incineration temperatures might not be adequate as levels of PCBs were detected in the soil around the site. A complicating factor was the presence of a municipal incinerator within 100 m, but some authorities feel that incineration temperatures should reach 2700°C. The PCB fire (deliberately set) at St. Basile le Grand in 1988 highlights the dangers of unsecured storage. Public concerns over PCB hazards have led to the NIMBY response (Not In My Back Yard) and resulted in resistance to the establishment of proper disposal sites and to refusal to allow ships containing toxic wastes to unload. Such ships, wandering the seas in search of a berth, create a potential for a marine environmental disaster, with contamination of food fish that could be far worse than the hazards associated with well-run disposal sites.

Solvents

CCl_4, chloroform, and methylene chloride (dichloromethane) are still popular industrial solvents because they are not flammable. Exposure is mainly an industrial problem but these agents may still appear as cleaning fluids in the home. A source of some concern is the presence of trace amounts of chloroform in drinking water as a result of the chlorination process.

Toxicity

These chemicals are hepatotoxic, causing central lobular necrosis with fatty degeneration of adjacent areas. They also can cause renal damage, and chronic exposure has been linked to neoplasms of the lung and liver. Cardiac arrhythmias have been reported, as has nausea and vomiting. The cardiac arrhythmias result from sensitization of the heart to catecholamines such as adrenaline and noradrenaline. Other halogenated anesthetics like cyclopropane and halothane will also do this, and hepatotoxicity has been reported as a toxic effect of halothane in some patients as well as in anesthetists routinely exposed to the drug. Liver toxicity can be increased in mice exposed to inducers of microsomal enzymes, suggesting that a toxic metabolite might be involved.

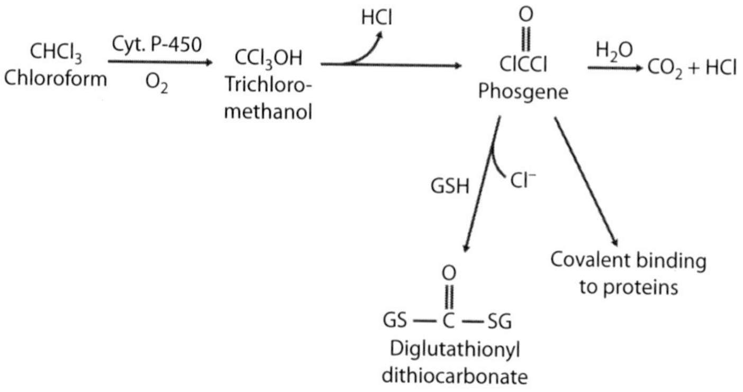

FIGURE 5.4
Biotransformation of chloroform.

Mechanism of Toxicity

These substances are metabolized by cytochrome P-450 enzymes. Chloroform is converted to chloromethanol, phosgene, and $CO_2 + HCl$. Phosgene is normally converted to glutathione for excretion, but if glutathione is depleted, covalent binding to proteins may lead to liver and kidney necrosis. The metabolic pathway for chloroform is shown in Figure 5.4. Methylene chloride is metabolized to $CO_2 + CO$. Carbon monoxide poisoning through the formation of carboxyhemoglobin can occur. Chloroform is an example of a trihalomethane.

Trihalomethanes

These are halogen-substituted, single-carbon compounds having the general formula CHX_3 where X may be chlorine, fluorine, bromine, iodine, or a combination of these. They are formed during the process of water chlorination from naturally occurring organic compounds. The most common agents found in drinking water are chloroform ($CHCl_3$), bromodichloromethane ($CHBrCl_2$), chlorodibromomethane ($CHClBr_2$), and bromoform ($CHBr_3$).

These are liquid at room temperature, rather volatile, and only slightly soluble in water. Their octanol–water partition coefficients range from 1.97 for chloroform to 2.38 for $CHBr_3$. All are sensitive to decomposition in air and sunlight. THMs are also released into the environment from industrial sources. Chloroform is the most common THM found in drinking water. Its major toxicity and metabolic transformation are noted earlier.

As a group, THMs are rapidly and well absorbed from the gastrointestinal tract, metabolized through di-halocarbonyl compounds via the cytochrome P450-dependent mixed function oxidases to CO_2 and CO and eliminated through the lungs. Because of their high lipid solubility they accumulate in adipose tissue > brain > kidney > blood.

Several epidemiological studies have shown a correlation between chlorination of surface or groundwater and the incidence of many cancers, but correlations with actual measured levels of THMs have been harder to demonstrate. Exceptions are pancreatic cancer in white males, rectal cancer in males only, and stomach cancer in both sexes. When population migration patterns were considered, however, the correlation with stomach and rectal cancer could not be demonstrated, and other studies have suggested that other water quality parameters may be involved.

In animal studies, chloroform has been shown to be carcinogenic in rats and mice. $CHClBr_2$ has been reported to be hepatotoxic in mice, but no evidence of carcinogenicity was obtained. These agents are probably mutagenic and teratogenic, as indicated in some studies.

In light of these facts, there are efforts directed to limiting the levels of THMs in drinking water. Standards vary widely throughout the world. In Canada, the current maximum standard is 350 µg/L, not to be exceeded. In the United States, the EPA has set a limit of 100 µg/L. This is an average based on quarterly samples, and is therefore more enforceable. The WHO sets a guideline of 30 µg/L, but with a warning that disinfecting efficiency should not be compromised in the pursuit of lower levels. The European Economic Community passed a directive that haloform levels in drinking water should be "as low as possible," which is unenforceable.

In is thus evident that the problem of THMs in drinking water is another example of how cost–benefit analysis must be performed to weigh the potential risks of cancer from the chemicals against the known risks of epidemic infections if water supplies are not treated.

Case Study 9

Part 1. Three members of a family became dizzy and nauseated within 1 h of eating snacks (taquitos) consisting of tortillas wrapped around a meat filling. Two of them subsequently had grand mal epileptic seizures. The snacks were commercially prepared and sold in sealed bags of 48. They had been purchased a few days earlier. In an unrelated case a few weeks later, a 17-year-old male had four closely spaced seizures 30 min after consuming taquitos from the same manufacturer and purchased from the same store. The boy was on long-term antiepileptic therapy because he had been diagnosed as an epileptic the previous year. Following the initial episode, the manufacturer had voluntarily removed from shop shelves and destroyed all existing packages of the product.

Q. What organ system seems to be the primary site of toxicity?

Q. Was the most likely cause of the reaction bacterial or chemical?

Q. Was the likely site of contamination the factory or the retail store?

Part 2. Analysis of some remaining taquitos from the first case revealed traces of endrin. No source or trace of endrin was found at the factory.

Q. To what class of compound does endrin belong?

Review the toxicity of this class of chemicals.

A statewide press release turned up several other cases of seizures including five persons who had experienced seizures within 12 h of consuming taquitos purchased from the same store.

Q. What preventive or remedial measures might you recommend?

Case Study 10

A maintenance employee in a factory died after acute exposure to solvent fumes. He had been using a mixture of chlorinated solvents to remove grease from machinery. The principal component was trichloroethane (methyl chloroform).

Q. What was the immediate cause of death?

Q. What steps might you suggest to prevent this type accident?

Q. This substance is similar to CCl_4. What would have been the nature of the toxic response if the exposure had been chronic rather than acute?

Review Questions

For Questions 1–6, use the following code:

Answer A if statements a, b, and c are correct.

Answer B if statements a and c are correct.

Answer C if statements b and d are correct.

Answer D if statement c only is correct.

Answer E if all statements (a, b, c, d) are correct.

1. Halogenated hydrocarbons are characterized by
   a. High lipid solubility
   b. Susceptibility to chemical breakdown
   c. Toxicity for microorganisms
   d. Decomposition at temperatures greater than 200°C

2. Dioxin (TCDD) toxicity is characterized by
   a. Chloracne
   b. Hepatotoxicity
   c. Porphyria
   d. None of the above (a, b, c, d)

3. Dioxin (TCDD)
   a. Induces the enzyme gamma-aminolevulinic acid (ALA) synthetase
   b. Interferes with feedback inhibition of ALA synthetase
   c. Inhibits porphyrin synthesis
   d. Induces aryl hydrocarbon hydroxylase

4. With regard to chloroform
   a. Phosgene is a major metabolite.
   b. Phosgene is detoxified by conjugation with glucuronide.
   c. Liver necrosis can occur if phosgene escapes the detoxification process.
   d. Phosgene is the only toxic metabolite of chloroform.

5. With regard to the detoxification of PCBs while in the environment (biodegradation)
   a. Sunlight may break down PCBs in surface water.
   b. No breakdown of PCBs occurs in bottom sediments.
   c. Bacteria may detoxify them by oxidation.
   d. Chemical dechlorination does not reduce the toxicity of PCBs.

6. With regard to the aromatic hydrocarbon (Ah) receptor
   a. TCDD attaches to it.
   b. Occupation of a certain minimum number of receptors is necessary for TCDD to be carcinogenic.
   c. The linear multistage assessment model for carcinogens may not be appropriate for TCDD.
   d. No other chemical is known to attach to the Ah receptor.

For Questions 7–11, match the chemical listed in the following to the appropriate use.

   a. Hexachlorophene
   b. Polybrominated biphenyls (PBBs)
   c. Polychlorinated biphenyls (PCBs)
   d. 2,4-Dichlorophenoxyacetic acid (2,4-D)
   e. Dichlorodiphenyltrichloroethane (DDT)

7. Insecticide
8. Disinfectant
9. Transformer insulator
10. Fire retardant
11. Herbicide

For Questions 12–15 answer true or false.

12. Dioxin (TCDD) can cause behavioral abnormalities.
13. Porphyrins are by-products of heme synthesis.
14. Victims of TCDD poisoning at Seveso showed no evidence of hepatotoxicity.
15. Pentachlorophenol is used as a wood preservative. From its name, one would predict that it would cause chloracne if accidentally consumed.

Answers

1. B
2. A
3. E
4. A
5. B
6. A
7. e
8. a
9. c
10. b
11. d
12. True
13. True
14. False
15. True

Further Reading

Axelson, O., Editorial: Seveso: Disentangling the dioxin enigma? *Epidemiology*, 4, 389–392, 1993.

Boffetta, P., Mundt, K.A., Adami, H.O., Cole, P., and Mandel, J.S., TCDD and cancer: A critical review of epidemiologic studies, *Crit. Rev. Toxicol.*, 41, 622–626, 2011.

Collins, J.J., Acquavella, J.F., and Friedlander, B.R., Reconciling old and new findings on dioxin, *Epidemiology*, 3, 65–69, 1992.

Collins, J.J., Strauss, M.E., Levinskas, G.J., and Conner, P.R., The mortality experience of workers exposed to 2,3,7,8,-tetrachloro-p-dioxin in a trichlorophenol process accident, *Epidemiology*, 4, 7–13, 1993.

Denison, M.S. and Helferich, W.G. (eds.), *Toxicant-Receptor Interactions*, Taylor & Francis, Philadelphia, PA, 1998.

Fingerhut, M.A., Halperin, W.E., Marlow, B.S., Piacitelli, L.A., Honchar, P.A., Sweney, M.H., Griefe, A.L., Dill, P.A., Steenland, K., and Suruda, A.J., Cancer mortality in workers exposed to 2,3,7,8-tetrachlorodibenzo-p-dioxin, *N. Engl. J. Med.*, 324, 212–218, 1991.

Fingerhut, M.A., Steenland, K., Sweeney, M.H., Halperin, W.E., and Piacitelli, L.A., Old and new reflections on dioxin, *Epidemiology*, 3, 69–72, 1992.

Hankinson, O., The aryl hydrocarbon receptor complex, *Annu. Rev. Pharmacol. Toxicol.*, 35, 307–340, 1995.

Jamieson, D.J., Terrell, M.L., Aquocha, N.N., Small, C.M., Cameron, L.L., and Marcus, M., Dietary exposure to brominated flame retardants and abnormal Pap test results, *J. Women's Health*, 9, 1269–1278, 2011.

Johnson, E.F., A partnership between the dioxin receptor and a basic helix-loop-helix-protein, *Science*, 252, 924–925, 1991.

Landers, J.P. and Bunce, N.J., The Ah receptor and the mechanism of dioxin toxicity, *Biochem. J.*, 276, 273–278, 1991.

Pesatori, A.C., Consonni, D., Rubagotti, M., Grillo, P., and Bertazzi, P.A., Cancer incidence in the population exposed to dioxin after the "Seveso accident": Twenty years of follow-up, *Environ. Health*, 8, 39–50, 2009.

Roberts, L., Dioxin risks revisited, *Science*, 251, 624–626, 1991a.

Roberts, L., EPA moves to reassess the risk of dioxin, *Science*, 252, 911, 1991b.

Ryan, J.J., Gasiewicz, T.A., and Brown, J.F., Human body burden of polychlorinated dibenzofurans associated with toxicity based on the Yusho and Yucheng incidents, *Fundam Appl. Toxicol.*, 14, 722–731, 1990.

Small, C.M., Murray, D., Terrell, M.L., and Marcus, M., Reproductive outcomes among women exposed to a brominated flame retardant in utero, *Arch. Environ. Occup. Health*, 66, 201–208, 2011.

Walker, C.H., Hopkin, S.P., Silby, R.M., and Peakall, D.B., *Principles of Ecotoxicology*, Taylor & Francis Ltd., London, U.K., 1996.

Walsh, A.A., Hsieh, P.H., and Rice, R.H., Aflatoxin toxicity in cultured human epidermal cells: Stimulation by 2,3,7,8,-tetrachlorodibenzo-p-dioxin, *Carcinogenesis*, 13, 2029–2033, 1992.

6

Toxicity of Metals

Mad as a Hatter.

Introduction

The process of felting, employed in making hats many years ago, required the use of mercurial compounds and many hatters suffered from the CNS disturbances, including behavioral disorders, associated with mercury toxicity. Metal intoxication as an occupational disease may be 4000 years old. Lead was produced as a by-product of silver mining as long ago as 2000 BC. Hippocrates described abdominal colic in a man who worked as a metal smelter in 370 BC and arsenic and mercury were known to the ancients even if their toxicity was not. In 1810 a remarkable case of mass poisoning with mercury occurred. The 74-gun man-o'-war *HMS Triumph* salvaged 130 tons of mercury from a Spanish vessel wrecked while returning from South America, where the mercury had been mined. The mercury was contained in leather pouches, which became damp and rotten, allowing it to escape and vaporize. Within 3 weeks 200 men were affected with signs of mercury poisoning including profuse salivation, weakness, tremor, partial paralysis, ulcerations of the mouth, and diarrhea. Almost all animals onboard died, including mice, cats, a dog, and a canary. Five men died. When the vessel put in at Gibraltar for cleaning, all those working in the hold salivated profusely.

The common nineteenth-century practice of adulterating foods and beverages (wine, beer, etc.) to increase profit led Accum to publish a treatise on the subject in 1820. Lead, copper, and mercury were frequently detected. Methods were not yet in place to detect arsenic, which was found to be a widespread adulterant later in the century. In 1875, the British parliament passed the first Food and Drugs Act as a result of these investigations.

In the past it was common to refer to heavy metal toxicity, as it was those metals that first emerged as industrial hazards. Heavy metals are arbitrarily defined as those having double-digit specific gravities and they include platinum (21.45), plutonium (19.84), tungsten (19.3), gold (18.88), mercury (13.55), lead (11.35), and molybdenum (10.22). These are in contrast to iron (7.87), manganese (7.21), chromium (7.18), zinc (7.13), selenium (4.78), and aluminum (2.70). Intermediate are copper (8.96) and cadmium (8.65).

In general, it can be seen that metals with sp.gr. less than eight are mostly essential trace nutritional elements (copper also is one and therefore the exception, as is aluminum, which is not a nutritional element), whereas those having sp.gr. greater than eight are the more toxic ones. It must be stressed once again that dose is all-important. Aluminum, with a sp.gr. of 2.70, has toxic properties. Arsenic exists in two solid forms: yellow arsenic (1.97) and grey or metallic arsenic (5.73). Both are highly toxic.

Lead

The Latin word for lead is plumbum, hence the chemical designation Pb. This word also gave origin to such English ones as plumbob (a mason's line with a metal ball attached for establishing vertical trueness), plummet (to fall as if leaden), and aplomb (to be as calm and undeviating as a plumb line). Lead was obviously well known to the ancients. In fact, they spent a lot of time trying to turn it into gold (alchemy). Lead toxicity was also familiar to them. Diascorides described its CNS toxicity as delirium.

Despite early knowledge of lead's toxic effects, the low melting point of the metal, coupled with its density, made it popular and useful. Well into the 1940s and early 1950s it was possible to buy lead toys, and kits were available to cast lead soldiers and lead fishing weights. An 1885 description of chronic lead poisoning is as good as any to be found in a modern text:

> The chief signs of chronic poisoning are those of general ill health; the digestion is disturbed, the appetite lessened, the bowels obstinately confined, the skin assumes a peculiar yellowish hue, and sometimes the sufferer is jaundiced. The gums show a black line from two to three lines in breadth, which microscopical examination and chemical tests alike show to be composed of sulphide of lead; occasionally the teeth turn black. The pulse is slow and all secretions are diminished. Pregnant women have a tendency to abort. There are also special symptoms one of the most prominent of which is lead colic. This colic is paroxysmal and excruciating.

Modern-day sources of lead are numerous. In the eighteenth century the industrial West discovered what the Chinese had known for centuries, namely that lead glazes produce crockery with a richer, smoother look. From this source and from lead solder in cans and kettles and water pipes leached by soft (but not hard) water, we consume about 150 µg/day. In some areas the figure may reach 1–2 mg. Occupational exposures were common in the past from lead mining, smelting, and the manufacture of products employing lead such as lead-based paints, toys, fuels, etc. Such exposures still occur in developing countries. Children are more vulnerable as all dirt and dust contain lead, especially in cities where lead from auto exhaust (tetraethyl lead) settles out on the ground.

This will persist long after the conversion to lead-free auto fuel. Children may also consume old lead-based paint, common in older buildings and which may also be on cheap wooden toys. In children, CNS toxicity is the dominant feature. This starts with vertigo and irritability, progressing to delirium, vomiting, and convulsions. The mortality rate is about 25% if treated and about 65% if untreated. In infants, exposure produces progressive mental deterioration after 18 months, with loss of motor skills, retarded speech development, and hyperkinesis in some cases. In the United States, the Lead Paint Poison Prevention Program was introduced in 1970. Since that time, the mean blood lead level of U.S. children has fallen from over 1 µmol/L (20.7 µg/dL) to less than 0.25 µmol/L (5.2 µg/dL). Only two deaths in children from acute lead encephalopathy have been reported in the past 30 years.

Children are not the only victims of lead poisoning from lead paint. Sandblasting of old, lead-painted buildings may, over time, cause chronic lead poisoning in workers who inhale the dust. Proper respirators and protective clothing are required for sandblasters. Heating of lead pint to a sufficiently high temperature can release lead fumes that can be inhaled. Cutting torches can produce sufficient heat to do this.

In all exposed individuals, subchronic toxicity can involve interference with mitochondrial heme synthesis at several levels, with resultant hypochromic (pale) microcytic (small) anemia. The pathway involved in this is illustrated in Figure 6.1.

Toxicokinetics of Lead

Elemental lead is not absorbed by the skin or through the alveoli of the lungs. Inhaled particulate lead is returned to the pharynx by the bronchial cilia and swallowed. Tetraethyl lead, however, may be absorbed across the skin and alveoli and readily penetrates CNS. Most of it is destroyed in exhaust emissions but sniffing leaded gasoline can result in severe CNS damage.

Gastrointestinal absorption of lead probably occurs via calcium channels as lead is a divalent cation (Pb^{2+}). It first appears in red blood cells, then hepatocytes, and then the epithelial cells of the renal tubules. It is gradually redistributed to hair, teeth, and bones where 95% of it is stored harmlessly. The T1/2 in blood is about 30 days, in bone, 25 years. Little reaches the adult brain but much more enters the infant brain. Renal excretion is the main route of elimination.

Cellular Toxicity of Lead

Lead affects oxidative phosphorylation and ATP synthesis in the mitochondrion. It also increases red cell fragility and inhibits sodium/potassium ATPase. Kidney tubular cells become necrotic, and chronic exposure may lead to interstitial nephritis. Nuclear inclusion bodies, consisting of lead bound to a protein, may be formed in renal cells. This may be considered as a protective mechanism. Carcinogenesis has been demonstrated in experimental

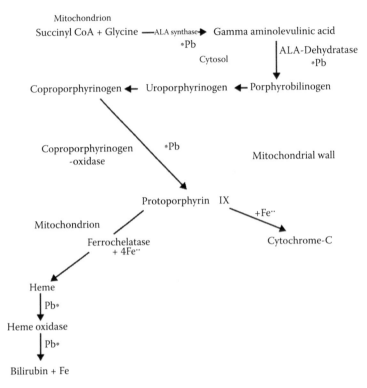

FIGURE 6.1
A simplified scheme showing points of interference of lead in heme synthesis (see also Figure 5.2) for ALA synthase and heme inhibition.

animals and chromosomal abnormalities have been observed, but evidence of tumor production in humans is scarce. Most of the toxic effects of lead and other heavy metals can be explained by their affinity for thiol groups. This is also the basis of chelation therapy.

Fetal Toxicity

A characteristic of all metals is their ability to penetrate the placental barrier, so that fetal toxicity can occur as a result of maternal exposure. Lead is considered to be a human carcinogen and pregnant women are generally removed from jobs where exposure may occur.

Treatment

Lead chelators are the treatment of choice. These bind lead (and other divalent cations) so that it can be excreted. Calcium/sodium ethylene diamine tetra-acetate (CaNa$_2$EDTA) and dimercaprol (British antilewisite, BAL) are given intramuscularly followed by oral penicillamine for several weeks. BAL was developed

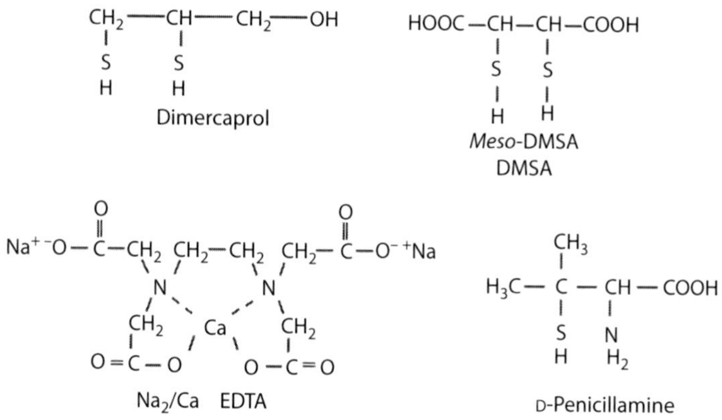

FIGURE 6.2
Chemical structures of some metal chelators.

during World War II as a treatment for lewisite, a vesicant arsenical poison gas. A newer chelator is meso-2,3-dimercaptosuccinic acid (DMSA). The chemical structures of these chelators are shown in Figure 6.2. In the case of EDTA, Pb is exchanged for Ca^{2+} whereas with the others, the Pb is bound to sulfhydryl (SH) groups. The complexes are excreted, mostly in urine. A disadvantage of chelation therapy is that it does not remove lead from the brain very efficiently. Cuprimine, another chelator, is D-penicillamine, 3-mercapto-D-valine.

Despite 50 years of use, objective evidence for benefit of chelation therapy for lead poisoning is scanty. It is widely agreed that it has drastically reduced the mortality from lead encephalopathy if diagnosis and treatment are started early. It also relieves lead colic, malaise, basophilic stippling, and it rapidly restores red-cell ALA dehydratase. It does not influence the residual manifestations of chronic lead poisoning such as peripheral neuropathy. A recent study showed that chelation with Cuprimine in lead-intoxicated patients significantly lowered blood Pb levels (initially 5.58 ± 2.02 µg/dL) and total Pb body burden as well as reducing blood zinc protoporphyrin. Cuprimine was given intramuscularly at 25–35 mg/kg of body weight per day in divided doses.

In nonindustrial cases of lead poisoning, it is essential to identify the source of the lead and remove it or remove the possibility of people contacting it. This is most important where children are involved.

Mercury

Mercury (Hg) exists in three forms: elemental mercury, inorganic compounds, and organic compounds. Elemental mercury causes toxicity when the mercury vapor is inhaled, as exemplified by the episode described at the beginning of this chapter. The major source of elemental mercury in the environment is the natural degassing of the earth's crust. Estimates of the level of mercury reaching the atmosphere range from 25,000 to 150,000 tons/year, and the atmosphere represents a major mechanism for global transport of metallic mercury. Conversely, anthropogenic sources account for only 10,000 tons/year, but because industrial effluent tends to be concentrated, these are the sources usually associated with toxicity. Metallic mercury and its vapor can be an industrial hazard. Mercury is used in the manufacture of chlorine and sodium hydroxide by the mercury cell process, in paint preservatives, and in the electronics industry. It is a by-product of smelting processes (most mineral ores contain mercury) and it is released during fossil fuel combustion.

Elemental Mercury Toxicity

In vapor form, elemental mercury is well absorbed across both the alveoli of the lungs and the blood–brain barrier. Acute poisoning usually occurs within several hours. Weakness, chills, metallic taste, salivation, nausea, vomiting, diarrhea, labored breathing, cough, and tightness in the chest may ensue. If the exposure is more prolonged, interstitial pneumonitis may develop. Recovery is usually complete except that residual loss of pulmonary function may persist. Chronic exposure to mercury vapor results in CNS disturbances including tremor and a variety of behavioral changes that can include depression, irritability, shyness, instability, confusion, and forgetfulness. Mercury vapor from mercury nitrate formerly used in the felting process accounted for the "mad hatter" syndrome. The behavioral abnormalities of the "Mad Hatter" in Lewis Carroll's *The Adventures of Alice in Wonderland* were really quite mild, compared to the other characters, which is in keeping with the topsy-turvy world that Carroll created. Thyroid disturbances may be present.

Inorganic Mercurial Salts

Inorganic salts such as mercuric chloride can cause severe, acute toxicity. The proteins of mucous membranes are precipitated, giving them an ash-gray color in the mouth and pharynx. Intense abdominal pain and vomiting are common. Loss of blood and fluid from the gastrointestinal tract results from sloughing of the mucosa in the stool and may lead to hypovolemia and shock. Renal tubular necrosis occurs after acute exposure and glomerular damage is more common after chronic exposure. A phenomenon called "pink disease" or acrodynia commonly follows chronic exposure to mercury ions. It is a flushing of the skin that is believed to have an allergic basis.

Organic Mercurials

Methylmercury is the commonest cause of organic mercurial poisoning and the most important one environmentally. It is extremely well absorbed from the gastrointestinal tract (90%) and deposited in the brain. Because of its high affinity for SH groups, methylmercury binds to cysteine and this may then substitute for methionine and be incorporated into proteins. This can result in the formation of abnormal microtubules required for cell division and neuronal migration. Many enzymes, such as membrane ATPase, are SH dependent and mercury may thus interfere with their function. Renal toxicity is a common manifestation of poisoning. Renal uptake of mercury is rapid and it accumulates in the kidneys. Mitochondrial dysfunction appears to be involved in cell damage in the proximal tubule.

The main signs and symptoms are neurological and consist of visual disturbances, weakness, incoordination, loss of sensation, loss of hearing, joint pain, mental deterioration, tremor, and in severe cases, paralysis, and death. Infants exposed *in utero* may be deformed and intellectually challenged. Experimentally, methylmercury has been shown in cell cultures to mobilize Ca^{2+} from intracellular stores that are sensitive to inositol 1,4,5-triphosphate.

Mercury is a waste product of many industrial processes. Hg is methylated in sediment by bacteria and cyanocobalamin. Several outbreaks of methylmercury poisoning have occurred. The most widely known began in Minimata, Japan, in 1953 near a plant that manufactured acetaldehyde and which discharged methylmercuric chloride ($MeHg^+Cl^-$) into Minimata bay. People who ate mollusks and large fish from the bay developed the symptoms that came to be known as Minamata disease. Nine hundred cases developed and there were 90 fatalities. Because of the high fetal toxicity of mercury many deformed infants were born. Another source of mercury toxicity is the consumption of seed grains treated with methylmercuric chloride as a fungicide. Several mass poisonings have occurred around the world. In Iraq in 1972, one such episode resulted in over 6500 cases of poisoning and 500 deaths.

Mechanism of Mercury Toxicity

Mercury toxicity can be explained entirely by its ability to bind with the hydrogen of SH groups to form mercaptides (i.e., X-Hg-SR and $HgSR_2$ where X = an electronegative radical and R = a protein). Organic mercurials such as methylmercury form mercaptides, R-Hg-SR'. The term *mercapto* means "to capture mercury" and refers to sulfur-containing groups. Since SH groups are important components of many enzymes, mercury acts as an enzyme poison and interferes with cell function at many levels. Mercury can also combine with other physiologically important ligands such as phosphoryl, carboxyl, amide and amine groups. Metallic Hg vapor may be oxidized by catalase enzyme in red blood cells to the less toxic divalent form. Alcohol competitively inhibits this process. Mercury was an important pharmaceutical agent for centuries, and its pharmacological properties also depend on its affinity for SH groups. It was used as an antibacterial agent (Ehrlich's 606 or Salvarsan) for syphilis, as a laxative, in skin creams, and in diuretics.

Mercurial diuretics were still in use in the 1960s. They were replaced eventually by safer agents. Aminomercuric chloride may still appear in freckle-removing creams, and daily application for years may result in increases in 24 h urine mercury excretions from 10 µg to 1 mg and the development of symptoms such as excessive salivation and insomnia.

Treatment of Mercury Poisoning

Mercury poisoning continues to be a significant industrial problem. A 2012 report from Iran compared 46 workers exposed to low levels of Hg vapors in a chlor-alkali plant to 65 healthy, unexposed employees. Atmospheric Hg levels were 3.98 ± 6.28 µg/m^3. Urinary Hg levels were threefold higher in the exposed workers. Symptoms reported included fatigue, anorexia, memory loss, erethism (a state of agitation), blurred vision, and dental problems.

Chelation therapy is recommended for elemental, inorganic mercury poisoning. Dimercaprol and penicillamine are SH-containing chelators. Dimercaprol is given intramuscularly and penicillamine, orally. Hemodialysis may also be used, and vomiting may be induced if there has been recent ingestion of mercury. These treatments are of little use in methylmercury poisoning, however. Dimercaprol actually increases brain levels of methylmercury and penicillamine and hemodialysis do not relieve symptoms. Some success has been achieved with binding resins taken orally. Since there is a significant enterohepatic recirculation of methylmercury (i.e., it is excreted in the bile and reabsorbed from the intestinal tract), binding it to a polythiol resin prevents its reabsorption because it is excreted in the feces.

The Grassy Narrows Story

In 1969 Norvald Fimreite, a PhD candidate in the Department of Zoology at the University of Western Ontario, first made public his findings on the mercury contamination of fish in Canadian and border lakes. The highest levels were recorded from a small lake, Pinchi, in British Columbia (10 ppm) and from Lake St. Clair (7.03 ppm) in the Great Lakes waterway. The (Canadian) federal standard for export and consumption was 0.5 ppm. His report was a bombshell coming on the heels of reports of Minamata disease from Japan. Fimreite estimated that Canadian industry was releasing 200,000 lb of mercury annually into the environment. Most of it came from chlor-alkali plants and from pulp and paper mills that used mercurials as antisliming (antialgal) agents, and the chlorine and alkali as bleaching agents. The question of mercury discharge from the Dow (Canada) Chemical plant had been raised 6 years earlier in the Ontario Legislature but nothing had been done. In 1970, the Ontario Water Resources Commission took steps to reduce Dow's output, but in Dryden, near the Manitoba border, the Dryden Pulp and Paper Co. (owned by the British Reed Group) had been emitting mercury vapor since 1962, and some workers developed bleeding gums and muscle twitches.

By 1970 it had pumped an estimated 20,000 lb of mercury into the surrounding environment, including discharges described as a brown froth into the Wabigoon River. Raw sewage also was discharged into the Wabigoon, providing a rich source of anaerobic bacteria to methylate elemental mercury. The Wabigoon is part of the English River system, and about 50 km downstream lie the Grassy Narrows and Wabaseemong Anishanaabe First Nations reserves. The residents gleaned a slim but adequate living as fishing guides and lived largely off the land, eating fish, deer, and moose supplemented with garden vegetables. In March of 1970, contamination of fish in Lake Erie was detected and the Lake St. Clair and Lake Erie fisheries were closed. Chloralkali plants and pulp mills were ordered to stop using mercury by the end of May after a concerted attack in the Ontario Legislature by opposition parties. Mercury, however, is not biodegradable, and it is only when it is buried by uncontaminated sediment that it ceases to be a threat. In June of 1970, the Lamms, owners of Ball Lake Fishing Lodge, hired Fimreite to conduct a survey of mercury levels in the fish of the English–Wabigoon system. The findings were appalling. Levels ranged from 13 to 30 ppm, as high as those from Minamata Bay. The government lifted Fimreite's license to collect specimens for scientific purposes and ignored his appeals to test the residents of the reserves until his data were made public, when it conceded that it had similar findings. A ban was placed on eating fish from the contaminated area but otherwise the government continued to downplay the problem. Tourist fishing dried up and the Indians went on welfare. Blood levels of mercury were not seriously studied until 1973, and ranged from 45 to 289 ppb (normal is about 20 ppb for a city dweller). Some residents were showing signs of mercury poisoning and the incidence of stillbirths was going up.

The social costs of this tragedy were perhaps even greater than the direct effects of mercury. In the years surrounding the discovery of mercury in the Grassy Narrows area, the death rate rose to 1 in 50, three times the national average. Most were alcohol related. Many of the deaths were newborn or very young infants. Violence became rampant. Dr. Peter Newbury, also a graduate of U.W.O., conducted a study for the Society of Friends (Quaker) and the National Indian Brotherhood and felt that the CNS effects of mercury were a contributing factor in the violence. Gasoline sniffing became common amongst young people. (It remains a problem on many reserves.) The Grand Council of Treaty Three District, which includes Grassy Narrows and Kenora, completed a study in 1973. They found that in the preceding 42-month period, there had been 189 violent deaths of native people. They reported 38 from gunshot, stabbing, or hanging; 30 in fires; 42 drownings; 25 from exposure; and 16 from car accidents. In the same year, members of the Ojibwa Warrior Society occupied Anicinabe Park on the outskirts of Kenora. Barricades were erected and manned by armed warriors. The park was claimed as First Nations land. The standoff lasted for several weeks but achieved little.

No follow-up of mercury contamination of fish in the river system was done since the 1970s until a report was published in 2007. This report was done at the request of the Grand Council treaty #3. Several species of fish were assayed for mercury including Walleye, Northern Pike, Large-mouth Bass, and Whitefish. A total of 851 fish samples were collected along the length of the river system. They found that mercury levels had declined over the preceding 25 years but a gradient was still observed with the highest levels detected in specimens collected closest to the original source of contamination. There also was a correlation between the length of the fish (an indicator of age) and the level of mercury. Moreover, the predatory species (the piscivores large-mouth bass, walleye, and northern pike) had much higher levels of mercury than the whitefish that survive on zooplankton. From Clay Lake the levels in the predators were all in excess of 2 mg/kg (ppm), whereas Whitefish levels were 0.2 mg/kg or less. It was recommended that fish consumption from Clay Lake should be discouraged. Whitefish from all other areas could be consumed regularly but consumption of the other species should be limited to smaller fish on a restricted basis. This study clearly illustrates how mercury contamination can persist for decades.

Cadmium

Cadmium (Cd) is present naturally in the environment in very low levels, being solubilized during the weathering of rock (levels are about 0.03 μg/g of soil, 0.07 μg/mL of freshwater, and 1 ng/m^3 of air). Dissolved cadmium may form a number of soluble and insoluble organic and inorganic compounds. Cadmium is chemically similar to zinc and it is present in zinc ore in a ratio of about 1/250. Most cadmium is produced as a by-product of electrolytic zinc plants. It is used in metal plating, in the manufacture of nickel-cadmium batteries, in the manufacture of pigments, in plastic stabilizers and small amounts are used in photographic chemicals, in catalysts, and in fungicides used on golf courses. Environmentally significant emissions come mainly from smelting operations for copper, lead, and zinc, from auto exhaust, and manufacture of pigments and alloys (most nickel-cadmium batteries are imported into Canada). Cadmium is readily taken up by plants and stored in the leaves and seeds. It is present in sewage sludge fertilizers (recommended maximum 20 ppm). Water pollution with cadmium may result in high levels in fish and especially in mollusks. The main sources in the human diet are organ meats (cadmium accumulates in liver and kidney), cereal grains, shellfish, and crustaceans.

Cadmium Toxicokinetics

Cadmium intake in Canada averages 50–100 μg/day from inhaled and ingested sources. Inhaled, unpolluted air may contribute up to 0.15 μg/day, whereas breathing air near a smelter can raise the level to 10 µg/day. Cigarettes contain cadmium and smoking increases exposure still further. About 50% of inhaled cadmium is absorbed. Only about 6% of ingested cadmium is absorbed, but it contributes most of the daily load. The FAO/WHO recommends a maximum weekly intake of 500 µg. Absorbed cadmium is bound to plasma albumin and cleared rapidly from the plasma. It is found in red cells only after high exposures. It is rapidly distributed to the liver, pancreas, prostate, and kidney, with slow redistribution to the kidney until, over time, it contains most of the cadmium. Renal levels increase up to age 50 and depend on the cumulative exposure. The T1/2 in humans is about 20 years. Cadmium is trapped in the kidney and liver by a cysteine (i.e., SH-rich) protein called metallothionein with a high affinity for cadmium and zinc. Cadmium normally binds to metallothionein, the synthesis of which is induced by the presence of the cadmium. High doses, however, exceed the binding capacity of the protein and the cadmium is free to bind to other essential cell components such as the basement membrane of the renal glomerulus.

Cadmium Toxicity

The kidney is the major organ of toxicity. About 200 µg/g wet weight of kidney appears to be the critical concentration in the renal cortex for damage to occur in the form of proximal tubule dysfunction. Once renal disease develops, cadmium is lost from the kidney. Nutritional deficiencies of zinc, iron, and calcium may predispose cadmium toxicity by increasing absorption from the gastrointestinal tract. Calcium deficiency increases the synthesis of calcium-binding proteins and cadmium absorption. Cadmium causes cellular damage probably by binding to SH groups of essential cell proteins. The production of reactive oxygen species can cause oxidative stress. The complexing of cadmium with metallothionein can increase renal toxicity because the complex is more readily taken up by the kidney than the free metal. It protects, however, against testicular damage, which can occur after a single exposure that can cause necrosis, degeneration, and loss of spermatozoa. This appears to be related to cadmium's effect on testicular vascular supply.

Workers in metal refineries may be exposed to high levels of cadmium fumes and develop respiratory difficulties. Chronic exposure may lead to obstructive pulmonary disease and emphysema. A major exposure occurred in Japan in the late 1940s. Effluent from a lead processing plant washed into adjacent rice paddies over decades and the rice accumulated high levels of cadmium. Because the people were calcium deficient due to a poor diet, they developed acute cadmium toxicity with severe muscle pain, malabsorption, anemia, and renal failure. The outbreak was named "Itai-Itai" (ouch-ouch) disease. The fetus appears to be protected from cadmium toxicity by placental synthesis of metallothionein, but heavy exposures can overwhelm this defense.

Animal studies have shown cadmium to be carcinogenic and there is a suggestion that it may increase the incidence of prostate cancer in elderly men. Other metals, notably arsenic, chromium, and lead, also have been implicated as carcinogens. A recent study provides convincing evidence that cadmium dietary intake is a significant risk factor for breast cancer in postmenopausal women. Cadmium is a common contaminant of food stuffs especially cereal grains. This Swedish study examined the incidence of estrogen receptor-positive (ER+) and estrogen receptor-negative (ER–) breast cancer in about 56,000 postmenopausal women. Dietary cadmium intake, adjusted for confounders such as whole grains and vegetables, which, although they are a source of cadmium, also contain anticancer phytochemicals, was associated with an increased incidence of total breast cancers that was statistically significant as was the association with ER+ ones.

Treatment

Chelation therapy is not effective. Treatment consists of removing the patient from the source of exposure and supportive measures.

Arsenic

Arsenic (As) pollution of water courses and groundwater is a major health issue in some parts of the world and in some countries it is the most important chemical pollutant in groundwater and drinking water. Up to 200 million people in 70 countries are exposed to arsenic from contaminated drinking water. They are at risk from debilitating disease and fatal cancers. In the Bengal delta region, it is estimated that about 35 million people have been drinking arsenic-contaminated water for decades and 15% had arsenic-related skin lesions. Arsenic, at a concentration of 50 µg/L in drinking water, can cause health problems after drinking it for 10–15 years. The Mekong River delta is another area of concern affecting Cambodia and Vietnam. Groundwater arsenic concentrations ranged from 1 to 3050 µg/L (average 159 µg/L) in Cambodia, and 1–845 µg/L in southern Vietnam. The Red River delta area also was affected (2–33 µg/L). Deep (12–40 m) core sediment samples contained arsenic 2–33 µg/g.

Arsenic is an age-old pharmaceutical preparation. It was believed to be a tonic because it causes facial flushing (rosy cheeks) and fullness (edema). It is still used in the treatment of trypanosomiasis. Ehrlich studied organic arsenicals and developed the first effective treatment for syphilis (Ehrlich's 606). The chemistry of arsenic is exceedingly complex since it can exist as a metallic form and as trivalent and pentavalent compounds. Trivalent forms include arsenic trioxide, arsenic trichloride, and sodium arsenite.

Pentavalent forms include arsenic pentoxide, arsenic acid, lead arsenate, and calcium arsenate. Organic arsenicals also may be trivalent or pentavalent. Arsenic in the environment arises from weathering of rock and from emissions from smelting of gold, silver, copper, zinc, and lead ores, combustion of fossil fuels, and the use of arsenicals in agriculture as herbicides and pesticides. Airborne particles may travel considerable distances and penetrate deeply into the lungs. Arsenic is taken up by plants and the degree of uptake varies with the soil type. Fine soils high in clay and organic material inhibit uptake. Arsenic also enters the water system through runoff and fallout. Wells drilled through rock containing arsenic will yield water high in arsenic. Chronic poisoning may result, and this is a problem in some parts of Nova Scotia (see above regarding water contamination in other countries). Tobacco contains arsenic. The average daily intake of arsenic in North America is about 25 µg.

Toxicokinetics of Arsenicals

Arsenic may be absorbed from the gastrointestinal tract, the lungs, and across the skin and mucous membranes. It penetrates intracellularly by an uptake mechanism used in phosphate transport. Like mercury, arsenic binds to SH and disulfide groups to poison numerous cell enzymes and respiration. Chromosomal breakage has been observed experimentally. In general, the order of toxicity is organic arsenicals > inorganic arsenicals > metallic arsenic. The trivalent arsenite has a high affinity for SH groups and interferes with the enzyme pyruvate dehydrogenase. Plasma pyruvate levels will increase. Some biotransformation between trivalent and pentavalent forms may occur. The T1/2 of arsenic is about 10 h and excretion is mainly by the kidneys. Arsenic is an effective uncoupler of oxidative phosphorylation. Numerous arsenite-oxidizing bacteria exist in the environment, forming arsenic trioxide. Arsenic trioxide has been reintroduced recently as a chemotherapeutic agent.

Toxicity of Arsenicals

As noted, the most poisonous forms are arsenic trioxide (As_2O_3) and sodium arsenite ($NaAsO_2$). Arsenic tends to accumulate in the liver, kidney, heart, and lung. It is also deposited in bone, teeth, hair, and nails and these become important tissues for diagnostic and forensic analysis. The average human intake is about 300 µg/day but it may be much higher if fish is a large part of the diet, as they accumulate the poison through biomagnification. In 2012 the U.S. Food and Drug Administration (FDA) released a report on arsenic contamination of fruit juices. Importation of several sources of pear juice and concentrate was blocked because the arsenic content was greater than 23 ppb, their level of concern. Contaminated apple juice and concentrate were also found.

Acute arsenic poisoning causes severe abdominal pain and it is rare. Chronic poisoning causes muscle weakness and pain, skin pigmentation, gross edema, gastrointestinal disturbances, kidney and liver damage, and peripheral neuritis with eventual paralysis. The fingernails develop white lines, called Mee's lines, which can be used to determine when exposure occurred.

Arsine is the gaseous form of arsenic resulting from electrolytic processes and it is extremely toxic, producing rapid and often fatal hemolysis. It has a garlic-like odor.

When Napoleon Bonaparte died in exile on the island of St. Helena in 1821 suspicions immediately fell on royalists fearful of his return to France. This theory was reinforced when tests in 1961 revealed high levels of arsenic in a sample of his hair. Subsequent studies, however, found that virtually everyone in the early nineteenth century had elevated arsenic in hair samples. Arsenic was used in dyes, cosmetics, as a pharmaceutical, and in many manufacturing process. Even the wallpaper in Bonaparte's damp, shabby home-in-exile contained it. In the hot climate the action of microorganisms could have caused off-gassing of arsine.

Treatment

Chelation therapy is used for arsenic poisoning. Both dimercaprol and penicillamine have been used.

Environmental Effects of Arsenic

Arsenic is toxic to a wide range of plants and animals including marine species. Of the plants, beans, peas, and rice are especially sensitive. Algae are sensitive, as well as some protozoa such as *Daphnia magna*. Finned fish are quite susceptible.

Chromium

Chromium (Cr) is used in the production of stainless steel, chrome plating, pigments, and in the chemical industry. Chromium has two oxidation states: Cr^{+3} and Cr^{+6}. The latter is much more toxic, causing severe respiratory irritation when inhaled, and possibly lung cancer after long chronic exposure. Kidney damage also occurs. The trivalent form binds readily to electron-donating ligands such as macromolecules like RNA but it does not readily cross the cell membrane. Conversely, the hexavalent form readily crosses cell membranes and is reduced to the trivalent form intracellularly. Toxicity is thus normally related to the presence of the hexavalent form in the environment. As the oxyanion it is taken up by the cell, probably by a sulfate transport system as shown later. Air levels as low as 10 $\mu g/m^3$ of Cr^{+6} can produce respiratory irritation. Chromium is distributed in the biosphere much like arsenic and can have similar effects.

Other Metals

Virtually any metal, if taken in excessive amounts or by an unusual route, can manifest toxicity. Thus, the inhalation of any metal dust can cause pulmonary fibrosis.

Aluminum

The kidney does not clear aluminum very efficiently and in renal insufficiency it may accumulate to toxic levels. Minute levels of aluminum are present in food and water, and if aluminum-free water is not used in dialysis machines, patients may accumulate high blood levels over time. This can cause microcytic anemia. This can be treated with the chelator desferoxamine, which removes the aluminum from the blood. Before the problem with aluminum and dialysis was identified, some patients developed very high blood levels and signs and symptoms of dementia similar to Alzheimer's disease.

Manganese

Public concern has developed over a potential source of inhaled manganese, the fuel additive MMT. Methylcyclopentadienyl manganese tricarbonyl is an octane booster employed by the petroleum industry and manufactured by the Ethyl Corporation. Automobile manufacturers object to it because it defeats antipollution devices. Environmentalists object to it because they believe it constitutes a threat to human health. The petroleum industry claims that it would be too costly to eliminate its use. A WHO group conducted a study of workers in Quebec who had been exposed to manganese fumes and dust during the manufacture of metal alloys. The workers demonstrated problems with motor coordination and mentation. Manganism is a condition in which Parkinson-like signs and symptoms appear; tremor, memory loss, irritability, insomnia, difficulties with speech, and, in severe cases, insanity. The workers were believed to be suffering from "micro-manganism." Attempts to ban MMT use in both Canada and the United States have foundered on the absence of definitive evidence that it constitutes a hazard and on concerted legal assaults by its defenders. Uncertainty regarding the health risk associated with the use of MMT as a fuel additive persists. However, a study released in 2012 indicates that new pharmacokinetic data for MMT will allow the prediction of tissue manganese levels for several portals of entry, exposure levels, and durations of exposure. This should permit a more accurate risk assessment.

Uranium

Uranium is nephrotoxic. It binds to albumin and to bicarbonate anion that is filtered by the kidney where it dissociates. The free uranyl cation binds to proteins in the proximal tubule and damages them. Toxic metals may also substitute for physiological ones, as when lead and strontium 90 are deposited in bones and teeth like calcium.

Antimony

Antimony is an industrial contaminant with distribution and toxicity essentially similar to that of arsenic's.

Nutritional Elements

Even essential metals like iron can be very toxic, especially to young children who may ingest iron-containing vitamin preparations, mistaking them for candy. Vomiting occurs and vomitus and stool may contain blood. Acidosis and shock develop. Kidney and liver damage can occur. In adults, iron overload sometimes occurs and hemosiderin is deposited in tissues including muscle. Desferroxime is a specific iron chelator with a low affinity for calcium that is used to remove systemic iron in both types of poisoning.

Metallothioneins

Metallothioneins (MTs) are low-molecular-weight (6000–7000 Da) proteins rich in SH groups and they are found in most mammalian cells. There are four classes of MTs based on their amino acid sequences. MT-I and MT-II are the most widely distributed, while MT-III and MT-IV are restricted to neurons, and squamous epithelial cells. MTs have a high affinity for many metals including Ag^+, Cu^+, Cd^{2+}, Hg^{2+}, and Zn^{2+}. MTs are found primarily in the cytoplasm and their function is to serve as storage depots and buffers for copper and zinc. The MT-I gene can be induced by cadmium, copper, mercury, and zinc. Such induction makes them important defenses against heavy metal poisoning, and experiments with knockout mice have shown that the absence of MT-I and MT-II genes makes them more vulnerable to cadmium toxicity. The Agency for Toxic Substances and Disease Registry maintains an excellent website at www.atsdr.cdc.gov/tfacts22.html

Carcinogenicity of Metals

Arsenic has long been recognized as being associated with an increased risk of skin and respiratory cancer. In 1930, workers in a factory making an arsenical sheep dip were identified as having an excessive incidence of skin cancer. Arsenic levels in air >54.6 $\mu g/m^3$ were associated with an increase in lung cancer incidence. High water content of arsenic also has been associated with increased cancer risk.

In 1976, a NIOSH study of 300 workers in a cadmium smelter revealed a significantly higher incidence of cancer. The incidence of lung cancer was twice normal and prostate cancer also was high. Some of these workers, however, were also exposed to arsenic. Inhalation of chromium dust by workers has been associated with an increased incidence of lung cancer. Nickel is also a respiratory carcinogen but it lacks other chronic toxic effects. There is some evidence suggesting that lead may be carcinogenic, or perhaps a co-carcinogen owing to its persistence in tissues. Case reports and epidemiological studies are difficult to sort out because exposures frequently involve more than one metal. This is true in the steel industry where increased cancer frequencies have been observed.

Many cationic metals will form complexes with thiol groups of cell components and the complex will mimic natural substrates to interfere with cell processes, mostly transport systems. Thus, methylmercury–cysteine complex mimics methionine and the complex is taken up into the brain by a transport system for neutral amino acids. Inorganic and organic mercury complexes with glutathione and is transported from liver cells into bile. Arsenic and copper do the same thing. Lead can substitute for Ca^{2+} in a number of transport and receptor-mediated processes. Voltage-activated calcium channels will admit a number of metallic cations, a fact that is exploited in research. Cadmium (Cd) and lanthanum (La) act in this way.

The case for mercury as a carcinogen is less clear-cut. Studies in experimental animals have confirmed carcinogenicity. Methylmercury chloride caused kidney tumors in male mice and mercury chloride was carcinogenic in male rats. Epidemiological evidence from dentists and dental nurses, chlor-alkali workers, and nuclear weapons workers is equivocal but suggests an increased risk of lung, kidney, and central nervous system tumors.

Unusual Sources of Heavy Metal Exposure

In 1988, the Texas Department of Health investigated illegal sales of drugs manufactured in Hong Kong. The tablets, sold as "chuifong tokuwan," contained a veritable pharmacy of drugs, including diazepam, indomethacin, hydrochlorothiazide, mefenamic acid, dexamethasone, lead, and cadmium! This potpourri of tranquilizer, diuretic, anti-inflammatory agents, corticosteroids, and heavy metals was repackaged and marketed as "The Miracle Herb–Mother Nature's Finest." Twenty-four percent of 93 persons who took this preparation had elevated urine cadmium levels; 1.8 µg/mL compared to 0.5 µg/mL for random controls. The upper limit of normal is considered to be 2.5 µg/mL. No elevated (>25 µg/dL) blood lead levels were detected, but 42% of these individuals had elevated urine levels of retinol-binding protein, indicative of renal tubular damage. Ayurvedic medicines also have been implicated in metal contamination.

Some of the "health" supplements such as bone meal (for calcium) contain high amounts of lead, and some zinc supplements are contaminated with cadmium. See also Chapter 8 for more on herbal remedies.

In Ohio, several members of a household were hospitalized with a diagnosis of acrodynia, a form of metallic mercury poisoning in which neurological and psychological disorders occur as well as hypertension, rash, sweating, cold intolerance, tremor, irritability, insomnia, anorexia, and diminished performance at school. Twenty-four-hour urine collections revealed mercury levels of 850–1500 µg/mL (normal <20 mg/mL). Careful inquiry of neighbors indicated that a previous tenant had spilled a large jar of elemental mercury in the apartment. Treatment was instituted with the oral chelating agent 2,3-dimercaptosuccinic acid (DMSA). Some neurological disorders persisted.

Case Study 11

In early spring, two of five workers employed in demolishing an old iron bridge visited the company's consulting physician complaining of muscle pain (myalgia), joint pain (arthralgia), headache, and nausea. These workers had been cutting up sections of the bridge using oxyacetylene torches. Large sections were lowered to a barge moored in the river below the bridge to be cut into smaller sections for hauling away. When the remaining three workers were questioned, it was discovered that they too had been suffering from similar symptoms as well as memory loss and irritability. A supervisor and a secretary who worked in the construction shack on shore were not affected, nor were four men involved in loading trucks or operating a small boom crane.

Q. What organ systems are involved in the affected workers?

Q. Could this be a toxicant causing the illness and, if so, what are its possible sources?

Q. What is the likely portal of entry?

Q. What diagnostic tests might be useful?

Q. What specific treatment might be appropriate?

Case Study 12

A 67-year-old man consulted his physician because of severe abdominal pain, weight loss, and fatigue. The doctor initially suspected gastric carcinoma but the patient was severely anemic, his red cells had basophilic stippling, and he had a blood lead level of 70 µg/dL. Six other household members also were affected, including an 8-year-old child. All had elevated blood lead levels. The home was located in a suburban residential area not near any industrial site.

Q. What is the probable portal of entry of the lead in these people?

Q. A search of the home revealed a ceramic jug as the offending agent. Why was it suspect?

Case Study 13

During a routine preemployment medical examination, a 46-year-old male was found to have a blood lead level of 50 µg/dL. He was subsequently investigated by a university hospital toxicology clinic that confirmed the same blood lead level 1 month later. Symptoms included numbness of the fingers and palms, tinnitus (ringing in the ears), and an apparent decrease in ability to do mental arithmetic and mild memory deficits. He had been taking ranitidine for indigestion. This is a histamine H2 receptor blocker that suppresses gastric acid secretion.

A detailed personal and employment history was obtained. He had spent 20 years as an electronics technician in the army and in civilian life but had had little exposure to lead from soldering or welding. He had no hobbies that could serve as a source of lead, no history of bullet or birdshot wounds and he denied drinking bootleg alcohol or using lead additives in his car.

Q. Is the source of lead likely to be work related or from the home environment? How could this be determined?

Q. What is the probable significance of the gastric distress?

Q. What therapeutic approach would be appropriate?

Review Questions

For Questions 1–6 answer true or false:

1. The term *heavy metal* usually refers to those with double-digit specific gravities.
2. Aluminum, which has a specific gravity of only 2.7, has no toxic properties.
3. All forms of arsenic are equally toxic.
4. Lead poisoning may be manifested as both gastrointestinal and central nervous system toxicity.
5. Acute, paroxysmal, colicky pain is common in severe lead poisoning in adults.
6. Mental retardation does not occur in lead poisoning in children.

For Questions 7–11 use the following code:

> Answer A if statements a, b, and c are correct.
> Answer B if statements a and c are correct.
> Answer C if statements b and d are correct.
> Answer D if statement d only is correct.
> Answer E if all statements (a, b, c, d) are correct.

7. a. Subchronic lead poisoning involves interference with mitochondrial heme synthesis at several levels.
   b. Anemia may accompany lead toxicity.
   c. Elemental lead is not absorbed through the skin or lungs.
   d. Elemental lead is not absorbed from the intestinal tract.
8. a. All metals can cross the placenta and cause fetal toxicity.
   b. Examining a blood sample is the only way of detecting lead in the body.

   c. Chelation treatment with dimercaprol may be used in lead poisoning.

   d. Chelation therapy is of no use in elemental mercury poisoning.

9. a. Elemental mercury poisoning never occurs.

   b. Methylmercury is the most important environmental source of mercury poisoning.

   c. Chelation therapy is useful in the treatment of methylmercury poisoning.

   d. Methylmercury poisoning may involve a variety of central nervous symptoms as well as signs of arthritis.

10. a. Cadmium accumulates in the liver and kidney.

   b. Cadmium induces the synthesis of the cadmium-binding protein metallothionein.

   c. The kidney is the major organ of toxicity for cadmium.

   d. Chronic inhalation of cadmium fumes may lead to pulmonary disease.

11. a. Like mercury, arsenic binds to sulfhydryl and disulfide groups.

   b. Trivalent forms of arsenic are the most toxic.

   c. Arsenic is readily absorbed from virtually all portals of entry.

   d. Arsenic does not accumulate up the food chain.

For Questions 12–16, select, from the following list, the correct mechanism for the stated metal.

(An answer may be used only once):

   a. Chelates with organic ligands containing SH groups

   b. Complexes with cysteine and competes with methionine in protein synthesis

   c. Mimics calcium and is deposited in bone

   d. Carried to the kidney by bicarbonate where it dissociates to cause renal damage

   e. The hexavalent form enters cells and is converted to the trivalent form that binds to, and poisons, macromolecules

12. Lead, arsenic.

13. Lead only.

14. Uranium (uranyl cation).

15. Methylmercury.

16. Chromium.

Answers

1. True
2. False
3. False
4. True
5. True
6. False
7. A
8. B
9. C
10. E
11. A
12. a
13. c
14. d
15. b
16. e

Further Reading

Angle, C.R., Childhood lead poisoning and its treatment, *Annu. Rev. Pharmacol. Toxicol.*, 32, 409–434, 1993.

Berg, M., Stengel, C., Trang, P.T.K., Viet, P.H., Sampson, M.L., Leng, M., Samreth, S., and Fredricks, D., Magnitude of arsenic pollution in the Mekong and red River deltas-Cambodia and Vietnam, *Sci. Total Environ.*, 372, 413–425, 2007.

Blyth, A.W., *Poisons; Their Effects and Detection*, Wm. Wood & Co., New York, 1885.

Boffetta, P., Merler, E., and Vainio, H., Carcinogenicity of mercury and mercury compounds, *Scand. J. Work Environ. Health*, 19, 1–7, 1993.

Centers for Disease Control (CDC), Cadmium and lead exposure associated with pharmaceuticals imported from Asia, *MMWR Morb. Mortal. Wkly Rep.*, 38, 612–614, 1989.

Centers for Disease Control and Prevention (CDC), Elemental mercury poisoning in a household, *Morb. Mortal. Wkly Rep.*, 39, 424–425, 1990.

Clarkson, T.W., Molecular and ionic mimicry of toxic metals, *Annu. Rev. Pharmacol. Toxicol.*, 32, 545–571, 1993.

D'souza, H.S., D'souza, S.A., Menezes, G., and Venkatesh, T., Diagnosis, evaluation and treatment of lead poisoning in general population, *Indian J. Clin. Biochem.*, 26, 197–202, 2011.

Fimreite, N., Mercury contamination in Canada and its effects on wildlife, PhD thesis, Western University Library, London, Ontario, Canada, 1970.

Hutchison, G. and Wallace, D., *Grassy Narrows*, Van Nostrand Reinhold, Toronto, Ontario, Canada, 1977.

Jaworski, J. (ed.), Effects of chromium, alkali halides, arsenic, asbestos, mercury and cadmium in the Canadian environment. NRC Executive Reports, Publ # NRC 17585, 1980.

Julin, B., Wolk, A., Bergkvist, L., Bottai, M., and Akesson, A., Dietary cadmium exposure and risk of postmenopausal breast cancer: A population-based prospective cohort study, *Cancer Res.*, 72, 1459–1466, 2012.

Kinghorn, A., Solomon, P., and Chan, H.M., Temporal and spatial trends in mercury in fish collected in the English-Wabagoon river system in Ontario, Canada, *Sci. Total Environ.*, 372, 615–623, 2007.

Klassen, C.D. (ed.), *Casarett and Doull's Toxicology: The Basic Science of Poisons*, 7th edn., McGraw-Hill Medical, New York, 2008.

Klassen, C.D. and Watkins, J.B. III. (eds.), *Casarett and Doull's Essentials of Toxicology*, McGraw-Hill Medical, New York, 2010.

Marquardt, H., Schafer, S.G., McCellan, R., and Welsch, F., *Toxicology*, Academic Press, New York, 1999.

McPhee, S.J. and Papadakis, M.A. (eds.), Rabow, M.W. (assoc. ed.), *Current Medical Diagnosis and Treatment*, 51st edn., Lange Medical Books/McGraw-Hill, New York, 2010.

Neghab, M., Norouzi, M.A., Choobineh, A., Kardaniyan, M.R., and Zadeh, J.H., Health effects associated with long-term occupational exposure of employees of a chlor-alkali plant to mercury, *Int. J. Occup. Saf. Ergon.*, 18, 97–106, 2012.

Santini, J.M. and Ward, S.A., *The Metabolism of Arsenic*, Taylor & Francis Group, Boca Raton, FL, 2012.

Timbrell, J.A., *Biochemical Toxicology*, 4th edn., Informa Healthcare, New York, 2009.

Walker, C.H., Hopkin, S.P., Sibly, R.M., and Peakall, D.B., *Principles of Ecotoxicology*, Taylor & Francis Group, London, U.K., 1996.

Zuber, S.L. and Newman M.C. (eds.), *Mercury Pollution: A Transdisciplinary Treatment*, Taylor & Francis Group, Boca Raton, FL, 2012.

7

Organic Solvents and Related Chemicals

Tis a sordid profit that's accompanied by the destruction of health.

Treatise on the Diseases of Tradesmen, B. Ramazzini, 1705

Introduction

Organic solvents are common in the workplace where they may constitute an occupational hazard, but they also occur in the home in the form of cleaning solutions, paint strippers, and brush cleaners, and they can thus be a source of household poisoning as well as being an environmental hazard. Since they are all fat solvents they may cause local defatting of tissue and local irritation on contact. Many also are systemic toxicants affecting the CNS-like volatile anesthetics or, in some cases, the hematopoietic (blood-forming) system. Commercial solvents are frequently complex mixtures having nitrogen- and sulfur-containing elements (e.g., gasoline and other petroleum-based products).

Industrially, the uses of solvents are many and varied. They are used in extraction processes in the food and pharmaceutical industries (e.g., ethyl alcohol and acetone), for the removal of impurities, for degreasing and vapor cleaning (e.g., trichloroethylene, 1,1,1-trichloroethane), as vehicles for paints, carriers for pesticides, for printing inks (toluene, ethyl acetate), in adhesives (hexane, toluene, methylethyl ketone). In short, volatile solvents are used whenever a fast-drying property is desired in order to leave a coating on a surface. They are also used in the chemical industry for a variety of manufacturing processes.

Classes of Solvents

Aliphatic Hydrocarbons

Aliphatic hydrocarbons are straight chain or branched carbon–hydrogen compounds. They are often present in complex mixtures in common commercial products such as gasoline, mineral turpentine, and kerosene.

These mixtures can also contain smaller amounts of unsaturated and cyclic carbon–hydrogen substances. While the vapors of these agents are generally less toxic than those of most organic solvents, inhaling the fumes can not only cause disorientation and stupor through effects on the CNS but they can sensitize the heart to adrenaline. The result may be ventricular fibrillation that can be fatal if not corrected quickly. Fatalities have occurred in workman cleaning storage tanks and rail tank cars. Chemical pneumonia often occurs from aspiration of the low-viscosity liquid (an occupational hazard for those who siphon gasoline from others' cars!). One member of this solvent group of special note is n-hexane, the main ingredient of petroleum ether. With a boiling point of 60°C–80°C it evaporates readily and produces an insidious form of poisoning. An outbreak of industrial poisoning occurred in Japan in 1973 when workers exposed to fumes from a glue used in making sandals lost sensation and function in fingers and toes from impaired nerve conduction. When the exposure was terminated they all recovered slowly over several months. Hexane neuropathy has both a central and peripheral component resulting from demyelination of nerve fibers. The actual toxicant is a metabolite of n-hexane, hexane-2,5-dione. An intermediate, 2-hexanone, is also more toxic than the parent substance. If the neurotoxicity of n-hexane is rated as 1, then 2-hexanone would be 10 and hexane-2,5-dione would be 40.

$$CH_3(CH_2)_4CH_3\text{-oxidation} \rightarrow CH_3\text{-}\overset{O}{\overset{\|}{C}}\text{-}(CH_2)_3CH_3\text{-oxidation} \rightarrow CH_3\text{-}\overset{O}{\overset{\|}{C}}\text{-}(CH_2)_3\text{-}\overset{O}{\overset{\|}{C}}\text{-}CH_3$$

n-hexane	2-hexanone	Hexane-2,5-dione

Metabolites are produced in the liver by cytochrome P450 oxidases. They appear to condense with lysine in the myelin, causing disorganization of the membrane.

Halogenated Aliphatic Hydrocarbons

This group of substances containing chlorine substituents includes methylene dichloride, chloroform, carbon tetrachloride, and chlorinated ethylenes (e.g., trichloroethylene) as well as a number of chemicals that contain other halogens such as bromine and fluorine. They are used as anesthetics and refrigerants. Hepatotoxicity and nephrotoxicity characterize this group. The liver converts carbon tetrachloride (CCl_4) to the free radical CCl_3 that attacks the endoplasmic reticulum, causing protein synthesis to cease. Recovery will usually occur from a single toxic exposure but repeated exposures lead to cirrhosis. Chloroform is converted to phosgene ($COCl_2$), which has similar toxicity to CCl_4. Phosgene was used as a poison gas in World War I. The chemical reactions involved in this process are illustrated in Chapter 5 (Figure 5.4).

Trichloroethylene was used as a light anesthetic in childbirth. It and similar agents are used as industrial solvents. Fatalities have occurred when workers were overcome in enclosed tanks. Death usually results from aspiration of vomit. It is also hepatotoxic. Dichloromethane (methylene chloride) was used for many years as a paint stripper. It has a boiling point of 40°C and thus readily forms vapor. It is converted to carbon monoxide (CO), which forms carboxyhemoglobin in the red blood cells (see Chapter 5). Bis (chloromethyl) ether is discussed later under "Cancer in the Workplace."

Other halogenated hydrocarbons are discussed in Chapter 5, and the chlorofluorocarbons (CFCs) and their impact on the environment are covered in Chapter 5. Some anesthetics are halogenated hydrocarbons. One such is halothane, chemically a haloalkane, which is a volatile liquid. Although most halothane is eliminated in the expired air, some undergoes hepatic biotransformation to a reactive metabolite, an alkylating radical, which causes lipid peroxidation and which is hepatotoxic. It does not deplete glutathione as do some hepatotoxic agents. The reaction is mediated by cytochrome P450 and toxicity is increased by hypoxia. The incidence of this toxic reaction is low, perhaps one per 38,000 patients, and it appears to be genetically determined. Other halogen-containing general anesthetics are methoxyflurane, enflurane, and isoflurane. Methoxyflurane is rarely used in North America now because it releases free fluoride during biotransformation. Free fluoride is toxic to the kidney.

Aliphatic Alcohols

Ethanol (ethyl alcohol) is present in alcoholic beverages and also in some lotions, perfumes, mouth washes, cough syrups, etc. Although there is a tendency to trivialize ethanol's toxicity, 3–6 mL/kg of pure ethanol may be fatal. Since 70 proof liquor contains 40% ethanol, this fatal adult dose equates with as little as 525 mL of liquor. In severe poisoning, the classical signs of inebriation (difficulty with balance, locomotion, and talking) progress to coma, metabolic acidosis, hypothermia, hypotension, and severe respiratory depression. Hypoglycemia may be present, especially in young children. Treatment is largely supportive, with correction of acidosis and hypoglycemia (5% glucose i.v.) and hemodialysis to remove ethanol and its metabolites. Other simple alcohols and glycols produce early effects resembling ethanol intoxication, which is why they are often abused by alcoholics who are at the bottom of the socioeconomic scale.

Methanol is by far the most toxic of the alcohols. In humans and other primates it is oxidized to formaldehyde, a very reactive substance that the eye cannot convert to the harmless formate anion. As little as 4 mL has caused blindness through retinal damage. The transformation is illustrated as follows:

$$CH_3OH \xrightarrow{\text{alcohol dehydrogenase}} HCHO$$

Methanol is present in paint thinners and removers, windshield washer fluids, and in fuels for small engines. It is also present in pineapple, where the levels may exceed the recommended limit of 2 ppm. Outbreaks of poisoning have occurred as a result of deliberate adulteration of wine. Home-distilled liquor also has caused poisonings, and "skid road" alcoholics may consume substances containing methanol as a substitute for ethanol. Initial symptoms are those of ethanol intoxication. A fairly long latency period may occur (6–30 h), followed by acidosis, delirium, coma, visual disturbances, which may or may not progress to permanent blindness, cardiac disturbances, and death. Treatment consists of the administration of activated charcoal by mouth, ethanol (orally or i.v.) as a competitive substrate for alcohol dehydrogenase, and hemodialysis to remove methanol, formaldehyde, and formic acid (which causes acidosis). A fatal dose is 60–250 mL. An experimental approach to treatment involves the administration of inhibitors of alcohol dehydrogenase, such as 4-methyl pyrazole.

Isopropyl alcohol is found in rubbing alcohols, aftershave lotions, and window-cleaning fluids, and it is a widely used industrial solvent. It is oxidized to acetone and causes acetonemia but not acidosis. Its toxicity lies between that of ethanol and methanol being about twice as toxic as ethanol, with coma and respiratory arrest resulting from CNS depression.

Higher alcohols are generally less toxic. N-butanol vapors can produce eye irritation (conjunctivitis, keratitis) and inhalation may cause pulmonary edema.

Acetone (C_3H_6O), although a ketone rather than an alcohol, produces similar toxic symptoms to those of ethanol and the treatment is the same. Acetone, like the aliphatic hydrocarbons, dissolves lipids from the skin and can be extremely damaging to the cornea by virtue of defatting the epithelial cells.

Note: The hepatotoxicity of the halogenated hydrocarbons can be potentiated by numerous alcohols including ethanol, methanol, and isopropyl alcohol, and acetone. The mechanism is unclear.

Glycols and Glycol Ethers

Ethylene glycol is present in antifreeze. A dose of 100 mL orally may be fatal for an adult. Early signs and symptoms resemble ethanol poisoning. After 24 h there may be pulmonary edema and myocardial depression and after 48 h renal tubular necrosis and renal failure. Hypoglycemia and hypocalcemia may occur. Treatment is as for methanol.

The toxicity of ethylene glycol is due to its metabolites. It is metabolized by alcohol dehydrogenase, which is why ethanol is given orally or i.v., as it is for methanol poisoning. When hemodialysis is used in conjunction with i.v. ethanol, ethanol is often added to the dialyzer fluid to maintain blood ethanol levels. Otherwise, the ethanol would be dialyzed because of the concentration gradient. One metabolite of ethylene glycol is oxalate, which

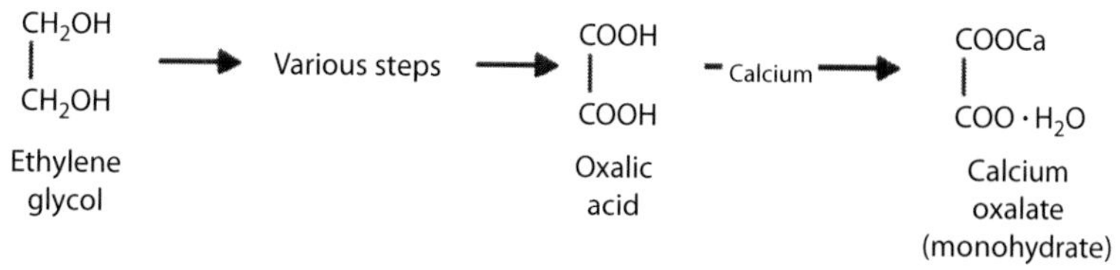

FIGURE 7.1
Conversion of ethylene glycol to calcium oxalate.

chelates calcium (hence the hypocalcemia; see also Chapter 11) and which precipitates as calcium oxalate crystals in the kidney, causing tubular necrosis. Other metabolites are aldehyde, glycolate, and lactate, which cause acidosis. A simplified scheme for the metabolism of ethylene glycol is shown in Figure 7.1.

Propylene glycol is fairly nontoxic and it is used in lotions and as a vehicle for injectable drugs. Too-rapid injection may cause cardiac depression.

A subgroup of this class of compounds is the glycol ethers. Monomethyl and monoethyl ethylene glycol are used extensively in industry, being present in latex paints and as solvents in the manufacture of lacquers, varnishes, and dyes. There is some evidence that they are reproductive toxins, causing teratogenesis in experimental animals. They are not well absorbed orally, but transdermal absorption and inhalation are the important portals of entry. Human toxicity has not been well established, but there have been reports of kidney damage and bone marrow depression. Precautions are taken to limit industrial exposure of women.

Aromatic Hydrocarbons

Benzene is one of the simplest and most toxic of these cyclic, special hydrocarbons. It is highly volatile, and exposure in the workplace is mainly by inhalation. Benzene is unique in this group because of its bone marrow toxicity. Chronic exposure causes a progressive reduction in all formed elements of the blood including red cells, white cells, and platelets. Aplastic anemia, resulting from almost complete destruction of the marrow, may occur and the mortality rate is high, the only treatment being bone marrow transplantation. Bone marrow depression is dose-time dependent. Leukemia also can develop. Several reports have noted a higher-than-normal incidence of acute myelogenous leukemia in workers exposed to benzene. Although leukemia has not been demonstrated in animals exposed to benzene, solid tumors have been observed.

Chromosomal abnormalities have been seen in animals exposed to benzene. The toxic agent may be a metabolite of benzene since animal studies have shown that toluene, which competes with benzene in its metabolic pathway, reduces its toxicity. A number of metabolites have been identified, including phenol.

Alkylbenzenes are a group of related compounds consisting of the aromatic ring with one or more aliphatic side chains. As a group, they lack the serious toxic effects of benzene because they are detoxified to metabolites with low toxicity, including hippuric acids, which are readily excreted by the kidney. They are, however, potent CNS depressants, acting like general anesthetics. No mutagenic properties have been demonstrated. The group includes toluene (1-methylbenzene) and the xylenes (1,2-, 1,3-, and 1,4-dimethylbenzene) as well as ethylbenzene and cumene (isopropylbenzene).

Dinitrobenzene is a related substance that is used extensively in the manufacture of plastics, dyes, pigments, explosives, insecticides, and in many other processes. The opportunity for an industrial exposure to toxic levels is thus fairly high. Meta, para, and ortho forms exist, all are crystals in the pure form and all volatilize with steam and are soluble in boiling water and organic solvents. They are absorbed across the skin and promote the formation of methemoglobin (MetHb), in which the normal ferrous (Fe^{++}) iron is converted (oxidized), to ferric (Fe^{+++}) iron. Methemoglobin binds irreversibly to O_2 and reduces the oxygen-carrying capacity of the red blood cells. If MetHb exceeds 1% of total hemoglobin, mild hypoxia results. If it exceeds 15%, a severe condition known as cyanosis-anemia syndrome results. Dinitrobenzene is the second-most common industrial cause of methemoglobinemia. The chemical structures of some of the aromatic hydrocarbons are shown in Figure 7.2.

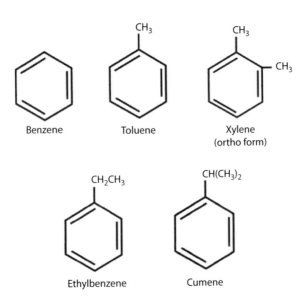

FIGURE 7.2
Chemical structures of some aromatic hydrocarbons.

Solvent-Related Cancer in the Workplace

Benzene

As noted earlier, benzene is capable of causing leukemia. The first case of "benzene leukemia" was observed in 1928 in a worker who was so heavily exposed that others could not work in the same environment without becoming acutely ill, which probably saved them from a similar fate. By 1980 about 200 cases of benzene-related leukemia were reported. In Italy, which banned it in 1963, workers in the shoemaking and rotogravure industries were estimated to have a risk of leukemia 20 times that of the general population. The latency period may be 15 years or more. In Japan, studies of the survivors of Hiroshima and Nagasaki revealed that the risk of leukemia was increased 2.5 times in those who had worked in jobs involving benzene exposure. In the United States, benzene was used extensively in the rubber industry until it was banned. Overall, workers in this industry had a threefold increase and those with high exposures a fivefold increase in their risk of leukemia.

Benzene metabolites formed in the liver and bone marrow damage hematopoietic cells by multiple mechanisms. These include chromosomal aberration and covalent binding, oxidative stress, altered gene expression, apoptosis, error-prone DNA repair, epigenetic regulation, and disruption of tumor surveillance. Individual genetic variations may affect susceptibility. *In utero* exposure to benzene may cause fetal DNA damage and result in increased risk of childhood cancer.

Bis(Chloromethyl) Ether

Bis(chloromethyl) ether (BCME) is a potent alkylating agent and carcinogen used in a variety of industrial syntheses. An American study conducted by NIOSH (U.S. National Institute of Occupational Safety and Health) of 136 men exposed to high levels of BCME indicated a 10-fold increase in risk of lung cancer. Exposures averaged 5 years or more, and cancer rates were highest in those exposed for 10 years or more. Sputum cytology tests indicated a 34% incidence of abnormal lung cells in exposed worker versus 11% in those not exposed. A much larger study involving 1800 workers in New York found a 2.5-fold increase in risk and in a Philadelphia chemical plant, heavily exposed workers had an eightfold increased risk. The discovery that fumes of HCl and formaldehyde could react spontaneously to form BCME led to regulations prohibiting the use of these agents in the same area without special ventilation.

Dimethylformamide and Glycol Ethers

These agents are widely used in the tanning industry for finishing hides. There have been three clusters of testicular cancer identified in this population of workers. Three cases were identified in a tannery in Fulton County, New York, between 1982 and 1984. Collection and analysis of air samples by NIOSH identified significant levels of the glycol ethers 2-ethoxyethanol, 2-ethoxyethyl acetate, and 2-butoxyethanol. The workers had previously been exposed to high levels of dimethylformamide (DMF) on the job, although none was detected at time of sampling. Two other clusters of testicular cancer in workers in the tanning industry have been identified. DMF has been shown to be a testicular toxin in animal studies, but not a mutagen, and glycol ethers are known to be reproductive toxins in animals. At the present time, the precise cause of these tumors has not been identified.

Ethylene Oxide (CH_2CH_2O)

Although not strictly speaking a solvent, it is appropriate to consider this chemical here. It exists as a sweetish, ether-like gas at room temperature, and becomes a liquid at 12°C. It is used as an industrial chemical in the manufacture of plastics, as a fumigant in agriculture, and as a sterilizing agent in health care facilities for heat-sensitive materials. All of these uses relate to the fact that this chemical is highly reactive with other organic substances, including proteins, because it contains an epoxide. The structural formula is

$$\begin{array}{c} CH_2 \\ | \quad \backslash \\ | \quad \; O \\ | \quad / \\ CH_2 \end{array}$$

It exhibits the same toxicity as other organic solvents including CNS depression, local irritation including respiratory irritation, frostbite due to the rapid evaporation of the liquid, and multiorgan toxicity following chronic exposure. It mixes with both organic solvents and water, and it floats on the latter. When exposures are minimal, as they should be when safe handling practices are observed, ethylene oxide poses little risk. Poor safety procedures or accidental exposure carry the risk of a toxic reaction. Populations at risk include workers in health care institutions, exterminators, fumigators, and chemical plant workers. The question of carcinogenicity of this agent remains unanswered. It is highly suspect because of its structure (see Chapter 1 and Figure 1.4), and animal studies have shown clear evidence of increased incidences of cancers of the adrenal gland, spleen, kidney, skin, lung, stomach, and brain, as well as mononuclear cell leukemia. A study of workers exposed to 5–10 ppm for an average of 10.7 years did not reveal any increased frequency of cancer.

Factors Influencing the Risk of a Toxic Reaction

Virtually all of the factors (discussed in Chapter 1) that can influence the response of experimental animals to a toxic agent can also affect toxicity to humans. The risk of a toxic reaction is a function of the toxicity of the chemical and the duration of the exposure; hence the need to establish exposure limits like the short-term exposure limit (STEL). As noted earlier, toxicity may also be affected by the presence of other agents such as ethanol. N-hexane and benzene will enhance the nephrotoxicity of chlorinated hydrocarbons. This may be especially important when exposure to mixed solvents occurs, as it often does. The uptake and distribution of a toxic agent will be influenced by its air/water (blood) partition coefficient and by its water (blood)/oil (fat) partition coefficient. Body depot fat may thus serve as a reservoir for highly lipid-soluble solvents, just as it does for some anesthetics (including sodium thiopental; see Chapter 1) and thus prolong their CNS-depressing effects. Age, the presence of hepatic disease, and general state of health, all may influence individual risk.

Nonoccupational Exposures to Solvents

Solvents are present in many household products including cleaning agents, waxes and polishes, glues, paints, automotive products (cleaners, polishes, etc.), paint removers, thinners and brush cleaners, even hair sprays, and, of course, in gasoline. Exposures may be accidental, usually involving skin contact or inhalation in an enclosed space or ingestion by small children; deliberate, as in the ingestion or inhalation of a substance in a suicide attempt; or as a consequence of substance abuse, as in gasoline sniffing. A great tragedy of our time is the high incidence of gasoline sniffing, which has resulted in several deaths, amongst native Canadian and American youths living on Indian reserves. Glue sniffing also can be a form of substance abuse practiced by adolescents in large cities and where access to other drugs of abuse is restricted.

A somewhat unexpected source of exposure is products used in arts and crafts. These include adhesives, paints and lacquers, and cleaning solvents. Cigarette smoke also contains volatile solvents, including benzene.

In 2009, the Government of Canada amended the Environmental Protection Act to regulate the sale of paints and other "architectural coverings" containing volatile organic compounds (VOCs). Retail outlets can sell existing stock until the fall of 2012 after which the VOC content must conform to the greatly reduced government standards. Essentially this means that water-based paints will become the industry standard and special formulations are being developed to allow the application of water-based paints over existing solvent-based ones. The move is intended to reduce air pollution by VOCs degassing from newly painted surfaces as well as eliminating the disposal problem associated with old cans of paint. Human health benefits should accrue from this move.

Paint technology has taken another major advance. A company called Boysen has developed a paint that actually cleans the air. It contains extremely fine particles of titanium dioxide, which serves as a catalyst to convert nitrogen oxides in air to calcium nitrate. Since catalysts are not changed by a chemical reaction, the calcium nitrate can be washed off the painted surface and the catalyst can continue doing its job. The paint is being used at high traffic intersections in Manila where very large murals are painted nearby. They have produced a measurable reduction in air content of NO_x (see http://www.knoxoutpaints.com).

Case Study 14

Five steam press operators in a rubber plant became ill with signs and symptoms including blue discoloration of the lips and nail beds, headache, nausea, chest pain, dizziness, confusion, and difficulty in concentrating. One worker suffered a seizure. Blood methemoglobin (MetHb) levels ranged from 3.8% to 41.2% (normal <1%). The product they were using was a solvent-borne adhesive.

Q. What information would you want to solicit concerning their working environment?

Q. What industrial chemicals could be responsible for the high MetHb levels?

Several days later, the steam press was operated by a supervisor for 2 h so that an industrial hygienist could collect air samples. The supervisor's MetHb level was 12.5% at the end of the 2 h. The air samples were unrevealing, so that technical assistance from the National Institute for Occupational Safety and Health (NIOSH) was solicited.

Q. What analysis might next be called for?

Q. What steps might you recommend in the interim?

Case Study 15

A 52-year-old woman was hospitalized for routine hemodialysis for chronic kidney failure. She became somnolent after dialysis and developed metabolic acidosis and shock within 12 h. She died 24 h after dialysis. The dialysis fluid had been made up with tap water.

Case Study 16

Two children, 4 and 7 years of age, were admitted around 7:00 p.m. to a small rural hospital with an acute onset of marked somnolence, vomiting, and weakness. They developed hematuria (blood in the urine) and were transferred to a nearby city hospital with a pediatric intensive care unit. Urinalysis showed calcium oxalate crystals. They were treated with intravenous fluid therapy and recovered uneventfully.

The children had attended a picnic that afternoon held in a park adjacent to a local firehouse. A survey of the nearly 400 guests revealed similar but less serious symptoms in 28 of them, two of whom required hospitalization. Nineteen of the 28 cases, including the two who required hospitalization, were children under 10 years of age. The most common symptoms were sleepiness, fatigue, dizziness, and unsteadiness when walking. A common denominator in those who became ill appeared to be consumption of a noncarbonated beverage made from crystals with tap water taken from the fire hall. A few individuals who had not consumed the beverage, but had drunk water, also became ill. Severity of symptoms correlated with the amount of beverage or water consumed.

Q. Do you think the offending agent in these cases is

   a. A pathogenic organism

   b. A toxic chemical

Give reasons for your choice.

Q. Do the calcium oxalate crystals provide a clue?

Q. What organ system is most affected?

In both of these cases, there was a direct connection between the general water supply and chilled-water air-conditioning systems. Could this have been a factor in these illnesses?

Review Questions

For Questions 1–8 answer true of false:

1. Aliphatic solvents are straight or branched chain carbon–hydrogen compounds.
2. Inhaling gasoline fumes can sensitize the heart to adrenaline.
3. Hexane is oxidized to a hepatotoxic metabolite.
4. Monomethyl ethylene glycol is less toxic acutely than ethylene glycol.

5. Propylene glycol is highly toxic.
6. Chloroform is hepatotoxic without being metabolized.
7. Ethylene oxide is liquid at room temperature.
8. Workers in the tanning may show an increased frequency of testicular cancer.

For Questions 9–12 use the following code:

Answer A if statements a, b, and c are correct.
Answer B if statements a and c are correct.
Answer C if statements b and d are correct.
Answer D if statement d only is correct.
Answer E if all statements (a, b, c, d) are correct.

9. Halogenated aliphatic solvents
    a. Cause CNS depression
    b. Are hepatotoxic
    c. Are nephrotoxic
    d. Never require biotransformation to exert their toxicity

10. Methanol
    a. Is converted to formaldehyde in the body
    b. Can cause blindness
    c. Poisoning is treated with ethanol
    d. May be present in fairly high concentrations in pineapple

11. Ethylene glycol
    a. Is metabolized to a toxic substance or substances
    b. Is toxic in the parent state
    c. Is metabolized to oxalate, which may precipitate as calcium oxalate in the renal tubules
    d. Poisoning is not treated with alcohol

12. Benzene
    a. Depresses the CNS
    b. May cause anemia and reduced platelet count
    c. Is carcinogenic in animals
    d. May cause fatal aplastic anemia

13. Ethylene oxide
    a. Contains an epoxide bond
    b. Is a suspected human carcinogen
    c. Depresses the CNS
    d. May cause multiorgan toxicity following long-term exposure

Answers

1. True
2. True
3. True
4. True
5. False
6. False
7. False
8. True
9. A
10. E
11. B
12. E
13. E

Further Reading

Centers for Disease Control and Prevention (CDC), Methemoglobinemia due to occupational exposure to dinitrobenzene, *Morb. Mortal. Wkly Rep.*, 37, 353–355, 1988.

Centers for Disease Control and Prevention (CDC), Testicular cancer in leather workers-Fulton County, New York, *Morb. Mortal. Wkly Rep.*, 38, 105–114, 1989.

Glenn, W., Ethylene oxide: Another question mark, *Occup. Health Saf. Can.*, 5, 28–33, 1989.

Klassen, C.D. (ed.), *Casarett and Doull's Toxicology: The Basic Science of Poisons*, 7th edn., McGraw-Hill Medical, New York, 2008.

Klassen, C.D. and Watkins, J.B. III (eds.), *Casarett and Doull's Essentials of Toxicology*, McGraw-Hill Medical, New York, 2010.

Marquardt, H., Schafer, S.G., McClellan, R., and Welsch, F., *Toxicology*, Academic Press, New York, 1999.

Niesink, R.J.M., de Vries, J., and Hollinger, M.A., *Toxicology: Principles and Applications*, CRC Press, Boca Raton, FL, 1996.

Tung, E.W., Philbrook, N.A., Macdonald, K.D., and Winn, L.M., DNA double-strand breaks and DNA recombination in benzene metabolite-induced genotoxicity, *Toxicol. Sci.*, 126, 569–577, 2012.

Wang, L., He, X., Bi, Y., and Ma, Q., Stem cell and benzene-induced malignancy and hematotoxicity, *Chem. Res. Toxicol.*, 25, 1305–1315, 2012.

8

Food Additives, Drug Residues, and Food Contaminants

The food that to him now is as luscious as locusts, shall be to him shortly as bitter as coloquintida.

W. Shakespeare

Food Additives

Food and Drug Regulations

All developed countries have regulations governing the ingredients and additives that can legally be present in foodstuffs. The nature of the regulations may vary from country to country, but those of Canada are fairly typical.

In Canada, regulations governing foodstuffs and food additives are part of the Food and Drug Act and are enforced by the Health Protection Branch of Health Canada (http://www.hc-sc.gc.ca). The following are some important definitions:

Food: Any substance, whether cooked, processed, or raw, that is intended for human consumption including drinks, chewing gum, and any substance used in the preparation of a food, but not including cosmetics, tobacco, or any substance used as a drug. This definition includes food additives.

Food additives: Any substance, including any source of radiation, the use of which results in, or may reasonably be expected to result in, it or its by-products becoming a part of, or affecting, the characteristics of a food. This definition does NOT include

1. Any nutritive material that is commonly recognized or sold as a food
2. Vitamins, minerals, and amino acids
3. Spices, seasonings, natural flavorings, essential oils, oleoresins, and natural extracts
4. Accidental contaminants such as pesticides, or drugs administered to farm livestock
5. Food packaging materials or components thereof

The regulations lay out which food additives are permitted, in what foods they may be used, and what are the maximum allowable amounts. A food additive must do at least one of the following:

1. Maintain nutritive value
2. Extend shelf life
3. Prevent spoilage during shipment
4. Enhance appearance or palatability
5. Assist in the preparation of the food or in the maintenance of its physical form

The Health Canada website lists nearly 400 substances that can be used as food additives. It has been estimated that about 1 billion lb of food additives are consumed annually in North America, or about 3.5 lb (1.6 kg) per person. The vast majority of these are harmless, but demonstrated toxicity in some experimental animals for some and public concern about harmful effects of man-made chemicals have created pressure on governments to tighten up regulations controlling their use and establish stricter limits on allowable levels in foodstuffs. As is the case for any xenobiotic, the use of such agents should only be undertaken on the basis of a cost/benefit analysis. If the advantage is trivial, such as enhancement of texture, then any associated risk would be unacceptable. On the other hand, prevention of spoilage or of the growth of pathogenic microorganisms might justify the acceptance of a slight risk. It is this area that generates the greatest conflict between environmental groups and growers' and manufacturers' lobbies, who may differ markedly on the definition of acceptable risk.

Some Types of Food Additives

Food additives are used for a variety of purposes. The following are some of the major ones:

Acidifiers or *acidulants* provide tartness and act as preservatives by lowering pH. They may also improve viscosity.

Adjuvants for flavor facilitate the action of the principal flavoring agent.

Aerating agents (propellants, whipping agents) are used to produce a foam as in whipped toppings, etc.

Alkalies control pH, neutralize high acidity foods (tomato products, some wines), and may improve flavor.

Antibiotics are used to prevent bacterial spoilage during storage and transportation.

Anti-browning agents prevent oxidation on the surface of some foods such as lettuce may cause brown spots.

Anticaking agents are added to powdered or crystalline products (drink mixes, powdered spices, salt, cake mixes, etc.) to prevent caking (formation of lumps).

Anti-mold agents are added to foods (bread, baked goods, dried fruit, cheeses, chocolate syrup) to prevent mould growth. They are also called *antimycotic* or *antirope* agents (liquid or viscous products that become moldy are described as being "ropy").

Antioxidants prevent the oxidation of fatty acids that causes rancidity, and of vitamins that lose potency.

Anti-staling agents prevent bread, etc., from going stale.

Binders are substances used to maintain "body" and hold a product together (e.g., processed meat, snack foods).

Bleaching agents are used to whiten flour, some cheeses.

Buffers are used in many processed foods.

Chelators or *sequestrants* are used to bind metallic ions that can hasten oxidation of fats and shorten shelf life.

Coating agents (*glazing* or *polishing agents*) are used to coat the skins of fruits and vegetables to prevent bruising, drying or spoilage, and to coat candies and tablets.

Defoaming agents (*antifoaming agents, surfactants*) are used to prevent excessive foaming in beverages when bottle filling.

Emulsifiers disperse fat droplets in an aqueous medium, for example, salad dressings, milk shakes, whipped cream, and toppings in pressure cans.

Extenders (fillers) are natural substances (casein, starch, soybean meal) used to add bulk to a food product.

Fixatives maintain the color of meat and processed meat.

Flavor enhancers intensify the natural flavor in soft drinks, fruit drinks, jams, and gelatin desserts.

Flavors (artificial): Any flavoring that does not occur in nature, even if the ingredients are all natural, is defined as an artificial flavor. When something is described as "chocolatey" rather than as chocolate, it indicates that the flavor is artificial, not natural, chocolate. This advertising ploy gets around the regulations prohibiting false advertising and it is widely used in North America.

Food colors are added to many products, including some fruit (oranges) to restore color lost in processing or transportation. Most (90%) are synthetic. Food colors that are bound to aluminum hydroxide are known as lakes. All synthetic food colors are highly water soluble. Only vegetable dyes are lipid soluble. This, of course, affects their absorption from the gastrointestinal tract. See in the following regarding artificial food colors.

Fumigants are toxic gases used to kill pests in harvested, dried grains, and nuts.

Fungicides prevent fungal growth on the surface of some fruits.

Humectants (*hydroscopic agents*) retain moisture and prevent drying in some candies and in ice cream.

Maturing agents (*dough conditioners*): Flour is better for baking if it is aged. Bleaches and other agents speed up the process.

Plasticizers (*softeners*) are used in chewing gum, candies, and edible cheese coatings to maintain pliability.

Stabilizers (*suspending agents*) prevent cocoa, orange pulp, and solids in ice cream from settling out.

Sweeteners (nonnutritional, artificial): There are many applications in low-calorie and diabetic diets.

The aforementioned is a partial list of the uses of food additives. It is probably unrealistic to expect that the use of such agents, many of which are synthetic chemicals, can be completely eliminated. Some of them at least are essential to allow the shipment of fresh fruit and vegetables over long distances as is necessary if these foods are to be available in countries with a short growing season. As is so often the case, public perceptions of risk cloud the issue of artificial food additives. Consumer advocacy groups continue to campaign for tighter controls on such agents. But when saccharin was banned because animal tests had shown the development of bladder tumors in rodents fed high doses, public outcry, originating from a perceived need for this product (vanity is a powerful motivator), resulted in a partial removal of the ban (see the following). It is now generally accepted that saccharin is not a carcinogen for humans. The remainder of this chapter will concentrate on the more common, or more controversial, food additives.

Artificial Food Colors

The common public perception is that synthetic food dyes are inherently more toxic than natural ones. In fact, they are highly purified chemicals, most of which have received extensive toxicity testing. Moreover, they are highly water soluble, so that absorption from the gastrointestinal tract is minimal. In contrast, natural dyes are complex mixtures of compounds that are generally more lipid soluble and therefore better absorbed. Since most natural food additives have been in use for decades, they are listed on the U.S. "Generally Regarded as Safe" (GRAS) list and have not been extensively tested. Most are lipid-soluble carotenoids (reds, oranges, and yellows) and are present in carrots, squash, yams, etc., and can be regarded as harmless. The possibility of an allergy to any food component cannot be discounted, however. Natural colors tend to be more subdued and to fade more quickly than synthetic ones. Thus synthetic colors are often preferred. (In Highland lore, "Ancient" tartans are those dyed with the original, more muted, vegetable dyes.)

Some synthetic dyes have been banned. Orange No. 1 and Red No. 3 caused diarrhea in children who consumed large amounts of candy, carbonated beverages, and other confections where these colors were used extensively. More recently, Red No. 2 (Amaranth) was banned because embryotoxicity was demonstrated in rats.

Considerable controversy surrounds the question of whether synthetic food dyes contribute to hyperactivity in children. Dr. Benjamin Feingold postulated that artificial colors and flavors, together with "salicylate-like" natural substances (present in apples, oranges, peaches, raisins and many berries, and in cucumbers and tomatoes) contribute to behavioral problems such as shortened attention span, easy distractibility, compulsiveness, and hyperactivity. Several studies have been conducted to examine this problem. In one type, groups of children were fed the "Feingold" diet or a normal diet and crossed over to the alternate diet after several weeks. This is called a double-blind crossover study. Neither the observers nor the subjects are aware of the treatment group to which they were assigned. Behavior was rated subjectively by parents and teachers. In studies at the Universities of Pittsburg and Wisconsin, 25%–33% of hyperactive children showed improvement when shifted to the Feingold diet according to the parents' ratings. Teachers' ratings showed much fewer differences. Moreover, shifts from the Feingold diet to the normal one were not accompanied by behavioral changes. In other studies, children received the Feingold diet throughout the experiment, but doses of a blend of eight certified colors were added periodically. Doses were 26–150 mg. The latter corresponds to the intake of the 90th percentile of American children. Both subjective estimates of behavior and objective tests of behavior and learning performance were used. At levels above 100 mg in a Canadian study, there was a slight deterioration in learning in 17 of 22 hyperactive children but no change in behavior. Two of 22 were affected at 35 mg. An Australian study employing a parent questionnaire reported substantial improvement on the special diet whereas a similar U.S. study did not.

The debate over the role of artificial food colors in causing hyperactivity in children is ongoing. Most researchers are in agreement that attention deficit hyperactivity disorder (ADHD) has a complex etiology dependent on the interaction of several factors. Autism Spectrum Disorder (ASD) is a rapidly growing problem in the population. Both may benefit in some cases from dietary modifications. Research conducted in the 20 years following Feingold's initial study yielded mixed results with some positive and some negative findings. There has been a resurgence of public interest in recent years, part of a general rejection of food additives. Public pressure prompted the British government in 2009 to request that food manufacturers remove most artificial food colors from their products. This action was likely motivated by a large 2007 study by McCann et al. involving 153 three-year-old children and 144 eight to nine-year-old children who were randomized to three groups. Two groups received a drink containing an artificial food coloring plus sodium benzoate and the third received a placebo drink. The results were based on observations by parents and teachers and the group concluded that the younger children had adverse behavioral effects after drinking mixture A, while the older children had adverse behavioral effects after drinking either mixture A or mixture B.

Nigg et al. conducted a meta-analysis of 24 publications studying effects of artificial food colors on ADHD and found evidence of notable effects but when corrected for publication bias the finding did not survive. The McCann study and renewed interest in the food color debate led the U.S. Food and Drug Administration to review existing evidence and hold a public hearing. Their conclusion was that there was nothing that warranted agency action.

The most reasonable conclusion after this decades-long debate is that there is a subpopulation of children with ADHD that is adversely affected by consumption of artificial food colors. Thus, this becomes a factor that parents of such children might want to consider in dealing with the condition.

Banned or Restricted Artificial Food Colors

Citrus Red No. 2 is restricted to surface use on oranges not to be processed. It has been shown to be carcinogenic in animals. Orange B is restricted to use in the casings of sausages and hot dogs. It is related to amaranth (Red dye #2). The manufacturer has discontinued production because of evidence of a carcinogenic contaminant. Red No. 40 was imputed to cause cancer but the evidence was ruled inconclusive. Yellow No. 5 (tartrazine) has been associated with allergic reactions, sometimes severe, and it is listed on the U.S. ingredients list (candies, desserts, cereals, dairy products). Cross-allergenicity with aspirin is common. Canada does not have a compulsory ingredients list. This is a matter of great concern to individuals, and their parents in the case of children, who have life-threatening allergies to foodstuffs and additives.

Blue Nos. 1 and 2, Green No. 3, and Yellow No. 6 are considered safe but WHO has raised questions about the adequacy of testing.

Emulsifiers

Carrageenin (Irish moss) is extracted from several species of red marine algae. It contains a variety of calcium, sodium, potassium and ammonium salts, plus a sulfated polysaccharide. It is widely used as an emulsifier and thickener in ice cream, milk shakes, and chocolate drinks. It keeps milk proteins in suspension. Estimated daily intake is 15 mg/person. Only the undegraded form is permitted. Rats fed 2000 mg/kg showed fetal deaths and young with underdeveloped bones. Increased vascular permeability and interference with complement have been shown experimentally. Although potentially serious, especially in ill people (complement is essential to the immune system) carrageenin is not well absorbed and the FAO/WHO Committee on Food Additives has established an acceptable level of 500 mg/day. Long-term testing is probably indicated. Furcelleran is a similar substance derived from a red seaweed and it is similarly used. Recent public concerns over the safety of food additives have raised again the issue of carrageenin's safety, but reviews of the literature have confirmed its safety at the concentrations encountered. There is some justification in limiting its consumption by infants due to the fact that carrageenin acts as a bulk expander like all high-fiber substances. The safety issue is clouded by confusion over degraded carrageenin, now known as poligeenan, which has toxic properties but which is not approved as a food additive. Studies have shown that there is no significant degradation of Carrageenan in the gastrointestinal tract.

Brominated vegetable oil is also used as an emulsifier to keep flavoring oils in suspension in soft drinks. A study in 1976 indicated that the daily U.S. intake averaged less than 0.2 mg/person. These products have been in use for over 50 years but toxicology studies in the early 1970s showed that doses of 2500 mg/kg of cottonseed BVO caused, in rats, heart enlargement and fatty deposits in heart, liver, and kidney after a few days. Doses as low as 250 mg/kg caused fat deposition in the heart. Maximum daily intake for a child probably does not exceed 0.05 mg/kg but it may occur over a prolonged period. Corn oil BVO fed to rats and pigs at 20 mg/kg for weeks caused deposits of brominated fat in liver and other tissues. The UN joint FAO/WHO Committee on food additives recommended in 1971 that BVO not be used as an additive. The U.S. FDA removed it from the GRAS list pending further safety studies by the manufacturer. These were judged to be faulty by the FDA. It is still used as an emulsifier, however, in soft drinks in Canada and the United States. In 1997 Horowitz reported on an unusual case of bromism in California resulting from excessive consumption (2–4 L daily) of cola soft drinks. The patient presented with headache, fatigue, ataxia, and memory loss. Ptosis (drooping) of the right eyelid and loss of ability to walk followed. Hemodialysis was required to restore function. Bromism was common in the early twentieth century when bromides were used as sedatives.

Preservatives and Antioxidants

Butylated hydroxyanisole (BHA) and butylated hydroxytoluene (BHT) are synthetics used to prevent premature rancidity in oil- and fat-containing foods. Total daily intake for a child could approach 0.5 mg/kg, which is the maximum recommended by FAO/WHO. Animal studies have consistently shown that high doses (over 500× the average human consumption) caused liver enlargement and induction of microsomal enzymes. Less than this had no effect. Recent studies have revealed evidence of carcinogenicity in off-spring of mice fed BHA. These agents may act as promoters through the enzyme induction mechanism. As free radical scavengers, these agents actually may have anticarcinogenic properties. Both the U.S. FDA and Health Canada have ruled that BHA and BHT are safe when used as directed both as a food additive and in cosmetics where they are also commonly used as antioxidants. Nonetheless, consumer resistance to food additives continues to generate objections to these agents. Limiting consumption in young children certainly will do no harm and may be a reasonable precaution. As is so often the case, it is a question of how much relevance should be attached to findings that suggest toxicity at high doses.

Sodium nitrate, sodium nitrite, and potassium nitrate are used as curing agents in meats such as bacon and smoked meats. They are always used in combination with salt. Nitrite also inhibits the growth of *Clostridium botulinum*, the organism responsible for botulism. Nitrate is converted to nitrite by bacterial and enzymatic action in the intestine. The average U.S. daily intake from food additives is about 11 mg. The major concern is that nitrites can combine with amines to form nitrosamines that are carcinogenic. This process is accelerated by cooking. However, it is important to note that nitrites from food additives account for less than 20% of daily intake, the rest coming from nitrates in drinking water and in vegetables such as celery, spinach, and other leafy vegetables. This is partly due to the use of nitrogen fertilizers and partly from natural sources. Saliva contains nitrate, perhaps providing over 100 mg/day to be converted to nitrites in the lower gastrointestinal tract through bacterial action. Thus the total nitrate load is about 90 mg from natural dietary sources, 100 from saliva and only 11 from food additives. The use of nitrates and nitrites is restricted to the minimum levels required to inhibit the growth of *C. botulinum*.

A particular concern has been the poisoning of infants by nitrates in well water. There are now thousands of such cases, and many deaths, reported because the formation of methemoglobin from nitrites from nitrates impairs the oxygen-carrying capacity of the blood. A high percentage of rural dwellers in North America and Europe draw their water supply from shallow wells supplied by groundwater. These are vulnerable to contamination by surface runoff and hence by nitrates from fertilizer. Typically, newborn infants in these rural areas would develop, after days or weeks, a syndrome that included cyanosis (blue baby), hypotension (nitrates and nitrites are potent vasodilators) and, eventually, coma and death. Most recovered when they were removed from the home and hospitalized. Invariably, these infants were being fed formula made with well water that contained 20–1000 ppm of nitrates. The latency period varied according to the degree of contamination. The conversion of nitrates to nitrites occurred as a result of bacterial action either in the well, or in the gastrointestinal tract of the infant, which is virtually neutral (pH 7±) and hence favorable to bacterial growth. With the decline in popularity of formula feeding and its replacement by breast-feeding, nitrite poisoning in infants has almost disappeared in North America.

In a rather perverse swing of the pendulum, there is now mounting evidence that dietary intake of inorganic nitrates and nitrites can have a protective effect on the cardiovascular system, reducing the risk of heart attack and stroke. It is proposed that the conversion of nitrite to nitric oxide, a potent vasodilator, helps to maintain vascular flow in the heart and brain. A Belgian study calculated the average daily intake of nitrate of 1.28 mg/kg bodyweight/day. The FAO/WHO recommends an acceptable daily intake (ADI) of 3.65 mg/kg bodyweight/day.

Artificial Sweeteners

Sodium saccharin is several hundred times sweeter than sucrose but leaves a bitter aftertaste. For many years it was the only sweetener in common use. The average daily intake is about 6 mg/kg but some individuals habituated to soft drinks may consume much more. Theodore Roosevelt first required a review of its safety in 1912. In the early 1970s, two studies, one by the FDA, reported that high doses (2500 mg/kg) caused an increase in the incidence of bladder cancer in rats. It was not known whether this was due to an impurity. In 1977 the Canadian Health Protection Branch confirmed that the saccharin was the causative agent, but only when rats were exposed *in utero* to high doses. It was concluded that saccharin is a weak carcinogen and probably a co-carcinogen, but public outcry blocked its recall in the United States. In Canada, it is available in tablet form in pharmacies but it cannot be used as a sweetener in prepared beverages or as a sweetener in restaurants. It was originally usually in combination with cyclamate (see the following), which originally was suspected as a carcinogen. It is also available in Great Britain. The International Agency for Research on Cancer, an agency of the United Nations, has not deemed it necessary to place saccharin on its list of proven carcinogens. It is doubtful if either agent poses a real risk.

Xylitol is a natural ingredient of many fruits and berries. It has the same caloric value as glucose but it does not affect blood glucose levels and so it can be used by diabetics. It does not cause caries because it is resistant to fermentation by plaque microorganisms. It has been used in "sugar-free" gum. Some evidence of carcinogenicity has been obtained in rats fed very high doses. Xylitol has been given i.v. to humans as a source of energy. Kidney, liver, and brain disturbances have occurred and some fatalities. Use in the United States has been voluntarily stopped pending a review. It has been replaced by aspartame. Sorbitol is another sugar substitute that is equal in calories to sucrose but which will not raise blood glucose levels. Its uses are the same as for xylitol. It is also used as a humectant in jellies, baked goods, and in canned bread to prevent browning. Nausea, cramps, and diarrhea have occurred in some individuals. Acesulfame potassium (an oxathiazinon-dioxide) was introduced in 1988 as a noncaloric sweetener. It is chemically similar to saccharin. No adverse reactions have yet been reported, but long-term safety is unknown.

Cyclamate was introduced as a sweetener after saccharin. It has the advantage of not leaving a bitter aftertaste. Both were included in the 1959 "Generally Regarded as Safe" (GRAS) list of the U.S. Food and Drug Directorate. Because cyclamate was fed along with saccharin in one study showing increased bladder tumors, it was removed from the GRAS list and it is not available in the United States or Great Britain, but it is in Canada because a Canadian study had shown saccharin to be the culprit. In May of 2000, the U.S. National Institutes of Health released its Ninth National Toxicology report on Carcinogens in which it dropped saccharin from its list as suspected carcinogens in light of recent studies that failed to show a clear association between the sweetener and cancer.

Aspartame is a dipeptide consisting of the amino acids aspartic acid and phenylalanine. Extensive testing has not revealed any carcinogenic potential and it has no aftertaste. It has largely replaced other sweeteners in soft drinks, gum, and other dietetic foods. It should not be used by people with phenylketonuria (PKU), a hereditary defect of the enzyme phenylalanine hydroxylase, which converts phenylalanine to tyrosine (Figure 8.1). These people use an alternate pathway that causes the accumulation of phenylpyruvic acid, which accumulates and deposits in the brain and causes mental retardation in infants. All newborns are tested for this and, if positive, are placed on a diet free of phenylalanine.

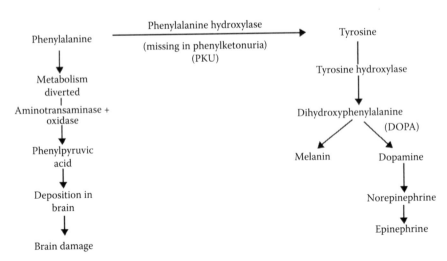

FIGURE 8.1
Phenylalanine metabolism in normal and PKU infants.

Since tyrosine is a precursor in the synthesis of catecholamine neurotransmitters, there was some speculation that high doses could cause behavioral changes but no evidence of this has been forthcoming in either children or adults.

An addition to this group is sucralose (trade name Splenda). It is a chemical modification of sucrose in which a number of H and OH groups are replaced by Cl. Sucralose cannot be broken down by digestive enzymes and it is not absorbed. Thus, it cannot enter metabolic pathways for glucose, it will not elevate blood glucose, and it provides no calories. It is rated as 600 times as sweet as sucrose. It has been pronounced safe by WHO and the Joint Expert Committee on Food Additives. In use for over a decade now, no adverse effects have been reported to date. A wide variety of products is presently in the laboratory testing stage of development, many of "natural" origin, and some of these will undoubtedly be making their appearance in the future.

One natural sweetening substance is stevia, from *Stevia rebaudiana*, known as sweetleaf or sugarleaf. It has been used as a sweetener in baking and elsewhere in South America for centuries. It is 200–300 times sweeter than sucrose and has been approved as an artificial sweetener in many countries including the United States and Canada. The United States has included it on the GRAS list. It has a negligible effect on blood glucose.

Flavor Enhancers

Monosodium glutamate (MSG) has been identified as the offending agent in the "Chinese Restaurant Syndrome." Subjective symptoms of numbness and tingling of the mouth and tongue have been reported, but double-blind studies with doses up to 3 g failed to confirm an effect. It appears that certain individuals are highly sensitive. Because glutamate is an excitatory amino acid, MSG has been given in doses up to 45 g daily to mentally retarded patients with no behavioral changes or ill effects. Animal studies have shown hypothalamic lesions and infants under 6 months of age may be especially susceptible to MSG toxicity.

By far the most commonly used flavor enhancers and the ones responsible for massive amounts of ill health are sugar and salt. The total, adverse health effects of all synthetic food additives combined is trivial in comparison. Childhood obesity is epidemic in North America and type 2 diabetes is becoming so in adults. Cardiovascular disease is the leading cause of death if smoking-related cancer is removed from the comparison with cancer deaths. Children of the twenty-first century are believed to be the first generation whose life expectancy will be shorter than that of their parents. The finger of guilt for all this is pointed at the typical North American diet, which is overly rich in sodium and calories. Preprepared foods and fast foods are particular culprits. The American Heart Association recommends a maximum daily sodium intake of 1500 mg. There is convincing evidence that normal subjects who limit their sodium intake can delay or prevent the need for antihypertensive therapy.

The Northern Manhattan Study examined the role of sodium intake on the incidence of ischemic stroke. Patients were analyzed in groups according to their daily sodium intake in mg. These were: ≤1500, 1501–2300, 2301–3999, and ≥4000. Findings were adjusted for sociodemographics, diet, behavioral/lifestyle, and vascular risk factors. Follow-up period was 10 years. There were 235 strokes among the 2657 subjects. The relative risk of stroke was 2.59 times greater in the ≥4000 mg/day group as compared to the ≤1500 mg/day group and the finding was statistically significant. For every 500 mg/day increase in consumption there was a 17% increase in risk of stroke. The North Manhattan Study also examined the effect of dietary fat intake on stroke risk and found that an intake greater than 65 g/day increased the risk of stroke by 1.6-fold and the finding was statistically significant. Canada is no different. Health Canada calculates that the average Canadian consumes 3400 mg/day, more than double the recommended daily intake of 1500 mg. Table 8.1 lists the content of some common food items that could be found in most homes. It is worth noting that the values from product labels are given as a percentage of the maximum recommended daily intake from the Canada Food Guide. If one were to consume one serving of all of the products listed, a not-impossible feat, they would take in 2220 mg of sodium, almost the total maximum recommended limit. It is further worth noting that some of these food items would be considered to be part of a healthy diet. Another little deception practiced by the food industry is to round fractions down instead of up. Thus, a sodium content of 20.9% becomes 20% on the label, rather than 21%. No better illustration of the hidden nature of our sodium consumption could there be.

It should be noted that sodium is sodium, regardless of the source. MSG is a source and sea salt offers no advantages over table salt.

The consumption of sugar-sweetened beverages (SSBs) is believed to be a significant factor in the development of childhood obesity, which in turn, sets the

TABLE 8.1

Sodium (Na) Content of Some Foodstuffs as Total mg and as a Percentage of the 2300 mg Maximum Daily Allowance*

Food Item	Mg Na	% Na*
Two slices of ancient grains bread	300	13.0
30 g of old cheddar cheese	200	8.7
1 tablespoon of tomato ketchup	140	6
16 pieces (85 g) of frozen French fries	170	7.4
3/4 cup (188 mL) mixed frozen vegetables	5	0.2
1 tablespoon (15 mL) calorie-wise dressing sauce (salad)	140	6.0
1/2 cup (125 mL) frozen yogurt	50	2.2
1 cup (250 mL) canned black beans	270	12.2
2/3 cup (150 mL) beef broth	550	20.9
1 cup (250 mL) "heart-healthy" soup	480	20.9

* % of maximum daily allowance.

child up for the future development of type 2 diabetes. Several studies have shown a positive association between the consumption of SSBs and weight gain in both children and adults. Since about 1985, the consumption of SSBs (mostly soft drinks) increased by about 300% in the United States with 56%–85% of schoolchildren consuming at least one drink daily. One can of soft drink contains about 65–70 g of sugar. It has been calculated that for every daily SSB consumed, the odd ratio for becoming obese is increased by 1.6. A school-based program in the United Kingdom designed to reduce consumption of SSBs was successful in preventing a further increase in obesity. In adult women, the consumption of SSBs has been associated with an increased risk of diabetes.

Of all the dietary factors that might concern a parent, including synthetic food additives and genetically modified foods, excessive sugar and salt intake by children are by far the ones that should give the greatest concern. Fortunately, there is some evidence that the consumption of SSBs in the United States has decreased in the last decade.

Drug Residues

The high-risk nature and narrow profit margin that are typical of the livestock industry have led to the extensive use of antibiotics, sulfa drugs, hormones, and other pharmaceuticals to improve productivity by increasing the rate or extent of weight gain per unit of food consumed. These pharmaceuticals are also used to prevent or treat disease. Many of the medications are also used in human medicine, others are unique to the agricultural field but they may have pharmacological or toxicological consequences for people as well. There are three main concerns about the possibility that traces of these substances might enter our food supply:

1. They may serve as a source of allergic sensitization.
2. Anti-infectives may contribute to the development of resistant strains of pathogenic bacteria.
3. They may exert direct toxic manifestations such as teratogenesis and carcinogenesis.

While there is ample evidence that 1 and 2 occur, the occurrence of 3 is mostly speculative at least as far as drug residues in meat are concerned.

Antibiotics and Drug Resistance

In the late 1950s, Thomas Jukes of Berkeley University reported that the antibiotic tetracycline at 50 ppm in animal feed significantly improved

TABLE 8.2

These Antibiotics Are Approved by Health
Canada as Growth Promotants in Livestock[a]

Bacitracin	Monensin
Bambermycins	Penicillin
Chlortetracycline	Salinomycin
Lasalocid	Tylosin
Lincomycin	Tylosin

[a] Other countries may employ other antibiotics.

the rate of weight gain and the gain-to-food consumption ratio in live-stock. Subsequent studies showed that this effect was not related to the prevention of disease and occurred even under optimal conditions of husbandry and hygiene. The effect was confirmed later for other antibiotics and the mechanism remains elusive although it is believed that alteration of the bacterial flora in the gastrointestinal tract is involved. There is by now a long list of antibiotics and other anti-infective agents that has been employed as growth promotants. Antibiotics approved by Health Canada as growth promotants for livestock are shown in Table 8.2. Others may be used in other countries.

Antibiotics are also used to prevent the outbreak of disease. Prophylactic use involves higher levels than those used for growth promotion and a typical mixture for preventing dysentery in swine in the 1960s contained chlortetracycline 100 g/T of feed, sulfamethazine 100 g/T, and penicillin G 50 g/T. Concern over the potential dangers of drug residues in food mounted over the next two decades. It began with a report in 1959 from Japan of an outbreak of *Shigella* dysentery in a hospital nursery. The outbreak was unique in that the infecting strain of the bacteria was resistant to several antibiotics and sulfa drugs, some of which had never been used in that hospital. The term *multiple drug resistance* (MDR) was coined for this phenomenon and in the next few years many reports emerged of MDR in livestock, and in 1965 an outbreak of Salmonellosis in Great Britain resulted in six deaths. It was traced to the consumption of veal from calves that had been treated with several antibiotics and the organism demonstrated MDR. In 1971, the Swann Commission in the United Kingdom recommended greater controls over the use of antibiotics in livestock. Government legislation was passed and many other countries, including Canada, followed suit.

Infectious Drug Resistance

During this period it was discovered that the pattern of resistance typical of one type of bacteria could be passed to other, unrelated, bacteria and species, and the term *infectious drug resistance* (IDR) came into use. It is also called *transferrable drug resistance*.

For many years it was thought that IDR involved Gram-negative enteric organisms exclusively. *Gram-negative* refers to the histochemical staining characteristics of the bacteria (one of the major means of classifying them, it is related to the composition of their outer cell wall) and "enteric" refers to the fact that they are common inhabitants of the intestinal or enteric tract of both animals and humans. This group includes strains of *Escherichia coli*, *Salmonella* spp., *Shigella* spp., and *Klebsiella* spp. It is now known, however, that most, and possibly all, bacteria are capable of developing MDR through transference of genetic information from one cell to another. This includes the transfer of resistance genes from nonpathogenic organisms to pathogenic ones.

The mechanism of IDR hinges on the fact that bacteria possess extra-chromosomal units of genetic information called plasmids. These are rings of DNA that are capable of replication independent of the chromosomes and that can be passed intact from one bacterial cell to another. Genes can be inserted into these plasmids from other sources, including bacterial chromosomes, and this is dependent upon the existence of discrete sequences of 800–1800 base pairs of amino acids called insertion sequences. When a gene for drug resistance is included between two insertion sequences, the unit is called a transposon or, more often, an R (for resistance) factor. The existence of plasmids has provided the means for genetic engineering and the bacterial synthesis of human insulin and other substances. There are several methods by which plasmids can be transferred from one bacterial cell to another. These are as follows:

1. *Transformation.* The lysis of a cell may release plasmids into the environment that may subsequently be absorbed by other cells. This is a highly species-specific phenomenon.
2. *Transduction.* This involves the participation of phage viruses that incorporate bacterial genetic information and transfer it to other cells. This also is very species-specific.
3. *Conjugation.* The plasmids of many bacteria possess a gene, called a fertility or "F" factor, which regulates a form of sexual reproduction. A fine tubule or "pilus" is formed between cells and intact plasmids may then be passed, along with their complement of genetic information, from one to the other. This process is not species-specific and it is the basis for IDR since R factors will be passed along as well. In this way, resistance genes may be shared amongst all enteric organisms that contact each other, and a multiple resistance pattern acquired. This process is illustrated in Figure 8.2.

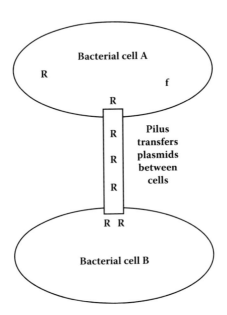

FIGURE 8.2
Bacterial conjugation and multiple (infectious) drug resistance. R = Resistance factor; f = Fertility factor.

The process of conjugation is clinically important because

1. Nonpathogenic organisms may serve as a reservoir of drug resistance to be passed on to more virulent ones. There are few species barriers to transfer.

2. The process of natural selection may, when an antibiotic is used therapeutically, result in the emergence of a strain of bacteria with MDR as susceptible strains are killed or inhibited.

3. Resistance patterns, or resistant organisms, may be passed from animals to humans. Many organisms are infectious for both.

4. The process is favored by exposure to low (nontoxic) levels of antibiotics.

The problem is attempting to strike a balance between the need to maximize production in the livestock industry, and thus keeping food costs down, while protecting the consumer from undue health risks. A study conducted in 1983 claimed that for an investment of $271 million in animal feed additives the American consumer saved $3.5 billion in food costs. Concern over the problems associated with antibiotics as food additives has resulted in a shift to hormonal ones (see the following). In 1984 the U.S. Centers for Disease Control issued a report on 52 outbreaks of Salmonellosis between 1972 and 1983. The source of the infection was identifiable in 38 of these and in 17 (44%) it could be traced to food animals (poultry, veal, and hamburger).

These accounted for 69% of all outbreaks involving strains with MDR. The mortality from these strains was 4.2% versus 0.2% from nonresistant strains. In March and April of 1985 the most massive outbreak of Salmonellosis in U.S. history occurred in Chicago, with 14,000 people eventually infected. There were only two deaths, yielding a mortality of 0.014%. This was markedly lower than the previous rates reported for either resistant or susceptible strains (the 4.2% mortality rate would have caused 588 deaths) and a follow-up revealed that over 16,000 people were eventually infected for a final mortality of 0.012%. Both fatalities, however, involved a tetracycline-resistant strain.

By 2012 there was less controversy as the link between animal-sourced drug-resistant infections by enteric organisms in people became more and more evident. Foodborne infections with pathogenic enteric organisms of animal origin have become a major public health concern. One such is the strain of *E. coli* O157:H7. This is the organism that caused the outbreak of enteritis in Walkerton because of contaminated well water (see Chapter 3). This strain frequently produces Shiga-toxin, transferred by plasmids from *Shigella* strains and a potent enterotoxin.

Early attempts to limit the development of resistant strains of bacteria centered on restricting their use as agricultural food additives to so-called *nontherapeutic antibiotics* or NTAs. This was done in Europe but it was soon discovered that NTAs could serve to select for antibiotic-resistant genes in pathogenic bacteria. As final confirmation of this, studies have shown that when the use of NTAs was abandoned the frequency of antibiotic-resistant bacteria dropped off dramatically. In one study, 73% of isolates from broiler chickens of *Enterococcus faecium* were resistant to avoparcin in 1995 when its use was discontinued. In 1996, the use of avoparcin was banned because of a link with vancomycin-resistant enterococci (VRE), an organism that has caused disastrous hospital infections. By 2000, the number of resistant strains had fallen to 5.8%. Other studies have found similar decreases after antibiotic use in livestock was discontinued.

Avoidance of meat is no guarantee that multiple resistance patterns can also be avoided. In one study, vegetarians had a higher frequency of MDR bacteria than did meat eaters. The use of manure as fertilizer for vegetables can serve as an efficient means of transfer when the vegetables are eaten raw. A 2012 study of soil that had been fertilized with chicken waste revealed that tetracycline resistance was detectable in soil organisms 2 years after the poultry farm had ceased operations. Meat packers and farm workers tend to acquire MDR patterns.

It must be remembered that the use of antibiotics in agriculture is not the only source of MDR strains. In one study, resistance patterns of coliform organisms in the sewage effluent from a general hospital, a psychiatric hospital, and a residential area were compared. The results are shown in Table 8.3. It is evident that the incidence of MDR closely parallels the use of antibiotics in a particular setting. Thus, it is higher in the general hospital than in either the psychiatric hospital or the residential area. Moreover, resistance

TABLE 8.3

Percent of Coliform Organisms Resistant to Chloramphenicol, Streptomycin, and Tetracycline

Effluent Source	Chloramphenicol	Streptomycin	Tetracycline
General hospital	48.8	0.4	24.3
Psychiatric hospital	9.5	0.03	0.04
Residential area	0.6	0.0007	0.1

to chloramphenicol, which is reserved for life-threatening situations where less toxic antibiotics will not work, was very low in all situations, but the correlation still held. In 1985 chloramphenicol was withdrawn as an approved drug for use in food animals.

Resistance to antibiotics is not the only type of genetic information that can be passed in this manner by plasmid transfer. Resistance to metals and bacterial virulence also is regulated by plasmid genes. Both vertical (within a species) and horizontal (amongst species) transfer is now thought to be commonplace in the microbial kingdom. Even transfer between Gram-negative and Gram-positive organisms has been demonstrated.

One of the concerns regarding MDR is that nonpathogenic organisms can pass their transposable resistance to pathogenic ones. Sweden banned the use of all antibiotics as growth promotants for livestock in 1986, and the European Union banned tylosin, zinc spiramycin, and virginiamycin in 1998. In 1996, the U.S. government created the National Antimicrobial Resistance Monitoring System (NARMS) to monitor changes in susceptibility of pathogens from human and animal specimens, healthy animals, and carcasses of food animals. The 2010 NARMS report (http://www.cdc.gov/narms) focused on the emergence of resistance to nalidixic acid of nontyphoidal *Salmonella*. Nalidixic acid is an elementary quinolone and a forerunner to the fluoroisoquinolones such as ciprofloxacin, a drug of first choice for treating *Salmonella* infections. Nalidixic acid resistance correlated well with increased resistance to ciprofloxacin. MDR was also noted in many *Salmonella* organisms.

The question of antibiotic use for growth promotion has already created international trade disputes over the import/export of agricultural products.

Infectious Diseases

No discussion of environmental hazards would be complete without passing reference to the hazard of infectious diseases. Not only has the emergence of antibiotic resistance led to clinical problems in treating old diseases such as tuberculosis, which has developed MDR and is on the rise, but many organisms seem to be developing more virulent strains and new ones are emerging. The AIDS virus, the *Hantavirus*, the bacteria of Lyme disease, and *Legionella*, and the development of more virulent strains of enteric pathogens such as O157:H7 are but a few examples. Perhaps this should not surprise us, given the rapidity of cell division of bacteria and the adaptability that has allowed them to survive for eons. Next to malnutrition, infectious diseases are the leading cause of death worldwide. There is an obvious relationship between the two.

Allergy

Legislation in most Western countries prescribes withholding periods during which livestock may not be shipped for food following the administration of an antibiotic or other drug by any route. This is designed to prevent the entry of drug residues into the human food chain. One reason for this is the possibility that such residues could result either in allergic sensitization or in an allergic reaction in individuals who are already sensitive. Significant levels of drugs could remain at the injection site, which is usually a large muscle mass that also constitutes a preferred cut of meat. Certainly, prior to the introduction of withholding times such reactions did occur. Of the drugs used in agriculture, those with the highest allergic potential are the penicillins. Penicillin G is still available without prescription in many countries for use in livestock and it is commonly employed for the treatment of mastitis (udder infection) in dairy cattle. It may thus be present in milk, although testing procedures are usually mandated by law. The test is not specific for a particular drug but employs a strain of bacteria that is extremely sensitive to inhibition by almost all antibiotics. More recently enzyme-linked immunosorbant assay (ELISA) has been used.

It may be significant that in 30% of patients suffering severe allergic reactions when treated with penicillin have no history of previous exposure, suggesting that sensitization occurred through environmental exposure. Although allergic reactions from contaminated food are rare and difficult to trace, they may still occur when drug residues escape detection. Sulfa drugs may also be offenders in this regard.

Hormones as Growth Promotants in Livestock

Diethylstilbestrol

Diethylstilbestrol or DES is one of a number of synthetic estrogens first synthesized by Charles Dodds in 1938. Since the work was conducted under the auspices of the British Medical Research Council, the chemicals were not subjected to patent restrictions and they rapidly became widely available. As a group, they revolutionized endocrinology, made possible the development of oral contraceptives and had a major impact on the practice of gynecology. The possibility of sex without pregnancy and the consequent sociosexual revolution of the sixties can be attributed directly to the development of these compounds and their chemical manipulation by the pharmaceutical industry.

DES mimics natural estrogens but it is much more potent and more toxic, and it is effective orally, which natural ones are not. It is very cheap to produce and by 1939 it was already popular in the United Kingdom, France, Germany, Sweden, and North America. It was used to treat menopausal complaints, amenorrhea, genital underdevelopment, and to suppress lactation in mothers who did not wish to nurse their infants. By 1942, agricultural use had begun on a limited scale. Early studies showed that rats fed DES gained weight more rapidly with no increase in food consumption and subsequent tests in cattle yielded similar results. Moreover, the "marbling" (diffuse fat distribution) was greater and this was considered to be a desirable flavor enhancer at the time. Athletes were not the first to take steroids to increase weight gain.

By 1945 DES became the first hormonal substance to be used as a growth promotant in livestock. It was being widely used in beef and poultry production. There were two methods of administration. One was a feed additive and the other a slow release pellet that could be injected under the skin of a steer's ear or the neck of a broiler chicken. The former method had the disadvantage of loss through spillage and the accidental ingestion by breeding stock caused some serious fertility problems. The necks and ears containing the remains of the pellet were supposed to be discarded at slaughter, but chickens are frequently sold with the necks attached, and in 1959, a report emerged of a 40-year-old male chef in New York who developed signs of feminization including loss of facial hair, mammary development, fat deposition of the female pattern, and a high voice. The problem was traced to the consumption of chicken necks taken from the restaurant and eaten as an economy measure. The use of pellet implants in poultry was banned that same year. The carcinogenic nature of estrogens has been known for some time. Experimentally, natural estrogens were shown to produce carcinomas of the vagina, cervix, and uterus in mice and carcinogenic potency was directly related to estrogenic potency. There were sporadic reports of DES inducing cancer in mice but in 1971 Herbst et al. reported a case of clear-cell adenocarcinoma in the vagina of a young woman who had been exposed to DES *in utero*. A DES registry was established in the United States and by 1984, 500 cases of clear-cell adenocarcinoma had been reported. It is now estimated that females exposed to DES *in utero* have somewhat less than a 1/1000 chance of developing it.

The use of DES in pregnancy was based on data indicating that it improved the vascularity of the uterus and promoted the synthesis of progesterone, both of which would help to maintain pregnancy in the face of a threatened abortion. Other conditions that have been causally associated with exposure to DES *in utero* include adenosis (a self-limiting condition involving the presence of cervical cells in the wall of the uterus), infertility, carcinoma of the cervix and vagina, and breast cancer. Cervical malformations occur in 18%–25% of DES daughters but most appear to correct themselves with age. Infertility is by far the most common problem experienced by DES daughters.

Pregnancy may be difficult to achieve and maintain. DES daughters have now entered middle age, the possibility that increased frequencies of other forms of cancer might emerge cannot be discounted; however, the association with breast cancer is becoming more tenuous.

The fetal risks of exposure to DES appear to relate to the fact that the mechanisms that detoxify natural estrogens cannot handle DES. In rats, a glycoprotein called alpha fetoprotein binds natural estrogens but not DES. Its presence has not been confirmed in humans but human fetuses have a very active process for sulfonating natural estrogens that may be less effective for DES. Active metabolites of DES may have a high affinity for estrogen receptors in the reproductive system and oxidative metabolites may damage DNA. Reactive metabolites may form after attachment to the receptor by the action of peroxidase enzymes present in estrogen-dependent tissues.

Attempts by the FDA to ban the use of DES in beef cattle were blocked by court actions launched by the beef producers until 1979 when liver and kidneys from beef cattle showed significant levels of DES. Illicit use continued thereafter, and in 1980 the USDA conducted 115 prosecutions against violators. It must be emphasized that all of the health problems associated with DES in North America (barring the chef noted earlier) are of iatrogenic origin, that is, they arose from medical treatment. None has been attributed to the consumption of meat from treated animals. This has not been the case elsewhere.

In the late 1970s and early 1980s, two epidemics occurred in which infants and children under 8 years of age displayed signs of abnormal sexual development such as breast development and precocious puberty. In Puerto Rico over 600 children were involved. DES was available without prescription and meat inspection controls were poor. In Italy, high levels of DES were detected in baby foods, especially those containing veal. Several hundred infants were affected. Occurrences declined after 1979 due to tougher controls and greater public awareness. In 1985, the DES Task Force of the U.S. Department of Health and Human Resources released its findings. These were as follows:

1. The risk of uterine carcinoma *in situ* was about twice as high in DES daughters as in nonexposed women (15.7/1000/year vs. 7.9/1000/year) although there were problems with the study.
2. The risk of genital herpes also was about twice as high (11.8% vs. 6.3%).
3. The risk of breast cancer was difficult to assess because of other predisposing factors.

The question of increased risk of breast cancer was addressed by an American multicenter epidemiological study. It compared over 3000 women who had DES prescribed during pregnancy with 3000 similar unexposed women. The risk of breast cancer/100,000 was 172.3 for exposed women and 134.1 for unexposed women. The rate increased markedly in both groups between

20 and 40 years post entry to the study. The results were statistically signifi-cant and it now seems conclusive that there is a real increase in the risk of breast cancer in DES-exposed women but probably not in their daughters.

A 2012 review of the DES problem by Harris and Waring reported signif-icantly higher incidences in DES daughters of a host of gynecological and reproductive problems including infertility, spontaneous abortion, preterm delivery, loss of second trimester pregnancy, ectopic pregnancy, pre-eclampsia, stillbirth, early menopause, as well as increased incidences of intraepithelial neoplasia and breast cancer.

The psychological costs to DES daughters are difficult to assess but prob-ably quite high. Increased incidences of psychiatric problems and suicide have been reported due to feelings of helplessness and frustration. Recently, one small report suggested a higher incidence of homosexual preference in DES daughters. It is well documented that experimental exposure to sex hormones before birth influences brain development and sexual behavior. In rats and guinea pigs, alpha-fetoprotein normally protects the brain from estrogens but it does not bind to DES and prenatal exposure resulted in abnormal patterns of sexual behavior.

There is statistical evidence that sons and daughters of DES daughters have an increased incidence of birth defects. Boys had a higher incidence of *hypospadia*, a condition in which the urethral opening is located on the underside of the penis. The story of DES is not over nor likely will be for some time as the consequences of its impact, both physical and psycho-logical, on the third generation continue. It is again worth reminding read-ers that none of these were the result of its agricultural use. The removal of such use is a clear example of the implementation of the precautionary principle.

Bovine Growth Hormone

Currently, a controversy is centered on the use of growth hormones in food animals. Bovine somatotropin, or bovine growth hormone (BGH), is used both to promote growth in beef cattle and to increase milk produc-tion in dairy cattle. It is produced by recombinant gene technology (rBGH). Studies indicate that it can increase lean content by at least 5% and feed efficiency by 8%. In dairy cattle, it can prolong the lactation period by several weeks, leading to greater profit. The consumer, however, remains understandably leery of such use. As long as profit margins in agriculture remain narrow, the pressure to use growth promotants will remain strong. The use of growth hormone in the United States has resulted in attempts to ban American beef in the European Economic Community. Canada has not approved its use and issued a notice of noncompliance in January of 1999. This decision was based largely on evidence of adverse effects in cat-tle including increased incidences of mastitis, lameness, and reproductive problems.

Concerns relate mainly to the fact that dairy cows injected with rBGH produce more of the natural hormone peptide insulin-like growth factor-1 (IGF-1) and IGF-1 is secreted in milk. Increased blood levels of IGF-1 have been associated with an increased risk of breast cancer in women. The debate is focused on whether IGF-1 in milk can be absorbed from the gastrointestinal tract. As a polypeptide the expectation is that it would be broken down by proteolytic enzymes. Some authorities claim that this prevented by the presence of casein to which IGF-1 is bound. While high molecular weight polypeptides and proteins are not generally absorbed from the gastrointestinal tract, there is some evidence that this can occur in some people. Experiments in rats showed that they can absorb IGF-1 from their intestinal tract, which is of course quite different from the human one (see Chapter 1).

A British study, published in 1998, of 397 women with breast cancer and 620 controls found that in premenopausal women under 50 with the highest levels of IGF-1, cancer risk was increased sevenfold over controls. The relative risk for other premenopausal women was 2.88. There was no significant increase in risk for any other subgroup. There is also evidence that elevated serum IGF-1 levels are associated with an increased risk of prostate cancer. A case control study within the (U.S.) Physician's Health Study involved 152 patients and 152 controls. There was a strong positive association between serum IGF-1 levels and risk for prostate cancer. Men in the highest quartile had a relative risk of 4.3. This association was independent of baseline prostate specific antigen (PSA) levels.

While it seems clear that increased IGF-1 levels constitute a risk factor for both breast cancer and prostate cancer, the extent to which the hormone can be absorbed from rBGH milk remains open at least for humans. Meanwhile, it would seem prudent to invoke the precautionary principle, especially as milk production in developed countries has a capacity already in excess of the demand.

Other Hormonal Growth Promotants

The removal of DES from agriculture was by no means the end of the use of anabolic female hormones as growth promotants. Currently approved for use in Canada, but only for beef cattle, are three natural ones, progesterone, testosterone, and estradiol-17β, and three synthetic ones, zeranol, melengestrol acetate, and trenbolone acetate (TBA). These may also be used in the United States.

Of particular interest is zeranol, a semisynthetic estrogen, which is related to the naturally occurring mycotoxin zearalenone and which may even occur as a breakdown product of it. Zearalenone itself is an endocrine disrupter (see Chapter 10 for more on mycotoxins) and a potential health hazard. This has complicated the identification of natural versus anthropogenic sources.

An Italian study published in 2008 looked at serum levels of zearalenone and its congener α-zearalenol (ZEAs) in 32 girls with central precocious puberty and found that 6 of them had high levels that correlated with height. Since the European Union (E.U.) banned zeranol as a growth promotant in livestock in 1981, the origin of the ZEAs presumably was natural.

A New Jersey study reported in 2011, examined urinary mycoestrogens in 163 girls 9 and 10 years of age (Jersey Girl Study) and looked for a possible relationship to body size and breast development. Mycoestrogens were detected in 78.5% of the girls and levels were associated with beef and popcorn intake. This is presumably (but not proven to be) because the *Fusarium* mold that produces zearalenone is a common contaminant of corn and zeranol is widely used in the United States as a growth promotant in beef cattle. Girls with urinary mycoestrogens tended to be shorter and less likely to have reached the onset of breast development. The authors concluded that ZEAs may exert antiestrogenic effects on prepubertal girls and that the subject merits a larger, longitudinal study.

Since ZEAs are anabolic substances they have the potential to be abused in the field of sports. A 2011 German study examined the problem of distinguishing natural versus self-administered substances in athletes. The authors noted that the illicit application of zeranol to livestock was a continuing problem in the E.U. Antidoping tests routinely monitor for zeranol and its major human metabolites zearalanone, 7β-zearalanol, as well as the mycotoxin zearalenone and its unique metabolic products α-zearalenol and β-zearalenol. They concluded that mycotic sources of ZEAs were a potential confounding factor in interpreting antidoping test findings.

Natural Toxicants and Carcinogens in Human Foods

Through the acquisition of folk knowledge, human beings have learned to avoid eating rhubarb leaves, daffodil bulbs, and other plants that contain toxic chemicals. Under some conditions, however, foods that are normally safe may become toxic, at least for some individuals, and carcinogens may be more prevalent in human foods than previously realized.

Some Natural Toxicants

Favism

The broad bean (*Vicia fava*) may induce acute hemolytic anemia in individuals with the hereditary defect glucose-6-phosphate dehydrogenase (G6PD) deficiency. This is especially true for those individuals with the Mediterranean phenotype. It is a problem primarily for males under 5 years of age. The exact nature of the mechanism remains obscure, but there is evidence that these individuals may be deficient in hepatic glucuronide conjugate and hence unable to detoxify the offending ingredient. Susceptibility in the same individual varies from time to time, and all those with the same defect are not necessarily affected.

Toxic Oil Syndrome

In 1981 in Spain, there was a remarkable epidemic that eventually affected over 20,000 people. In the first 12 months of the epidemic there were 12,000 hospitalizations and over 300 deaths. According to a review of the event 20 years later, the final count was over 20,000 people affected and over 1200 deaths in the affected cohort. It began in a small town on May 1, when a boy was admitted to hospital with acute respiratory failure that rapidly led to death. The epidemic peaked in June, at which time 2000 new cases were being reported every week. The patients usually presented initially with respiratory distress, cough, low-grade fever, oxygen deficiency, pulmonary infiltrates and pleural effusions, and a variety of skin rashes. Nausea and vomiting were sometimes present, as were enlarged liver, spleen, and lymph nodes. Virtually all patients had elevated eosinophil counts (>500 to >2000/mm³, normal value, 0–500/mm³). The condition often progressed to severe muscle pain, muscle and nerve degeneration, and even paraplegia.

An astute medical clinician traced the problem to the consumption of a cheap, unlabeled cooking oil sold as pure olive oil in the open markets in small towns and villages. The oil consisted of low-grade olive oil mixed with various seed oils including a rapeseed oil that was imported in a denatured form by mixing it with aniline to render it unfit for human consumption. Two significant clues were the facts that the condition affected low-income families almost exclusively, and nursing infants were never affected. Geographically, the epidemic was limited largely to central and northwestern Spain.

Chemical analyses of suspect oil samples, and comparison with nonsuspect samples, failed to reveal the presence of known toxicants such as heavy metals, but the suspect oil contained significant levels of aniline and fatty acid anilides, reaction products of the oil with the aniline. A dose–response relationship between the degree of contamination and the severity of signs and symptoms was also noted.

The toxicity of aniline is well known because of its heavy industrial use and although there are some similarities with the toxic oil syndrome, skin lesions for example, aniline toxicity involves CNS symptoms (vertigo, headache, mental confusion) and blood disorders including methemoglobinemia and anemia. It was felt, therefore, that the offending agent was likely a reaction product. To date, however, the syndrome has not been reproduced in an animal model. Twenty years after the event the exact etiology remains unknown. If fatty acid analides were indeed the source of toxicity they must have been extremely toxic as they were present in very small amounts.

Eosinophilia-Myalgia Syndrome

The toxic oil syndrome might be regarded as a historical curiosity if it were not for the fact that in 1989, a similar problem emerged in the United States. Called the eosinophilia-myalgia syndrome (EMS), it affected over 1500 people and it appeared to be caused by consuming contaminated L-tryptophan as a food supplement. Again, the eosinophil count was usually elevated above 2000/mm^3. Like the toxic oil syndrome, there was also inflammation of muscles and their nerve supply. There was a notable absence of acute respiratory symptoms, unlike the toxic oil syndrome.

The Centers for Disease Control determined that the product came from one supplier and that it had a contaminant, the di-tryptophan aminal of acetaldehyde. This was either the toxic agent or a marker thereof as there was a strong association between the level of this substance and the incidence of the disease. The FDA banned the sale of L-tryptophan as a precaution in 1991. The incidence of EMS fell rapidly. The FDA ban was lifted in 2002 and some new cases have emerged.

Efforts at identifying the toxic agent and developing an animal model for the condition are continuing. Elevated levels of serum quinolinic acid have been recorded clinically in patients with EMS. Self-injection of quinolinic acid over a 1 month period resulted in an almost threefold rise in the eosinophil count, a common observation in EMS. A 2007 paper showed that 3-(*N*-phenylamino) alanine or PAA is activated by human liver microsomes. PAA was identified as one of the contaminants in the batch of L-tryptophan associated with the EMS outbreak. The paper identified a common toxic metabolite, 4-aminophenol that could have been generated by a common pathway in both EMS and toxic oil syndrome.

Herbal Remedies

In recent years there has been a tremendous upsurge of interest in, and use of, herbal remedies. One in five Canadians and one in three Americans indicated in a recent survey that they used herbals, often in conjunction with prescription drugs and often without informing any health care professional. This has created a five billion dollar industry with poor controls for efficacy and potency. The phenomenon appears to be another manifestation of the simplistic "natural is good, synthetic is bad" philosophy that arose following public concern over the presence of toxic chemicals in the environment. This concern was fuelled by episodes such as the Love Canal situation and perhaps by the emergence of "New Age" spiritualism. The attraction to herbals is mostly based on the assumption that, because they are natural, they must be safer. This assumption ignores the fact that, if they possess enough of the active ingredient to have a pharmacological effect, they must also carry the potential to cause adverse reactions and to interact with other pharmacological agents, including other herbals. Of course, if they do not possess such potency, they are unlikely to be effective. Since these preparations are presently defined as food supplements they are not subjected to the rigorous regulations governing prescription and patent medicines. Leaving aside the question of potency, which often is an unknown quantity, there is mounting evidence of the potential for some herbals to do harm. A wide variety of herbals possess diuretic activity with associated potassium loss. They may thus interact with prescription diuretics used to treat hypertension and cause excessive potassium loss and low blood pressure. St. John's Wort, cherry stems, and parsley are but a few examples. Others may interfere with antihypertensive therapy and cause a dangerous elevation in blood pressure. Ephedra, the active ingredient of which is ephedrine, a known hypertensive agent, is present in natural weight loss products including fen-phen and these have been associated with over a dozen deaths and hundreds of cases of illness reported to the U.S. Food and Drug Administration. Licorice can cause pseudoaldosteronism in which sodium is retained, with a resulting increase in blood pressure. Numerous other products may depress blood pressure to dangerous levels in patients receiving antihypertensive therapy. Cat's claw, devil's claw, and garlic fall into this category. *Rauwolfia serpentina* is present in many natural products. Its active ingredient is reserpine, a potent depressor of blood pressure.

The agents listed earlier are but a small sample of the host of herbal agents with the potential to cause untoward side effects and drug interactions. Public demand for these products will force regulatory agencies to divert billions of dollars to analyze and test them, with the predictable result that many will be banned and others reclassified as drugs. This will no doubt lead to an even greater public outcry. For a more complete list of herbal remedies and their potential side effects and interactions with prescription drugs refer to Philp (2004).

Natural Carcinogens in Foods

Considerable evidence exists that bracken fern produces bladder cancer in cattle that eat excessive amounts when better fodder is unavailable, and in rats fed large amounts of it. Because the young shoots, called fiddleheads because of their curled shape, are eaten as a delicacy in many parts of the world, including Canada and Japan, there has been concern over the potential for carcinogenic effects in humans. Epidemiological studies, however, have failed to demonstrate such an association, and it is now felt that the eating of fiddleheads does not constitute a risk factor for cancer.

Natural carcinogens and precursors have been detected in many other foods. In their 1987 review, Ames et al. list the carcinogenic TD_{50} for nitrosamines present in many foodstuffs as 0.2 mg/kg for rats and mice. In contrast, PCBs, with a similar daily dietary intake of 0.2 µg, had Td values of

1.7–9.6 mg/kg in these species. Other substances shown to be carcinogens in animal tests, but for which evidence of a risk to humans is weak, include allyl isothiocyanate (in kale, cabbage, broccoli, cauliflower, horseradish, and mustard oil), safrol (nutmeg, cinnamon, black pepper) and benzo[a]pyrene (produced during cooking, especially charcoal broiling). By extremely conservative methods, the risk for benzo[a]pyrene has been estimated at 1.5/100,000 at high levels of consumption.

It has been stated that exposure to carcinogens is an unavoidable fact of life, but that the levels in foods are so low that further reductions would not have a significant effect on cancer incidence. This "bad news" is offset by the "good news" that there are probably many more anticarcinogens in natural foods than there are carcinogens. In one study of human dietary habits, individuals who ate meat but not vegetables on a daily basis had a colon cancer risk (per 100,000 population) of 18.43, those who ate vegetables but not meat on a daily basis had a risk of 13.67, whereas those who ate both had a risk of only 3.87. Vitamins A, C, and E and carotenoids have been shown to be protective against cancer, probably because of their antioxidant properties. Dietary fiber is protective, and even meat has been shown to have anticancer properties. Indoles and isothiocyanates present in cruciferous vegetables such as cabbage, cauliflower, broccoli, and Brussels sprouts have been shown to be anticarcinogenic.

Carcinogenic mycotoxins abound in nature and these will be dealt with in Chapter 10, and irradiated foods in Chapter 12.

Case Study 17

Three men were admitted to the emergency department of a large hospital, all suffering from similar symptoms. Inquiry revealed that they had just finished dining in a nearby Chinese restaurant. All had a severe headache, cardiac palpitations, facial flushing, vertigo (dizziness), and perfuse sweating. The symptoms commenced shortly after finishing their meal. Their faces and chests were reddened, blood pressures slightly below normal, and their pulses were quickened.

Q. What is the likely portal of entry of this apparent toxicant?

Q. Is this likely due to

    a.   Food poisoning of bacterial origin

    b.   Contamination with a pesticide

    c.   A food additive

Q. In a call to the restaurant in question, the attending physician inquired as to whether the trio had consumed mushrooms or fish. Why was this question asked?

The response to the aforementioned question was no. It was revealed that many other patrons had eaten similar food without trouble, the exception being pork chow yuk. These three were the only ones to consume this dish.

Q. How can this information be helpful?

Review Questions

For Questions 1 to 13 use the following code:

Answer A if statements a, b, and c are correct.

Answer B if statements a and c are correct.

Answer C if statements b and d are correct.

Answer D if statement d only is correct.

Answer E if all statements (a, b, c, d) are correct.

1. With regard to diethylstilbestrol (DES):
   a. It was used mainly to synchronize the estrus cycle of dairy cows.
   b. It was given to livestock as a growth promotant in feed or as subcutaneous pellets.
   c. In North America and Great Britain its use as a feed additive for livestock resulted in significant human health problems.
   d. It was given parenterally to pregnant women to prevent impending abortion.

2. Human health problems associated with diethylstilbestrol (DES):
   a. Have been reported in some countries in infants who were fed formula containing high levels of DES
   b. Affect women who were exposed in utero to high levels from their mother's blood
   c. Generally involve abnormalities of the genitourinary system
   d. Have caused widespread problems in people who eat meat containing pellet residues

3. With regard to Salmonella infection:
   a. It is commonly transferred to humans by eating or handling undercooked meat.
   b. It is most often a problem for the general public.
   c. It sometimes involves multiple drug-resistant strains.
   d. It has a high mortality rate in normal individuals.

4. Which of the following statements is/are true?

    a. Estrogens, including DES, have been shown to be carcinogenic.

    b. Women exposed to DES in utero have an increased incidence of cervico-uterine deformities.

    c. DES is not metabolized significantly by the human placenta or fetus.

    d. The human fetus cannot deactivate natural estrogens.

5. Regarding the use of antibiotics in agriculture:

    a. They are used exclusively for treating infections in animals.

    b. They may be used prophylactically to prevent infections in animals.

    c. There are no regulations governing their use in agriculture.

    d. Very low levels added to feed might have a growth-promoting effect.

6. Multiple drug resistance is

    a. Characteristic of Gram-negative enteric bacteria

    b. Possible even when the organism has not contacted all of the anti-infective agents to which it is resistant

    c. Determined by extrachromosomal DNA

    d. Seen only in farm livestock exposed to antibiotics

7. The process of bacterial conjugation

    a. Refers to the release of DNA from lysed cells

    b. Requires the participation of plasmids

    c. Refers to the transfer of DNA by phage viruses

    d. Requires the participation of an "F" (fertility) factor

8. Which of the following statements is/are true?

    a. Salmonella infection is rarely a problem in nursing homes.

    b. The rank order of frequency for resistant forms of intestinal infections is general hospital > psychiatric hospital > extrahospital environment.

    c. Multiple drug resistance occurs only in enteric organisms from animals.

    d. The frequency of MDR increases with antibiotic exposure.

9. The legal definition of food additives in most countries includes

    a. Vitamins

    b. Food colors

    c. Spices

    d. Nitrates

10. Which of the following substances used in or on foods have been associated with a high degree of allergic reactions?
    a. Tartrazine
    b. Carrageenin
    c. Sodium metabisulfite
    d. Saccharin

11. Which of the following statements regarding nitrosamines, nitrates, and nitrites is/are true?
    a. The major source of nitrosamines for people is the nitrates and nitrites used as meat preservatives.
    b. Nitrosamines have been shown to be carcinogens in animals.
    c. There are no natural food sources of nitrates and nitrites.
    d. Nitrates and nitrites inhibit the growth of *C. botulinum*, the cause of ptomaine poisoning.

12. Regarding artificial sweeteners:
    a. Individuals with phenylketonuria should avoid aspartame.
    b. There is no convincing evidence that aspartame in normal amounts causes behavioral problems.
    c. Saccharine has been shown to cause bladder cancer in rats exposed in utero to very high levels.
    d. Cyclamate or a metabolite is a proven carcinogen for humans.

13. Toxic oil syndrome
    a. Involves respiratory distress, eosinophilia, and myalgia
    b. Produced a high mortality rate in an outbreak in Spain
    c. Resembles a disease traced to tryptophan consumption in the United States.
    d. May be due to a reaction product of aniline and a component of a cheap, low-grade mixture of cooking oils

---

Answers

1. C
2. A
3. B
4. A
5. C

6. A

7. C

8. C

9. C

10. B

11. C

12. A

13. E

Further Reading

Aarestrup, F.M., Seyfarth, A.M., Emborg, H.-D., Pedersen, K., Hendriksen, R.S., and Bager, F., Effect of abolishment of the use of antimicrobial agents for growth promotion on the occurrence of antimicrobial resistance in fecal enterococci from food animals in Denmark, *Antimicrob. Agents Chemother.*, 45, 2054–2059, 2001.

Allen, J.A., Peterson, A., Sufit, R., Hinchcliff, M.E., Mahoney, J.M., Wood, T.A., Miller, F.W., Whitfield, M.L., and Varga, J., Post-epidemic eosinophilia-myalgia syndrome associated with L-tryptophan, *Arthritis Rheum.*, 63, 3633–3639, 2011.

Ames, B.N., Magaw, R., and Gold, L.S., Ranking possible carcinogenic hazards, *Science*, 236, 271–280, 1987.

Baily-Klepser, T., Doucette, W.R., Horton, M.R., Buys, L.M., Ernst, M.E., Ford, J.K., Hoehns, J.D., Kautzman, H.A., Logemann, C.D., Swegle, J.M., Ritho, M., and Klepser, M.E., Assessment of patients' perceptions and beliefs regarding herbal therapies, *Pharmacotherapy*, 20, 83–87, 2000.

Bandera, E.V., Chandran, U., Buckley, B., Lin, Y., Isulapalli, S., Marshall, I., King, M., and Zarbi, H., Urinary mycoestrogens, body size and breast development in New Jersey girls, *Sci. Total Environ.*, 409, 5221–5227, 2011.

Boden-Albala, B., Elkind, M.S., White, H., Szumski, A., Paik, M.C., and Sacco, R.L., Dietary total fat intake and ischemic stroke risk: The Northern Manhattan study, *Neuroepidemiology*, 32, 296–301, 2009.

Boullata, J.L. and Nace, A.M., Safety issues with herbal medicine, *Pharmacotherapy*, 20, 257–269, 2000.

Centers for Disease Control, Analysis of L-tryptophan for the etiology of eosinophilia-myalgia syndrome, *Morb. Mortal. Wkly Rep.*, 39, 581–591, 1990.

Chan, J.M., Stampfer, M.J., Ma, J., Gann, P., Gazianio, J.M., Pollak, M., and Giovannucci, E., Plasma insulin-like growth factor-I and prostate cancer risk, *Science*, 279, 563–566, 1998.

Cohen, S.M. and Ito, N., A critical review of the toxicological effects of carrageenan and processed eucheuma seaweed on the gastrointestinal tract, *Crit. Rev. Toxicol.*, 32, 414–444, 2002.

Colton, T., Greenberg, E., Noller, K., Resseguie, L., Van bennekom, C., Heeren, T., and Zhang, Y., Breast cancer in mothers prescribed diethylstilbestrol in pregnancy: Further followup, *JAMA*, 269, 2096–2100, 1993.

Comas, A.P., Precocious development in Puerto Rico, *Lancet*, 319, 1299–1300, 1982.

Dupont, H.L., The growing threat of foodborne bacterial enteropathogens of animal origin, *Clin. Infect. Dis.*, 45, 1353–1361, 2007.

Flamm, W.G., Pros and cons of quantitative risk analysis. In *Food Toxicology: A Perspective on the Relative Risks*, Taylor, S.L. and Scanlan, R.A. (eds.), Marcel Dekker, New York, 1989, pp. 429–446.

Frisoli, T.M., Schmeider, R.E., Grodziki, T., and Messerli, F.H., Salt and hypertension: Is salt dietary reduction worth the effort? *Am. J. Med.*, 125, 433–439, 2012.

Fujisawa, S., Atsumi, T., Kadoma, Y., Ishihara M., Ito, S., and Yokoe, I., Kinetic radical scavenging activity and cytotoxicity of 2-methoxy- and 2-butyl-substituted phenols and their dimers, *Anticancer Res.*, 24, 3019–3026, 2004.

Gans, D.A., Behavioral disorders associated with food components. In *Food Toxicology: A Perspective on the Relative Risks*, Taylor, S.L. and Scanlan, R.A. (eds.), Marcel Dekker, New York, 1989, pp. 225–254.

Gardner, H., Rundek, T., Wright, C.B., Elkind, M.S., and Sacco, R.L., Dietary sodium and risk of stroke in the Northern Manhattan Study, *Stroke*, 43, 100–1205, 2012.

Hall, R.L., Dull, B.J., Henry, S.H., Schleuplein, R.J., and Rulis, A.M., Comparison of the carcinogenic risks of naturally occurring and adventitious substances in food. In *Food Toxicology: A Perspective on the Relative Risks*, Taylor, S.L. and Scanlan, R.A. (eds.), Marcel Dekker, New York, 1989, pp. 205–224.

Harrington, S., The role of sugar-sweetened beverage consumption in adolescent obesity: A review of the literature, *J. Sch. Nurs.*, 24, 2–12, 2008.

Harris, R.M. and Waring, R.H., Diethylstilboestrol—A long-term legacy, *Maturitas*, 72, 108–112, 2012.

Holly, J., Insulin growth factor-I and new opportunities for cancer prevention, *Lancet*, 351, 1373–1375, 1998.

Holmberg, S.D., Osterholm, M.T., Senger, K.A., and Cohen, M.L., Drug-resistant *Salmonella* from animals fed antimicrobials, *N. Engl. J. Med.*, 311, 617–622, 1984.

Hord, N.G., Tang, Y., and Bryan, N.S., Food sources of nitrates and nitrites: The physiologic context for potential health benefits, *Am. J. Clin. Nutr.*, 90, 1–10, 2009.

Horowitz, B.Z., Bromism from excessive cola consumption, *J. Clin. Toxicol.*, 35, 315–320, 1997.

James, J. and Kerr, D., Prevention of childhood obesity by reducing soft drinks, *Int. J. Obes. (Lond.)*, 29(Suppl. 2), S54–S57, 2005.

Johnson, R.J., Gold, M.S., Johnson, D.R., Ishimoto, T., Lanaspa, M.A., Zahniser, N.R., and Avena N.M., Attention-deficit disorder: Is it time to reappraise the role of sugar consumption? *Postgrad. Med.*, 123, 39–49, 2011.

Kanarek, R.B., Artificial food dyes and attention deficit hyperactivity disorder, *Nutr. Rev.*, 69, 385–391, 2011.

Kaneene, J.B., Warnick, L.D., Bolin, C.A., Erskine, R.J., May, K., and Miller, R., Changes in tetracycline susceptibility of enteric bacteria following switching to non-medicated milk replacer for dairy calves, *J. Clin. Microbiol.*, 46, 1968–1977, 2008.

Loizzo, A., Gatti, G.L., Macri, A., Moretti, G., Ortolani, E., and Palazzesi, S., The case of diethylstilbestrol treated veal contained in homogenized baby food in Italy. Methodological and toxicological aspects, *Ann. Ist. Super. Sanita.*, 20, 215–220, 1984.

Lundberg, J.O., Cardiovascular prevention by dietary nitrate and nitrite, *Am. J. Physiol. Heart Res.*, 296, H1221–H1223, 2009.

Lundberg, J.O., Carlstrom, M., Larsen, F.J., and Weitzberg, E., Role of dietary inorganic nitrate in cardiovascular health and disease, *Cardiovasc. Res.*, 89, 5525–532, 211.

Malik, V.S., Schulze, M.B., and Hu, F.B., Intake of sugar-sweetened beverages and weight gain: A systematic review, *Am. J. Clin. Nutr.*, 84, 274–288.

Marshall, B.M. and Levy, S.B., Food animals and antimicrobials: Impacts on human health, *Clin. Microbiol. Rev.*, 24, 718–733, 2011.

Marti, L.F., Effectiveness of nutritional interventions on the functioning of children with ADHD and/or ASD. An updated review of research evidence, *Bol. Assoc. Med. P R*, 102, 31–42, 2010.

Martinez-Cabot, A. and Messeguer, A., Generation of quinoneimine intermediates in the bioactivation of 3-(*N*-phenylamino)alanine (PAA) by human liver microsomes: A potential link between eosinophilia-myalgia syndrome and toxic oil syndrome, *Chem. Res. Toxicol.*, 20, 1556–1562, 2007.

Mashour, N.H., Lin, G.L., and Frishman, W.H., Herbal medicine for the treatment of cardiovascular disease, *Arch. Intern. Med.*, 158, 2225–2234, 1998.

Massart, F., Meucci, V., Sagesse, G., and Soldani, G., High growth rate of girls with precocious puberty exposed to estrogenic mycotoxins, *J. Pediatr.*, 152, 690–695, 2008.

Mcann, D., Barrett, A., Cooper, A., Crumpler, D., Dalen, L., Grimshaw, K., Itchin, E., Lok, K., Porteous, L., Prince, E., Sonuga-Barke, E., Warner, J.O., and Stevenson, J., Food additives and hyperactive behavior in 3-year-old and 8/9-year-old children in the community: A randomized, double-blinded placebo-control trial, *Lancet*, 370, 1560–1567, 2007.

Munro, I.C., A case study: The safety evaluation of sweeteners. In *Food Toxicology: A Perspective on the Relative Risks*, Taylor, S.L. and Scanlan, R.A. (eds.), Marcel Dekker, New York, 1989, pp. 151–167.

Nigg, J.T., Lewis, K., Edinger, T., and Falk, M., Meta-analysis of attention deficit/hyperactivity disorder or attention deficit/hyperactivity disorder symptoms, restriction diet, and synthetic food color additives, *J. Am. Acad. Child Adolesc. Psychiatry*, 51, 86–97, 2012.

Noakes, R., Spelman, L., and Williamson, R., Is the L-tryptophan metabolite quinolinic acid responsible for eosinophilic fasciitis? *Clin. Exp. Med.*, 6, 60–64, 2006.

Oberrieder, H.K. and Fryer, E.B., College students' knowledge and consumption of sorbitol, *J. Am. Diet. Assoc.*, 91, 715–717, 1991.

Pariza, M.W., A perspective on diet and cancer. In *Food Toxicology: A Perspective on Their Relative Risks*, Taylor, S.L. and Scanlan, R.A. (eds.), Marcel Dekker, New York, 1989, pp. 1–10.

Philp, R.B., *Herbal-Drug Interactions and Adverse Effects: An Evidence-Based Quick Reference Guide*, McGraw-Hill Medical Publishing, New York, 2004.

Poirier, L., Lefebvre, J., and Lacouriere, Y., Herbal remedies and their effects on blood pressure: Careful monitoring required, *Hypertension Canada*, 65, 1–8, 2000.

Poseda de la Paz, M., Philen, R.M., and Borda, A.I., Toxic oil syndrome: The perspective after 20 years, *Epidemiol. Rev.*, 23, 231–247, 2001.

Report of the DES Task Force, *J. Am. Med. Assoc.*, 255, 1849–1886, 1986.

Saito, M., Sakagami, H., and Fujisawa, S., Cytotoxicity and apoptosis induced by butylated hydroxyanisole (BHA) and butylated hydroxytoluene (BHT), *Anticancer Res.*, 23, 4693–4701, 2003.

Sun, M., Use of antibiotics in feed challenged, *Science*, 226, 144–146, 1984.

Temme, E.H.M., Vandevijvere, S., Vinkx, C., Huybrechts, I., Goeyens, L., and Van Oyen, H., Average daily nitrate and nitrite intake in the Belgian population older than 15 years, *Food Addit. Contam. A*, 28, 1193–1204, 2011.

Thevis, M., Fussholler, G., and Schanzer, W., Zeranol: Doping offence or mycotoxin, *Drug Test Anal.*, 3, 777–783, 2011.

Titus-Emstof, L., Troisi, R., Hatch, E.E., Palmer, J.R., Hyer, M., Kaufman, R., Adam, E., Noller, K., and Hoover, R.N., Birth defects in the sons and daughters of women who were exposed *in utero* to diethylstilbestrol (DES), *Int. J. Androl.*, 33, 377–384, 2010.

Weiner, M.L., Toxicological properties of carrageenan, *Agents Actions*, 32, 46–51, 1991.

Weiss, B., Synthetic food colors and neurobehavioral hazards: The view from environmental research, *Environ. Health Perspect.*, 120, 1–5, 2011.

Welsh, J.A., Sharma, A.J., Grellinger, L., and Vos, M.B., Consumption of added sugars is decreasing in the United States, *Am. J. Clin. Nutr.*, 94, 726–734, 2011.

You, Y., Hilpert, M., and Ward, M.J., Detection of a common and persistent tet(L)-carrying plasmid in chicken-waste-impacted farm soil, *Appl. Environ. Microbiol.*, 78, 3203–3213, 2012.

9

Pesticides

> So naturalists observe, a flea has smaller fleas that on him prey; And these have smaller still to bite 'em, And so proceed *ad infinitum*.

Jonathan Swift

Introduction

The term pesticides refers to a large body of diverse chemicals that includes insecticides, herbicides, fungicides, rodenticides, and fumigants employed to control one or more species deemed to be undesirable from the human viewpoint. Pesticides are of environmental concern for two main reasons. Although considerable progress has been made with respect to their selective toxicity, many still possess significant toxicity for humans, and many are persistent poisons, so that their long biological T1/2 allows bioaccumulation and biomagnification up the food chain (see Chapter 3). There is, thus, the possibility that they may enter human food supplies as well as constituting an ecological hazard. By their very nature, pesticides must have an impact on any ecosystem since they are designed to modify it by their selective elimination of certain species. As is always the case in considering chemicals used in the service of humankind, there is a complex risk–benefit equation that must be taken into account in making decisions regarding the use of pesticides. There is no question that they have increased agricultural production when used properly, and they have, in the past, been highly effective in controlling the insect vectors of human diseases like malaria and yellow fever spread by mosquitoes, and African sleeping sickness, which affects both humans and animals and which is spread by the tsetse fly. As shall be seen, however, these gains have not been without their problems.

Efforts to control agricultural pests probably evolved in parallel with cultivation techniques. Early methods included manual removal of weeds and insects, rigorous hoeing to prevent weed growth, and the use of traps for animal and insect pests. The first chemical controls to be used against agricultural pests were the arsenical compounds. In 1910, Erlich discovered that arsphenamine was an effective treatment for syphilis. This was the first chemotherapeutic agent for a bacterial infection and the first example of a structure–activity relationship. It opened the door on the entire field of chemical control of both infections and of pests. Paracelsus had introduced the use of inorganic arsenicals, notably arsenic trioxide (As_2O_3, white arsenic) into medicine in the sixteenth century, but its use was limited by its extreme toxicity. Ehrlich's discovery revived interest in these compounds and in 1824 the Colorado potato beetle was discovered east of the Rockies and its eastward spread accelerated the search for an effective control. As_3O_3 was found to be effective and came into widespread use. Other arsenicals were developed, including Paris green (copper arsenite) which is still used as slug bait. Being a heavy metal, arsenic is persistent in the environment, the significance of which was not appreciated when it was being widely used.

Natural-source insecticides also evolved fairly early on. Certain plants have been employed as fish poisons in Southeast Asia and in South America for centuries, and in 1848 a decoction of derris root was used to control an insect infestation in a nutmeg plantation in Singapore. By 1920, large amounts were being imported into North America. The active ingredient is rotenone and it has the advantages of low mammalian toxicity and short T1/2 in nature. Pyrethrum flowers (chrysanthemums) have been known for their insecticidal properties for centuries. Commercial manufacture began in 1828. In 1945, the United States imported 13.5 million lb. By 1954 this had fallen to 6.5 million because of the widespread use of DDT, the banning of which has led to a resurgence of use of pyrethrin compounds. Nicotine sulfate (Blackleaf 40 is a 40% solution) from tobacco is used to control aphids and other insects. It has a short biological T1/2 but significant mammalian toxicity.

The mechanization of farming led to a second agricultural revolution by making possible the planting and cultivation of vast tracts of land. Pest control techniques also changed from the small-scale operations of the past to include mechanized spraying from the ground and the air (see also Chapter 4). This involved a marked increase in the use of pesticides and it coincided with the introduction of the first, modern, synthetic insecticide, DDT.

Dichlorodiphenyltrichloroethane, or DDT, was first synthesized in 1874, but its insecticidal properties were not recognized until 1939. Its structural formula is shown in Figure 9.1. Its first major use occurred in Sicily in 1943, where it was used to halt an epidemic of tick-borne typhus.

Sometimes called the grandfather of all chlorinated aromatic hydrocarbons, DDT was the first of such agents to arouse environmental concern. Rachel Carson's *Silent Spring* called attention to the ecological damage caused by DDT and led to its banning in the United States and

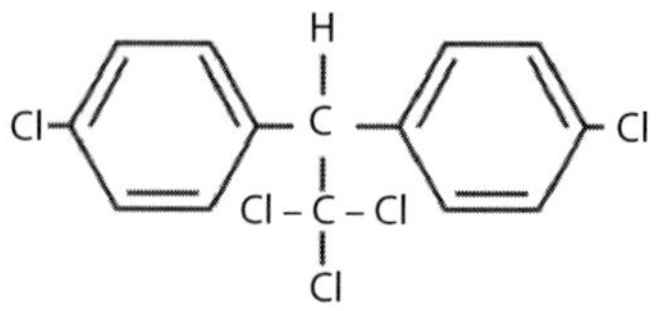

FIGURE 9.1
DDT, chemically 1,1,1-trichloro-2,2-bis (p-chlorophenyl) ethane.

Canada in 1972. Prior to that, however, its use had led to the eradication of malaria in 37 countries and dramatically reduced its incidence in a further 80, providing relief to 1.5 billion people. Its effectiveness in controlling agricultural pests, coupled with its low mammalian toxicity (oral LD_{50} 113 mg/kg, dermal LD_{50} 2.5 gm/kg), resulted in extensive use in North America. U.S. production reached 50,000 metric tons annually. The availability of cheap, surplus aircraft after World War II resulted in the spraying of huge areas to control not only agricultural pests but human ones as well. Organochlorines, including the cyclodienes, dominated the insecticide field until the early 1960s, when organophosphorus insecticides (organophosphates) and carbamates were developed. These, plus the development of more disease-resistant hybrid crops, led to the "Green Revolution" of the 1960s, with dramatic increases in food production.

Classes of Insecticides

Organochlorines (Chlorinated Hydrocarbons)

As already discussed, the parent compound of this group is DDT. Its human toxicity is extremely low. In one rather heroic experiment, volunteers were fed 35 mg/day for up to 25 months without obvious ill effects. Another study of 35 male workers who had DDT levels in fat and liver 80 times the American average, and who had worked in a manufacturing plant for up to 19 years, showed no ill effects. DDT is, however, a potent inducer of cytochrome P450 hepatic microsomal enzymes and may thus affect the rate of biotransformation of other chemicals and drugs. Extremely high doses cause neurological signs and symptoms including numbness of the tongue, lips and face, dizziness, hyperexcitability, tremor, and convulsions.

DDT has very high lipid solubility and it is sequestered in body fat. Virtually everyone who was alive after 1940 has DDT in body fat. In the 1960s, significant amounts were found in people all over the world from Sri Lanka to North America. In 1970, the mean concentration in human fat was 7.88 ppm. After the ban, it fell to 4.99 in 1975. There is no evidence that chronic exposure to DDT has resulted in any health problems. In insects, DDT opens up ion channels to prevent normal axonal repolarization. Disorganized neuronal function leads to death.

Other life forms are not as resistant as humans are. Fish are extremely vulnerable, and die-offs occurred after heavy rains washed DDT into streams and lakes. Deformities also occur. Predatory birds at the top of the food chain are very vulnerable as well. Reproduction is disturbed in a number of ways. DDT induces cytochrome P450 to increase estrogen metabolism and DDT itself has estrogenic activity that affects fertility. Ca^{2+}-ATPase is inhibited as is calcium deposition in eggshells. This effect is largely due to stable metabolites, notably DDE (dichlorodiphenyldichloroethane). Some bird species are only now recovering. The limited use of DDT against the tussock moth was reapproved in the United States in 1974 and its use in malarial areas has continued without interruption, so that DDT exposure on a worldwide basis still occurs.

The cyclodienes are a subgroup of the organochlorines. This group includes aldrin, dieldrin, heptachlor, and chlordane. Their mechanism of insecticidal action is the same as for DDT, but their toxicity for humans is much greater because of more efficient transdermal absorption. Signs of excessive CNS excitation and convulsions occur before less serious signs appear. Several deaths, mostly in those who handle the pesticide, have occurred. These agents too, are persistent in the environment. There is concern about their potential for carcinogenicity since this has been shown in some animals. However, Ribbens reported on a study of 232 male workers who had been exposed to high levels of cyclodienes in a manufacturing plant in Holland for up to 24 years (mean 11 years). Mortality and cancer incidence were compared to the means for the Dutch male population of the same age group. The observed mortality in the group was 25, which was significantly lower than the expected mortality of 38. Nine of the deaths were from cancer, as opposed to an expected incidence of 12. These workers had been exposed to very high levels of cyclodienes in the early days of manufacture, with recorded dieldrin blood levels of up to 69 µg/L at sometime in their history.

Other organochlorines include methoxychlor, lindane, toxaphene, mirex, and chlordecone (kepone). Mirex and kepone are extremely persistent, toxic to mammals (CNS toxicity), and carcinogenic in animals. They also induce cytochrome P450. They are no longer used in North America. Lindane shares the same toxicity but is much less persistent and it is used to treat head lice. Lindane (chemically 1,2,3,4,5,6-hexachlorocyclohexane) is the active isomer of benzene hexachloride. Toxaphene induces liver tumors in mice and is fairly toxic and its use is declining. Methoxychlor is similar to DDT but it is much less persistent and less toxic to mammals, which can metabolize it. It also is stored in fat to a much lesser degree. Its formula, along with that of lindane, is shown in Figure 9.2.

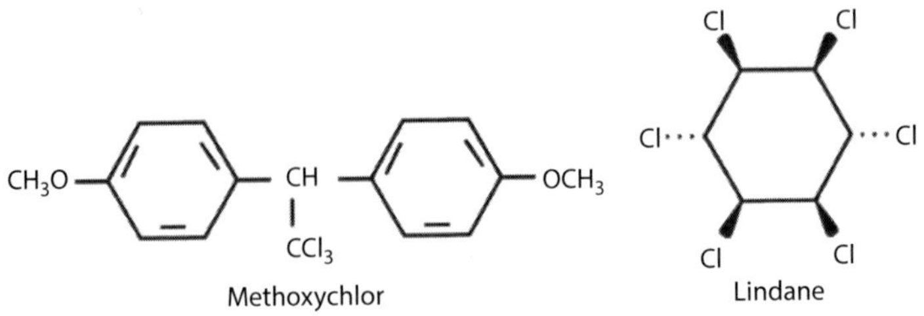

FIGURE 9.2
Methoxychlor and lindane.

Organophosphorus Insecticides

These insecticides, often referred to as organophosphates, are the most frequent cause of human poisonings. The group includes parathion, dichlorvos (present in Vapona strips), and diazinon. They all act as irreversible inhibitors of acetylcholinesterase, so that the neurotransmitter acetylcholine is not inactivated following its release from the nerve terminal. Signs and symptoms are those of a massive cholinergic discharge and include dizziness and disorientation, profuse sweating, profuse diarrhea, constricted pupils, and bradycardia (slowing of the heart) possibly with arrhythmias. Parathion has a dermal LD_{50} of 21 mg/kg and an oral LD_{50} of 13 mg/kg in male rats but the NOEL in both rats and humans is only 0.05 mg/kg. Parathion itself is not toxic but it is transformed in the liver to para-oxone, its oxygen analog (see Chapter 1, Figure 1.3).

The following is a typical case history of organophosphorus poisoning:

A 52 year old white, male farmer was admitted to a hospital emergency department following a highway accident in which his tractor collided with the rear of a motor vehicle about to make a turn. He incurred numerous lacerations and contusions and a fractured right humerus. He was restless, incoherent, and required physical restraint. His pupils were bilaterally constricted, his heart rate was 55 beats/min and he was sweating profusely. His clothing had a strong, chemical odor. His wife volunteered that he had several episodes of visual difficulty over the preceding 2 weeks. Further questioning revealed that he had been spraying organophosphorus insecticides during this period (organophosphorus poisoning is frequently delayed). Atropine was given intravenously in repeated small doses until the signs of cholinergic discharge abated. Another drug that can be used is pralidoxime, which complexes with the phosphate component of the organophosphorus and releases the cholinesterase. The principal advantage of the organophosphates is their short life in the environment. The sites of action of organophosphates, atropine, and pralidoxime are shown in Figure 9.3.

Carbamate Insecticides

Carbamates (e.g., Sevin) are also inhibitors of acetylcholinesterase, but they do not require metabolic activation and they are reversible. They are not persistent in the environment. Because they lack the phosphate group, pralidoxime cannot be used for treatment of poisoning. In fact, it is contraindicated because it may tie up more reactive sites on the enzyme and increase the degree of inhibition. This group includes aldicarb (Temik), carbaryl, and Baygon. The dermal LD_{50} for aldicarb in male rats is 3.0 mg/kg. It is also fairly

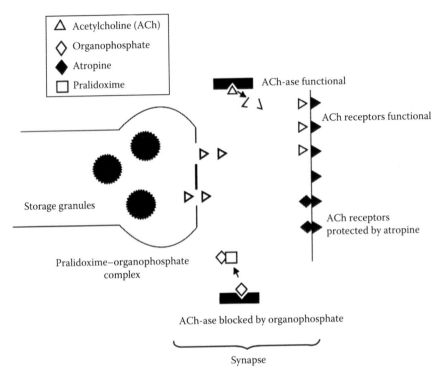

FIGURE 9.3
Sites of action of organophosphorus insecticides, atropine, and pralidoxime. Although the neurotransmitter site is labeled a synapse, muscarine is primarily a muscarinic receptor-blocking agent, acting at parasympathetic neuroeffector junctions. Acetylcholine is present there, as well as in all ganglia, at the neuromuscular junction, the brain, and the adrenal gland.

toxic for humans. Although these agents are generally not persistent in the environment, aldicarb may be an exception. Under certain conditions (sandy soil over aquifers) it may reach water supplies and persist for a considerable time. In Long Island New York it has been estimated that the levels of 6 ppb may persist for up to 20 years.

Botanical Insecticides

The more common botanical insecticides were discussed briefly earlier. While it is commonly felt that natural-source insecticides are safer than synthetic ones (another example of the "nature knows best" syndrome), this is not necessarily so. Pyrethrins and rotenone have oral LD_{50}s of about 600–900 and 100–300 mg/kg, respectively. Nicotine is quite toxic, with an oral LD_{50} of 10–60 mg/kg. The main problem with pyrethrins has been the rapidity with which they are destroyed in the environment. Newer ones have been isolated with longer T1/2s to permit more effective kills.

Herbicides

Chlorphenoxy Compounds

These agents, characterized by 2,4-D and 2,4,5-T, act as growth hormones, forcing plant growth to outstrip the ability to provide nutrients. They are employed as a variety of salts and esters. The acute toxicity of these agents is relatively low, with LD_{50}s of 300–>1000 mg/kg reported for several species of mammals. The dog may be more sensitive (LD_{50} 100 mg/kg). Ventricular fibrillation appears to be the immediate cause of death. Acute toxicity in humans is manifested largely as chloracne.

The main concern about 2,4-D and 2,4,5-T is the likelihood of their contamination with dioxin (TCDD). This subject is dealt with in Chapters 2 and 4. The chemical structures of these compounds are shown in Figure 9.4.

Dinitrophenols

Several substituted dinitrophenols are used as herbicides, the most common probably being Dinoseb (see Figure 9.5). It has been reported to have an LD_{50} of 20–50 mg/kg in rats. Dinoseb, first registered in 1947, is out of favor because handlers may be at considerable risk for teratogenic effects, cataracts, and male reproductive disturbances, even when protective clothing is worn. The U.S. EPA suspended all use in 1987.

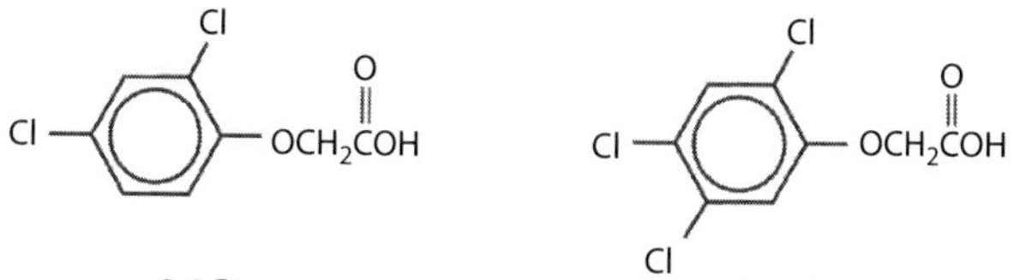

2,4-D
(2,4-dichlorphenoxyacetic acid)

2,4,5-T
(2,4,5-trichlorphenoxyacetic acid)

FIGURE 9.4
Chemical structures of 2,4-D and 2,4,5-T.

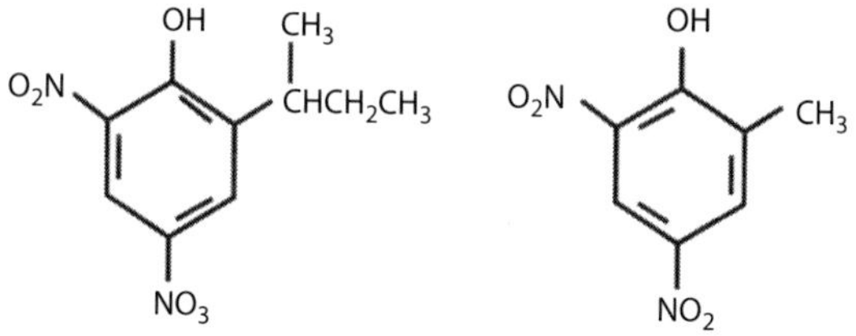

Dinoseb
(2-sec-butyl-4,6-dinitrophenol)

DNOC
(4,6-dinitro-o-cresol)

FIGURE 9.5
Chemical structures of Dinoseb (2-sec-butyl-4,6-dinitrophenol) and DNOC (4,6-dinitro-o-cresol).

4,6-dinitro-o-cresol (DNOC, Figure 9.5) has caused acute poisoning in humans with signs and symptoms including nausea, vomiting, restlessness, and flushing of the skin, progressing to collapse and coma. Hyperthermia may occur. Death may ensue in 24–48 h. Uncoupling of oxidative phosphorylation is probably the mechanism of toxicity. Atropine is contraindicated in DNOC poisoning because there is no anticholinesterase activity and the CNS effects of atropine may complicate the outcome. Treatment is symptomatic and includes ice baths to reduce fever, fluids intravenously, and the administration of O_2.

Bipyridyls

Paraquat and diquat are the most familiar members of this group (see Figure 9.6). Both are toxic but their toxicity differs. The principal organ of toxicity for paraquat is the lungs, although liver and kidney also may be damaged. Respiratory failure may be delayed for several days after the ingestion of paraquat. It appears to be selectively concentrated in the lungs by an energy-dependent system. Paraquat is believed to undergo conversion to superoxide radical ($O_2^{\bullet-}$), which causes the formation of unstable lipid hydroperoxides in cell membranes. Widespread fibroblast formation occurs, and O_2 transfer to capillary blood is impaired. Treatment consists of attempts to remove or neutralize any paraquat remaining in the gastrointestinal tract by gastric lavage, cathartics, and Fuller's earth as an adsorbant. In complete lung failure, double lung organ transplant offers the only hope for recovery.

In contrast, diquat toxicity is centered on the liver, kidney, and gastrointestinal tract. Superoxide anion formation is believed to play a role also in these organs. Poisoning with paraquat is far more common and it has been used as an instrument of suicide on numerous occasions.

Carbamate Herbicides

Unlike the insecticide carbamates, the herbicides do not possess anticholinesterase activity. They have low acute toxicity. Dithiocarbamates are used as fungicides and have similar low acute toxicity; LD_{50}s for these agents are in the gm/kg range for rodents.

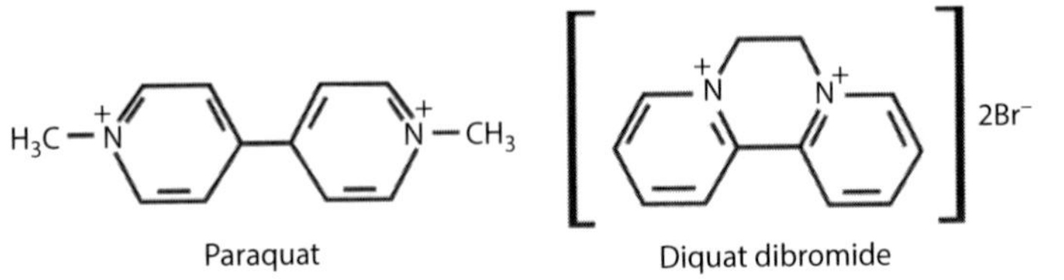

Paraquat   Diquat dibromide

FIGURE 9.6
Chemical structures of paraquat and diquat.

Triazines

This group, typified by Atrazine, also is characterized by low acute toxicity. Amitrole is a herbicide somewhat related to the triazines. It has similar low acute toxicity, but it has peroxidase-inhibiting activity and it has been associated with tumor formation in the thyroid in rats fed the chemical for 2 years.

Fungicides

A wide variety of agents has been used for their fungicidal properties, some of them quite toxic. Seed grains treated with mercurials have sometimes entered the human food supply with disastrous results (see Chapter 6). Pentachlorophenol and hexachlorobenzene are halogenated hydrocarbons with the toxicity typical of that group (see Chapter 4). Thiabendazole is a fungicide of low toxicity as evidenced by the fact that it is also used as an anthelmintic in domestic animals and humans for the eradication of roundworms.

Dicarboximides

Captan and Folpet are agents of some concern. Structurally similar to thalidomide, they have been shown to possess similar teratogenic properties in the chick embryo. Captan has been shown to be mutagenic, carcinogenic, and immunotoxic in animals. The EPA has judged Folpet to be a probable human carcinogen with a lifetime risk of cancer of 2 per million for lettuce and small fruits and a total of 5.5 per million when all food sources are combined.

Newer Biological Control Methods

The earliest form of biological control no doubt was the development of strains of plants and animals with a high degree of resistance to disease, through selective breeding. Observant farmers probably began this process soon after the domestication process began, and it continues today. Over 40 years ago, as a high school student, the author worked with Professor Waddell who developed, at the Ontario Agricultural College, the first strains of wheat to be resistant to wheat rust, a fungal infestation. Recently, a strain of American elms with a high degree of resistance to Dutch elm disease has been developed. Ladybugs have been bred and released to control the cottony cushion scale on oranges in California, and *Bacillus thuringiensis var. kurstaki* (Btk) is used to control forest pests (see also Chapter 4 regarding aerial spraying of Btk).

One of the earliest "high-tech" biological controls was developed in the 1950s and involves sterilization by radiation of millions of male insects that are then released to mate with the females. In species in which the female only mates once, this results in a high frequency of infertile unions with a resulting decline in the insect population. This method was first used successfully to control the screwworm fly in the southern United States. This fly lays its eggs in wounds in the skin of cattle and other livestock. The larvae then live on the flesh of the unwilling host. By 1966 the screwworm had been successfully eradicated in the United States and northern Mexico. It resurfaced in Libya, creating a political dilemma for the United States. Withholding technological assistance could result in massive infestations throughout Africa (the fly will also lay its eggs in wounds on humans), but the alternative at the time was to offer help to Quaddafi. Humanitarian considerations prevailed. This form of biological control has also been used more recently to control the Mediterranean fruit fly in California.

Analogs of insect hormones have been developed that are highly specific to a given species. These hormones trigger the molting metamorphosis in the larval stage so that the larva cannot develop normally and dies. Others, such as the pheromone for the light brown apple moth (see also Chapter 4), can be released to flood an area with this mating hormone so that male moths cannot home in on the females.

Pathogenic bacteria exist that can be cultured in commercial quantities and released to control specific pests. Some agents have been genetically modified for this purpose, but public concerns about "superbugs" have blocked approval of all but a few of these. Given that there are no, known bacteria that are infectious for both insects and mammals (as opposed to insects being vectors for infection), this fear seems unjustified. A more legitimate concern is that beneficial or harmless species may also be attacked by the organisms (see also Chapters 4 and 13).

Government Regulation of Pesticides

Most governments have regulations regulating the use of pesticides. The Canadian regulations are fairly typical of those in place in industrialized countries. The Pest Control Products Act, administered by Agriculture Canada, regulates the introduction of new pesticides. The risk–benefit principle is applied to decisions, that is, the degree of risk must be acceptable in light of the potential benefit to be derived from pest control with the new agent. Its relative safety and effectiveness compared to existing pesticides will influence the decision.

Table 9.1 compares the rodent LD_{50} values and the estimated lethal doses for humans of a number of pesticides. This is just a small sample of the

TABLE 9.1

Oral LD$_{50}$ (Rodent) Values and Estimated Lethal Doses for Some Pesticides for a 70–155 kg Human

Class and Chemical Names[a]	Oral LD$_{50}$ mg/kg	Adult Lethal Dose (mL)
Insecticides		
Aldicarb (a carbamate) Temik	5	0.3
Carbaryl (a carbamate) Sevin	500	30
Chlorpyrifos (organophosphate) Lorsban	92–276	3–30
Diazinon (organophosphate)	300–400 (tech)[t]	To 25
Methoxychlor (chlorinated hydrocarbon)	6000	300–900
Lindane (chlorinated hydrocarbon)	88–125	3–30
Permethrin (a pyrethrin) Ambush, Pounce	tech[t] (>4000)	30–300
Pyrethrum	1500	30–300
Herbicides		
Alachlor, Degree	tech[t] 930–1550	30–300
Diquat	600	30
Glyphosphate, Roundup, Touchdown	>5000	>300
Oxyfluorfen, Goal	>5000	>300
Paraquat, Gramoxone Max	150	3–30
2,4-D (acid)	375	3–30
2,4-DB (Butyrac)	>2000	150
Fungicide		
Captan	9000	500
Fluazinam, Omega	>5000	>300

Sources: This table was compiled from data on websites provided by the British Columbia Ministry of Agriculture, http://www.agf.gov.bc.ca/b_1.htm; Pennsylvania State University, http://pubs.cas.psu.edu/FreePubs/pdfs/uo222.pdf

[a] Chemical names are followed by trade names.
[t] Technical grade.

hundreds listed on the Pennsylvania State website. It is evident that the cholinesterase-inhibiting insecticides, the carbamates and organophosphorus compounds, are the most toxic agents on the list. The herbicide paraquat is also very toxic.

Problems Associated with Pesticides

Development of Resistance

Insects, like microorganisms, possess the most important characteristics for the evolution of resistant strains; an extremely short reproductive cycle and the production of vast numbers of progeny. Most species of insects can go through many generations in one season, producing millions of offspring.

There is, thus, the capacity for multiple, sequential mutations to occur, and a good chance that some of these will be resistant to one or more insecticide. The development of resistance requires the presence of the appropriate insecticide to select out the resistant strain (by killing off the susceptible ones) with the means of detoxifying the chemical or excluding it from absorption. Only 3 years after the introduction of DDT, houseflies and mosquitoes were showing signs of resistance and, by 1951, DDT, methoxychlor, chlordane heptachlor, and benzene hexachloride (of which lindane is the active isomer) no longer had any effect on houseflies, which proliferated abundantly. By the end of 1980, 428 species of insects and acarines (mites, ticks) were classified as resistant.

In the pre-DDT era, relatively few species developed resistance. This has been attributed to the multisite mechanisms of action of earlier pesticides (making single mutation resistance unlikely) and to their ionic nature, making detoxification by metabolism nearly impossible.

Closely related insecticides may be detoxified by the same mechanism and generally act at the same target site, so that if resistance evolves to one, either by the evolution of a detoxification process or by modification of the target molecules, cross-resistance to the others will occur. This type of resistance tends to be under the control of a single gene allele or closely linked genes. Cross-resistance to DDT and methoxychlor, aldrin, and heptachlor has developed in this manner. One study reported that 216 weeds in 45 countries were resistant to a variety of herbicides including the old 2,4-D and the newer glyphosate. This problem has led to the use of combinations of herbicides to make it less likely that adaptation to them all can develop. This is much like the use of combinations of antibiotics in an effort to thwart the emergence of resistant strains of bacteria.

Multiple Pesticide Resistance

In common with bacteria, protozoa, and cancer cells, insects can develop resistance to several insecticides. This is referred to as multiple resistance. It is the result of the existence of several, independent gene alleles producing resistance to unrelated agents (e.g., organochlorines and synthetic pyrethrins, called pyrethroids) with different modes of action and different detoxification pathways. It can be a very serious problem of insect control. The mechanism of multiple resistance is obviously quite different from that of bacteria, which is discussed in Chapter 8.

Multiple resistance to herbicides in weeds is also becoming a problem. Like the development of resistance in other organisms, from bacteria to rats, the greater the degree of exposure of the target organism to the pesticide, the greater the likelihood that a resistant mutation will be selected out. Although motivated more by concerns about adverse health effects, the ban or regulation of the use of pesticides for cosmetic purposes should help slow the emergence of resistant strains.

Nonspecificity

Broad-spectrum pesticides, as most are, make no distinction between true pests and species that are harmless or even beneficial. They, therefore, may disrupt the natural competition among species to permit the proliferation of one previously held in check or, as has been observed, they may kill off predator species and permit the expansion of a prey species, which then becomes a pest. This has happened with spider mites.

The problem of emerging resistance has actually been exploited in recent years to address the nonspecificity of herbicides. By developing genetically modified crops with resistance to a specific herbicide it allows the use of that herbicide for weed control in that crop. This is most commonly done with glyphosate (Roundup). Canola, in 1995, was the first crop plant to be genetically modified in this way but others have followed including corn. The approach is not without its own problems, however. The proliferation of naturally resistant strains and the emergence of resistance in previously susceptible ones have limited this usefulness of this approach. Recently, the U.S. Department of Agriculture decided not to allow the growing of glyphosate-resistant sugar beet and wheat because of these factors.

Environmental Contamination

A greater danger than direct toxicity to humankind is probably the contamination of the environment and the subsequent bioaccumulation and biomagnification which occur with persistent pesticides. These chemicals may end up in soil, water, air or all three, depending on their characteristics (see Chapters 3 and 5). The Great Lakes are accumulating hazardous chemicals as a result of agricultural runoff and industrial discharges, frequently accidental ones. It must be stressed, however, that the greatest source of chemical contamination is residential sewage. Even after treatment, phosphates and other household chemicals may enter the water system. The water table in many areas has been contaminated with the herbicide atrazine, commonly used in corn fields.

Callous disregard for the environment can be the result of greedy individuals attempting to increase their profit margin. There have been numerous cases in Toronto of trucks dumping toxic wastes in Lake Ontario after dark, in violation of provincial and municipal laws. In the United States, careless dumping of chlordecone in the James River by a chemical company resulted in a ban on fishing. This cyclodiene is used in ant and roach baits.

There is no doubt that persistent poisons can have a catastrophic effect on the environment and this is probably the most compelling argument for limiting their use. In 1996, over 4000 Swainson's hawks were found dead in a 50 km^2 area in Argentina. The deaths were attributed to the spraying of crops to kill grasshoppers. The hawks follow tractors that stir up the hoppers and the hawks then feast on them. Ecologists feared that losses may reach 20,000 birds out of a world population of about 400,000.

Balancing the Risks and Benefits

The widespread use of pesticides means that there are trace quantities present in or on almost all foodstuffs. Major advances in analytical techniques over the last 30 years mean that chemicals can now be detected at levels never before possible. Headlines proclaiming that dioxins (or PCBs, etc.) have been detected in Lake Erie fish seldom go on to say that the quantities were at the parts-per-billion level. The detection of dioxins in milk, leached from the carton paper, is an example of this. Actual levels were comparable to a drop in an olympic-sized swimming pool.

Nonetheless, there is a growing feeling in the public that the use of pesticides should be greatly curtailed because the risks are unacceptable or, what is almost as bad, unconfirmed. In recognition of this, the Ontario Ministry of Agriculture and Food established in 1983 the Food Systems 2002 program that attempted to reduce the use of pesticides in agriculture in Ontario by 50% by the year 2002. In 2008, the Ontario Ministry of Agriculture, Food, and Rural Affairs looked at the impact of the program on pesticide use in agriculture from 1983 to 2008. Pesticide use, excluding greenhouse of pesticides and growth regulators, was, overall, 40% lower in 2008 than in 1983. It was, however, higher than in 2003 when a 54% reduction in use was observed. The greatest reductions occurred in the use on corn and tobacco but it should be noted that tobacco farming had greatly declined in the intervening years. Pesticide use on fruits and vegetables fell by only 5% but the health risk fell much more, 23%, due to the use of lower risk pesticides.

The impact of pests on food production is so great that temporary approval is sometimes granted to new agents before all of the required tests have been completed. In 1977 it was discovered that an American testing company, Industrial Bio-Test Laboratories, had misrepresented toxicological data on numerous agents. Of the 405 pesticides registered in Canada, 106 had been approved partly on the basis of Bio-Test data. In 1983 Health and Welfare Canada announced that five of these were to be withdrawn, one being the fungicide Captan, which was found to be teratogenic and carcinogenic.

Other pesticides that have been banned or which are under investigation include the following:

Chlordane. This cyclodiene was withdrawn in 1986 because it was considered to be an epigenetic tumor promoter. It is still registered for use against termites.

Alachlor. This herbicide was withdrawn in 1985 because of evidence of carcinogenicity in animals. It was introduced in 1969 and widely used on corn and soybean crops.

Cyhexatin. Dow chemical voluntarily withdrew this insecticide because of evidence of teratogenicity (hydrocephaly) in rabbits.

All of these agents may still be being used in other parts of the world, legally or otherwise.

Toxicity of Pesticides for Humans

Pesticide applicators are at risk primarily from inhalation and dermal contact with pesticides. Protective clothing and equipment are important means of reducing risk. Nonoccupational poisonings occur largely from oral ingestion of contaminated food although dermal exposure has resulted in poisonings in infants (the pentachlorophenol treatment of hospital linens) and inhalation exposure from spray drift can occur. Household pesticides can cause poisoning by all three routes.

There have been a few isolated fatalities in North America from acute pesticide poisoning, but elsewhere in the world, many cases of mass poisonings have occurred. Consumption of seed grains treated with hexachlorobenzene and organic mercury has resulted in mass outbreaks of poisoning in Turkey, West Pakistan, Iraq, and Guatemala. Accidental contamination of foodstuffs like flour, sugar, and grain with parathion and other agents has occurred in several places around the world.

Effects of long-term exposure to very low levels of pesticides on human health remain conjectural but evidence is accumulating that they exist. In one study in Great Britain, a battery of neuropsychological performance tests was administered to two groups of males, 16–65 years of age. One was a group of 146 sheep farmers exposed to organophosphates in sheep dip over several years. The other was a group of matched quarry workers not exposed to organophosphates. The farmers performed significantly worse on tests of short-term memory and cognitive function. The tests included simple reaction time, symbol digit substitution, digit span, serial word learning, and others. There is also the possibility that contaminants may emerge as a greater risk than the pesticide itself, as was the case with TCDD.

One epidemiological study reviewed several reports of human exposures and found an association with cancers of the blood, neurotoxicity (e.g., Parkinson-like symptoms), and behavioral and reproductive problems. There have been indications that agricultural workers have an increased incidence of brain tumors. The U. S. National Institute of Occupational Safety and Health (NIOSH) is conducting a long-term study on brain cancer incidence and its association with a selected group of 134 pesticides of the 600 or so currently in use. One recent study found that farmers and farm workers who spent 55 years or more on the farm had an increased incidence of the brain tumor glioma, especially in association with the use of paraquat, bufencarb, and chlorpyrifos. A modest association between pesticide use and breast cancer incidence has also been shown.

Epidemiological studies linking cancers to pesticide exposures, and reviews of such studies, continue to appear regularly in the scientific literature. Cancers so linked include testicular cancer, breast cancer, leukemia, non-Hodgkin's lymphoma, prostate cancer, and most likely others. Most of these studies are vulnerable to criticism on several grounds including small group sizes, exposure to more than one pesticide, and the presence of confounding factors such as exposure to known carcinogens such as PCBs, PBBs, and dioxins. Indeed it seems probable that we live in a milieu of carcinogens and trying to identify a single one or even a group of them as a causative agent for a cancer type is nearly impossible. Nevertheless, there is a sufficiently high index of suspicion to warrant reducing pesticide use where possible and continuing to study the situation.

There is some hope that nature may be developing her own protective processes against pesticides. There is some new evidence that fields sprayed with the same chemicals year after year may develop a population of bacteria that break down the pesticides and may even adapt to the point that they utilize them as a food source. Tests on prairie soils indicated that the organism *Rhodococcus* breaks down thiocarbamate insecticides in the test tube within 2 h. It may be possible through gene splicing to develop plants that protect themselves against pesticide residues.

Case Study 18

A 43 year old male crop duster was admitted to the emergency department of a rural hospital following an accident in which the aircraft he was attempting to land on a grass strip hit hard, collapsed the undercarriage, and nosed over. The pilot suffered numerous lacerations and bruises but no serious injuries. He was restless and incoherent and he had to be physically restrained. A rapid breath alcohol test was performed and it was negative. His pupils were constricted, his heart rate was slowed and he was sweating profusely. His ground assistant volunteered the information that the pilot had complained of visual disturbances on several occasions during the previous few days. His clothing smelled strongly of a chemical.

Q. Some of these symptoms could be due to a head injury or to chemical intoxication. What facts point to the latter?

Q. What information would you want to seek from his assistant?

The ground crew revealed that the pilot had been spraying crops with parathion during the preceding 2 weeks.

Q. What class of pesticide is this?

Q. What drugs would be indicated for treatment?

Q. How would treatment differ if a carbamate insecticide such as Sevin had been used?

Q. What blood test might assist in confirming the diagnosis?

Case Study 19

During the months of June and August of 1993, 26 men, 19–72 years of age, were admitted to three different local hospitals with an array of symptoms that included nausea, vomiting, dizziness, visual disturbances, muscle weakness, abdominal pain, headache, sweating, and excessive salivation. The men all worked in apple orchards, 19 different ones in all.

Q. What do these symptoms suggest?

Q. What inquiries would you want to make of these men?

Q. What inquiries would you want to make of the orchard operators?

Review Questions

For Questions 1–8, use the following code:

Answer A if statements a, b, and c are correct.
Answer B if statements a and c are correct.
Answer C if statements b and d are correct.
Answer D if only statement d is correct.
Answer C if all statements (a, b, c, d) are correct.

1. Which of the following statements is/are true?
    a. Lindane is commonly used for the control of head lice.
    b. Toxicological testing is usually performed only on the active ingredient of an insecticide.
    c. Lindane is an organochlorine.
    d. A cause-and-effect relationship for progressive chronic disease resulting from prolonged pesticide use is well established.

2. Which of the following is/are true?

   a. The herbicide "atrazine" has contaminated the water table in many areas.

   b. Eggshell strength is adversely affected by contact with pyrethroids by the female bird.

   c. The fungicide "Captan" is a teratogen and carcinogen in experimental animals.

   d. Natural pesticides are always safer than synthetic ones.

3. The main reason for the carcinogenicity in animals of the herbicides 2,4,-D and 2,4,5,-T is

   a. Their action as growth hormones

   b. Their mutagenicity

   c. Their photosensitizing properties

   d. The presence of dioxin (TCDD) as a contaminant

4. The mechanism of TCDD carcinogenicity in humans may involve

   a. Its pleiotropic response to the Ah locus

   b. Its conjugation with glutathione

   c. Its lack of gene restriction

   d. Its ability to cause chloracne

5. The insecticide parathion

   a. Is an organochlorine

   b. Is biotransformed by a mixed function oxidase to its toxic form

   c. Biomagnifies in the environment

   d. Causes symptoms that can be treated with atropine

6. Which of the following statements concerning DDT is/are incorrect?

   a. It has a low acute LD_{50}.

   b. It is very persistent in the environment.

   c. It is very water soluble.

   d. It is still used for mosquito control in many places.

7. Which of the following statements is/are true?

   a. Organophosphates are very persistent in the environment.

   b. Organophosphates are inhibitors of acetylcholinesterase.

   c. Carbamate insecticide poisoning can be treated with pralidoxime.

   d. Carbamate insecticides act as reversible inhibitors of acetylcholinesterase.

8. Which of the following statements is/are true about insect resistance to insecticides?

   a. Over 400 species of insects, mites, etc., showed resistance to pesticides by 1980.

   b. Resistance to preorganic insecticides is especially prevalent.

   c. Cross-resistance to closely related chemicals may occur.

   d. Resistance is more likely to occur if an insecticide has several sites of action.

Answers

1. A
2. B
3. D
4. A
5. C
6. C
7. C
8. B

Further Reading

Alavanja, M.C. and Bonner, M.R., Occupational exposures and cancer risk: A review, *J. Toxicol. Environ. Health B Crit. Rev.*, 15, 238–263, 2012.

Arbuckle, T.E. and Sever, L.E., Pesticide exposures and fetal death: A review of the scientific literature, *Crit. Rev. Toxicol.*, 28, 229–270, 1998.

Baldi, M., Mohammed-Brahim, B., Brochard, P., Dartigue, J.F., and Salamon, R., Delayed health effects of pesticides: Review of current epidemiological knowledge, *Rev. Epidemiol. Sante Publique*, 46, 134–142, 1998.

Boada, L.D., Zumbado, M., Henriquez-Hernandez, L.A., Almeida-Gonzalez M., Alverez-Leon, E.E., Serra-Majem, L., and Luzardo, O.P., Complex organochlorine mixtures as determinant factor for breast cancer risk: A population-based case-control study in the Canary Islands (Spain), *Environ. Health*, 11, 28 (Epub ahead of print), 2012.

Carson, R., *Silent Spring*, Fawcett Crest Books, New York, 1962.

Clapp, R.W., Jacobs, M.M., and Loechler, E.L., Environmental and occupational causes of cancer: New evidence 2005–2007, *Rev. Environ. Health*, 23, 1–37, 2008.

Cory-Slechta, D.A., Studying toxicants as single chemicals: Does this strategy adequately identify neurotoxic risks? *Neurotoxicology*, 26, 491–510, 2005.

Daniels, J.L., Olshan, A.F., and Savitz, D.A., Pesticides and childhood cancers, *Environ. Health Perspect*, 105, 1068–1077, 1997.

Duke, S.O., Taking stock of herbicide-resistant crops ten years after introduction, *Pest Manage. Sci.*, 61, 211–218, 2005.

Flanders, R.V., Potential for biological control in urban environments, In *Advances in Urban Pest Management*, Bennett, G.W. and Owens, J.M. (eds.), Van Nostrand Reinhold, New York, 1986, pp. 95–129.

Georghiou, G.P. and Saito, T. (eds.), *Pest Resistance to Pesticides*, Plenum Press, New York, 1983.

Hall, S.H. and Dull, B.J., Comparison of the carcinogenic risks of naturally occurring and adventitious substances in food. In *Food Toxicology: A Perspective on the Relative Risks*, Taylor, S.L. and Scanlan, R.A. (eds.), Marcel Dekker, New York, 1989, pp. 205–224.

Klassen, C.D. (ed.), *Casarett and Doull's Toxicology: The Basic Science of Poisons*, 7th edn., McGraw-Hill Medical, New York, 2008.

Klassen, C.D. and Watkins, J.B. III (eds.), *Casarett and Doull's Essentials of Toxicology*, McGraw-Hill Medical, New York, 2010.

Lee, W.J., Colt, J.S., Heineman, E.F., McComb, R., Weisenburger, D.D., Lijinsky, W., and Ward, M.H., Agricultural use and risk of glioma in Nebraska, United States, *Occup. Environ. Med.*, 62, 786–792, 2005.

McEwen, F.L. and Stephenson, G.R., *The Use and Significance of Pesticides in the Environment*, Wiley Interscience Publication, New York, 1979.

McGlynn, K.A. and Trabert, B., Adolescent and adult risk factors for testicular cancer, *Nat. Rev. Urol.*, 9, 336–349, 2012.

Muir, K., Rattanamongkolgul, S., Smallman-Raynor, M., Thomas, M., Downer, S., and Jenkinson, C., Breast cancer incidence and its possible spatial association with pesticide application in two counties of England, *Public Health*, 118, 513–520, 2004.

Nelson, N.J., Studies examine whether persistant organic agents may be responsible for rise in lymphoma rates, *J. Nat. Cancer Inst.*, 97, 1490–1491, 2005.

Ontario Ministry of Agriculture, Food and Rural Affairs, Evaluation of the changes in pesticide risk- executive summary, 2008, http://www.omafra.gov.on.ca/english/crops/facts/pesticide- use-exec.htm (accessed on May 16, 2012).

Owen, S.O. and Zelaya, I.A., Herbicide-resistant crops and weed resistance to herbicides, *Pest Manage. Sci.*, 61, 301–311, 2005.

Palca, J., Libya gets unwelcome visitor from the west, *Science*, 249, 117–118, 1990.

Ribbens, P.H., Mortality study of industrial workers exposed to aldrin, dieldrin and endrin, *Int. Arch. Occup. Environ. Health*, 56, 75–79, 1985.

Sanderson, W.T., Talaska, G., Zaebst, D., Davis-King, K., and Colvert, G., Pesticide prioritization for a brain cancer case-control study, *Environ. Res.*, 74, 133–144, 1997.

Sankpal, U.T., Pius, H., Khan, M., Shukoor, M.I., Maliakal, P., Lee, C.M., Abdelrahim, M., Connelly, S.F., and Basha, R., Environmental factors in causing human cancers: Emphasis on tumorigenesis, *Tumour Biol.*, 33, 1265–1274, 2012.

Stephens, R., Spurgeon, A., Calvert, I.A., Beach, J., Levy, L.S., Berry, H., and Harrington, J.M., Neuropsychological effects of long-term exposure to organophosphates in sheep dip, *Lancet*, 345, 1135–39, 1995.

10

Mycotoxins and Other Toxins from Unicellular Organisms

Anything green, that grew out of the mould,
Was a wonderful drug to our fathers of old.

Anon

Introduction

Fungi have been a great boon to humankind, making possible the production of cheeses and alcoholic beverages, and for driving the fermentation process involved in many manufacturing processes, providing a source of food in the form of edible mushrooms, puffballs, and the much sought after truffles, making antibiotics such as the penicillins and others, and being involved in the production of citric acid, an important food preservative. Many species, however, produce mycotoxins that constitute a threat to human health and agriculture by contaminating grain crops and because of their toxicity to most mammals and avians.

Mycotoxins are a group of chemically diverse and complex substances present in a wide variety of filamentous fungi (molds). They are secondary metabolites of fungal metabolism of uncertain function. Some may have a survival advantage by virtue of their toxicity to competing organisms in the microenvironment. The biological function of others is unclear, but many have significant biological activity and several are toxic to mammals. They, therefore, have significant public health and economic implications. It is estimated that 25% of the world's annual food crops are contaminated by mycotoxins.

Some are of interest to pharmacologists and toxicologists because they have served as research tools to study cell function and to identify various types of neurotransmitters and blocking agents. Poisonous mushrooms are solid (not filamentous) spore-forming fungi that also constitute a health hazard and that have historic, pharmacological significance, but they shall be considered separately in the following chapter.

Some Health Problems Associated with Mycotoxins

Ergotism

Reports of toxicity from molds are as old as recorded history. The oldest recorded source of fungus poisoning is ergotism. Ergot is the common name for the fungus *"Claviceps purpurea"* that affects cereal grains, especially rye, and that produces a number of very potent pharmacologically active agents that cause toxic reactions when people eat bread made from contaminated flour. Periodic epidemics of ergotism have occurred throughout history, and in medieval times these were often attributed to supernatural causes. The earliest reference to ergot seems to have been on an Assyrian tablet circa 6000 BC that refers to "a noxious pustule in the ear of grain." The Parsees, an ancient religious community in India, referred in their writings to noxious grasses that caused women to abort and to die shortly thereafter. The ancient Greeks escaped the scourge of ergotism because they never developed a taste for rye bread, which they referred to by a phrase that translates roughly as "that filthy Macedonian muck." Since rye bread did not reach Europe until after the decline of the Roman Empire, few if any references to ergotism exist in Roman writings.

There was an early association of ergot with St. Anthony of Egypt, who lived sometime between 250 and 350 AD. He is considered to be the founder of Christian monastic life and spent long sojourns in the desert, experiencing visions and hallucinations. These frequently involved attacks by Satan in various guises (wild beasts, soldiers, and women) that either physically attacked him or tempted him. Contemporary witnesses claimed that he behaved like an individual being physically abused. Because the signs and symptoms of ergot poisoning resembled his attacks, sufferers in the Middle Ages attached religious significance to them and they came to be known as "St. Anthony's Fire" (or sometimes "Holy Fire"). In 1100, the Monastery of the Hospitallers of St. Anthony was established at La Motte in France. It became the site of pilgrimages by those afflicted with this malady. The signs and symptoms included intense, burning pain in the extremities followed by a blackened, necrotic appearance; hence the association with fire. Epidemics of madness in the Middle Ages may also have been the result of ergotism.

Ergot was employed as an abortifacient long before it was known to be the cause of St. Anthony's Fire, and this use continued well into the nineteenth century when it was finally abandoned because of its highly toxic nature. One of the components of ergot is still employed to stop *post partum* hemorrhage because it constricts blood vessels and contracts the uterus. To summarize, the symptoms of ergotism include drowsiness, nausea, vomiting, muscle twitch, staggering, gangrene, hallucinations, and abortion. Ergot contains a veritable potpourri of pharmacological agents that were characterized by Sir Henry Dale in the early 1900s and that contributed to the development of many new drugs.

The active components of ergot, the ergot alkaloids, are all derivatives of lysergic acid. They bear some molecular resemblance to adrenaline (epinephrine), dopamine, and serotonin, all central neurotransmitters. Methysergide is the precursor of LSD. Two derivatives still have medicinal application. Ergometrine (ergonovine) is the one used to control *post partum* bleeding. Ergotamine is a powerful vasoconstrictor responsible for the gangrene of ergotism. In very small doses it may be used to treat migraine. Ergotism may also affect farm livestock, causing similar signs and, in addition, loss of milk production in cattle and necrotic combs, feet, and beaks in poultry.

Aleukia

Aleukia (literally "absence of white cells") occurs when millet and other grains that have become moldy are consumed. The toxin is a trichothecene from species of *Fusarium* and it damages bone marrow. Because of severe food shortages in Russia during World War II, the eating of moldy grain resulted in several epidemics of aleukia including a very large one in 1944. The mortality rate of those with marrow damage was 60%. Other symptoms included hemorrhages in the skin and mucous membranes. The condition also is referred to as alimentary toxic aleukia.

Some Specific Mycotoxins

Aflatoxins

Aflatoxins are a family of heterocyclic, oxygen-containing compounds secreted by the molds *Aspergillus flavus* and *Aspergillus parasiticus*. These molds grow abundantly on many kinds of plants in very hot conditions. Peanuts are especially prone to infection. Peanuts and maize (corn) constitute the predominant food items in most developing countries. Taste is not affected and the toxins are heat stable. Aflatoxin B_1 (AFB$_1$) is the most frequently encountered member of the group and it is a potent carcinogen in experimental animals (rodents, birds, and fish). Rats fed a diet containing 15 ppb of AFB$_1$ develop hepatitis that is often followed by cancer of the liver. Epidemiological studies of populations in Uganda, Kenya, and Thailand have shown a close correlation between the incidence of liver cancer and the consumption of food containing aflatoxins. It is estimated that the risk factor for liver cancer is increased 10-fold by AFB$_1$. Infectious hepatitis also increases the risk 10-fold. If present together, the risk is increased 100-fold. Other aflatoxins include B_2, G_1, and G_2. Lactating cattle excrete hydroxylated metabolites M_1 ad M_2 in milk. They have lower toxicity than the parent compounds but are of concern because of the large amount of milk consumed by infants and young children.

Thus this natural carcinogen, at least as potent as TCDD in animal studies, is a confirmed cause of human cancer, unlike the latter synthetic agent, that has received much more media coverage. Experiments employing human liver microsomal preparations indicate that cytochrome P450 (Cyp) 3A4 is largely responsible for the formation from AFB_1 of the 8,9 endo-epoxide, a genotoxic metabolite. Cyp 1A2 also forms this epoxide but is much less active. It also forms other nongenotoxic metabolites and thus may play a role in detoxification. An inhibitor of Cyp 1A2, alpha naphthoflavone, increased the formation of the 8,9-epoxide. The exo-epoxide of AFB_1 (i.e., preformed before ingestion) and the endo-epoxide are detoxified by conjugation with glutathione. Experimental evidence suggests that species differences in toxicity are due to differences in the rate and extent of conjugation with glutathione. Whereas rat liver cytosol conjugated endo-epoxide, mouse liver cytosol conjugated the exo-epoxide almost exclusively and both were much more active than the cytosol of human liver cells. The mucus cells of the intestine also make the 8,9-epoxide, but because they are sloughed frequently, this does not comprise a risk factor for cancer.

A 2012 review by Turner et al. confirms that the metabolite 8,9-epoxide is highly reactive and forms covalent bonds with proteins and nucleic acids. The epoxides are probably responsible for the genotoxic carcinogenicity of aflatoxins. There is not likely any safe exposure level.

Other aflatoxins found together are B_2, G_1, and G_1. Acute toxicity also can occur in humans. In 1974, an outbreak in India resulted in about 100 deaths and in Kenya 12 died in an outbreak in the early 1980s.

The aflatoxin story began in 1960, with a serious and mysterious outbreak of a disease in turkeys in Great Britain. Turkey poults developed loss of appetite, feeble fluttering, lethargy, and they frequently died in a few days. Necropsies revealed hemorrhage and necrosis of the liver and kidney. The disease was dubbed "X" disease. Outbreaks also occurred in ducklings in Europe, the United Kingdom, and Africa. The common denominator was groundnut (peanut) meal used as feed and subsequently shown to be contaminated with *A. flavus*. Livestock (mammals) may also be affected. Death has occurred in humans consuming an estimated 6 mg/day of aflatoxin B_1. Growth faltering in West Africa cannot be fully explained by poor nutrition and mycotoxins are suspect.

In North America there is concern about the long-term effects of consuming low levels of aflatoxins and monitoring systems are in place. Fortunately, the mold is not adapted to the colder northern climate so that Canadian-grown peanuts are free of the toxins. The majority of peanuts and peanut butter are still imported from subtropical climes, however. The possibility exists that climate change may alter the geographic distribution of the mycotic organisms. Peanuts are not the only source of these toxins. Because of drought conditions in the United States in the summer of 1988, up to 30% of the corn

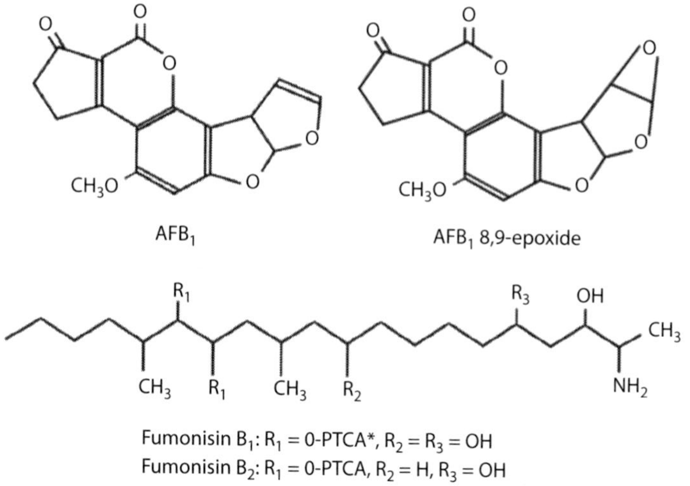

FIGURE 10.1
Chemical structures of aflatoxins and fumonisins. *PTCA, propanetricarboxylic acid.

crop may have been contaminated. The Quaker Oats Co. was turning away almost one truckload in five at its Cedar Rapids plant in the fall of 1988. The company tests six samples from each truck. Inspections of the 340,000 ton of corn crossing the Canada-U.S. border annually were stepped up in 1988–1989 by Agriculture Canada. It has been noted that climate change will undoubtedly shift the geographic distribution of mycotic organisms and hence of the health and agricultural risks of mycotoxins. The general structure of aflatoxins is shown in Figure 10.1.

Because of the ubiquity of aflatoxins, the presence of AFB_1 alone is not a good indicator of exposure level. Serum aflatoxin-albumin is useful in this regard as is the presence of metabolites in urine as well as AFB_1-DNA depurination product.

Fumonisins

Produced by *Fusarium moniliforme* and *Fusarium proliferatum*, these mycotoxins are ubiquitous in many parts of the world, including South Africa, where they were first identified, and many states bordering the Great Lakes. All of the fumonisins (FB_1, FB_2, and FB_3) are potentially carcinogenic. FB_1 appears to account for about 70% of all fumonisins with the order of frequency being $FB_1 > FB_2 > FB_3$. Fumonisins do not seem to undergo significant metabolism so that metabolites cannot be used as biomarkers.

Many experts feel that it will become the most significant mycotoxin for human health. FB_1 has been shown to be carcinogenic for animals (a promoter and initiator of liver cancer in rats), and there is a high degree of correlation between the incidence of esophageal cancer in humans and the presence of

FB$_1$ in corn in specific areas of South Africa. The fungus infects corn, millet, sorghum, and rice around the world. Fumonisins are toxic for many species of animals, especially horses, and outbreaks have caused numerous deaths in horses and swine in Texas, Iowa, and Arizona. Severely infected corn cobs may contain up to 900 mg/kg. Symptomatology in various species is as follows:

- *Swine*: Vomiting, convulsions, sudden death, abortion, pulmonary edema (porcine pulmonary edema syndrome). Symptoms may occur at levels above 20–50 mg/kg of FB$_1$.
- *Poultry*: Ataxia, paralysis, sudden death, stunted growth. Toxicosis occurs at levels of contamination of 10–25 mg/kg FB$_1$.
- *Cattle*: Poor weight gain, liver damage.
- *Horses*: Equine Leukoencephalomalacia (ELEM) is the disease caused by fumonisins. Brain degeneration occurs with focal necrosis. Symptoms may involve blindness, wild behavior, liver damage, staggering, ataxia, etc. and may occur at levels of 10 mg FB$_1$/kg for over 40 days.

FB$_1$, FB$_2$, and hydrolyzed FB$_1$ have been shown to be specific inhibitors of *de novo* sphingolipid synthesis and sphingolipid turnover. FB$_1$ has been shown to inhibit sphingosine (sphinganine) *N*-acetyltransferase, leading to the accumulation of sphyngoid bases. This has been shown to stimulate DNA synthesis and it is hypothesized that this interference with normal cell function could account for the toxicity of fumonisins.

The fumonisins are themselves analogs of sphingosine, with structural similarities to the phorbol esters, which are known carcinogens (see Chapter 11). The general structure of fumonisins is shown in Figure 10.1.

In one study of women in Mexico, urinary FB$_1$ levels were correlated with tortilla (corn) intake. Fifty-six (74.6%) of the 75 women in the study had detectable levels (>20 pg/mL) and it was most frequently detected in the high (96%) consumption group compared to the medium (80%) and low (45%) consumption groups. Means for the three groups were 145, 108, and 35 pg/mL and the findings were statistically significant. Other studies reviewed by Turner et al. supported the contention that urinary levels of FB$_1$ are a useful measure of fumonisin exposure. Fumonisins have also been implicated in growth faltering in infants exposed to high levels. Neural tube defects in experimental animals have raised concerns about exposure *in utero*.

Ochratoxins

Ochratoxins A, B, and C are formed by *Aspergillus* and *Penicillium* species. Ochratoxin A (OTA) is by far the most prevalent. It has been shown to be embryotoxic, carcinogenic, and teratogenic in several laboratory species of mammals (pig, dog, mouse, rat) and birds and it is therefore viewed as a potential and probable human teratogen and carcinogen. Acute effects in animals (including swine and poultry) include renal and hepatic destruction. Both humoral and cellular immune systems are adversely affected. Neuronal damage has been demonstrated experimentally. Contamination of bread and cereals has been documented in parts of central Europe (Yugoslavia, Bulgaria, Poland, and Germany) and levels have been detected in human milk, urine, blood, and kidneys. Foodstuffs deemed vulnerable to OTA contamination in regions where it is common include wine, beer, coffee, dried vine fruit, grape juice, pork, poultry, dairy, spices, and chocolate. Although levels in an individual food may be low, its ubiquity means that it is possible for exposed individuals to accumulate significant amounts. It is one of 20 mycotoxins monitored in foods.

OTA has a long serum half-life in humans due to binding to serum macromolecules such as albumin. Biotransformation by, and induction of, CP450 enzymes and enzymatic hydrolysis are believed to play a role in toxicity and DNA adducts occur and oxidative stress is also thought to be involved in toxicity. Several epidemiologic studies indicate the involvement of OTA in human nephropathies including Balkan endemic nephropathy (a severe, progressive, and ultimately fatal renal disease) and chronic interstitial nephropathy in North Africa. A study in Sri Lanka found that over 93% of patients with renal disease had urinary levels of OTA. In a study of serum levels, persons with kidney disease had levels ranging from 35 to 100 ng/mL (ppb) compared to European levels of 0.1 to 2.0 ng/mL. In one case, acute renal failure occurred in two individuals who had been exposed to inhaled OTA for 8 h. One of the couple, the woman, eventually required treatment for 40 days before recovery occurred. The postproximal nephron and proximal tubule are the targets of OTA toxicity. Renal cancer is also associated with OTA exposure.

Concerns for human health have focused mainly on dietary exposure to OTA. There is, however, evidence that inhalation also constitutes a health risk. Water-damaged buildings may become fertile grounds for the growth of the *Aspergillus* and *Penicillium* spp. that produce OTA. Elevated urinary levels of OTA have been reported in people exposed to such buildings. Dust and aerosol samples from Norwegian cowsheds have also contained OTA at levels ranging from 0.2 to 70 µg/kg (ppb). Systemic appearance of OTA after inhalation is rapid and suggests 98% bioavailability.

In conclusion it is clear that OTA constitutes a significant health threat in areas where it is prevalent. A study by Health Canada concluded that exposure to OTA from Canadian foods did not constitute a significant risk factor for renal tumors.

Patulin

Patulin is potentially a carcinogenic toxin produced by several species of fungi including some *Penicillium* spp. It is a highly potent inhibitor of RNA polymerase, having a strong affinity for sulfhydryl groups. It therefore inhibits many enzymes. Teratogenicity has not been demonstrated in mammals but embryolethality occurs at higher doses (2 mg/kg intraperitoneally). A common source of patulin is *Penicillium expansum*, a common spoilage microorganism in apples. Apple juice can sometimes contain significant amounts of patulin. Acute toxicity in rodents is manifested largely as gastrointestinal symptoms, including hemorrhage. Carcinogenicity studies in rats were negative, but clastogenic activity has been shown in some systems. T-2 toxin is produced by various *Fusaria* and is both potent and common. It inhibits protein and DNA synthesis and it is therefore potentially teratogenic and carcinogenic. The genotoxicity of patulin has been postulated to be due to structural DNA damage by cross-linking forming nucleoplasmic bridges. Patulin has a strong affinity for sulfhydryl groups and depletion of glutathione has been shown experimentally to increase toxicity.

Fungi tend to produce mixtures of toxins so that exposure to a single agent is unlikely to occur. There is some experimental evidence that ochratoxin A potentiates the teratogenic effects of T-2 (see in the following). It must be emphasized that any mycotoxin that has been shown to be toxic in several mammalian species (including various farm livestock) must be regarded as potentially toxic for humans.

Fusarium Species

Fusarium graminearum is the fungus responsible for maize ear rot in corn and head blight in wheat. Its life cycle is typical of all *Fusaria*. Spores survive in crop debris from the previous season (stubble, stalks, and seeds) to reinfect the next year's crop. Intensive farming practices that involve planting susceptible species in the same fields year after year thus favor the spread of infection. Birds such as starlings and red-winged blackbirds, which puncture the corn kernels to eat the milk, can also spread spores. Insect such as the picnic or corn-sap beetle seek out damaged kernels and also may spread spores.

Certain weather conditions favor the spread of infection. Fungus growth is favored by warmth (15°C–35°C) and by surface wetness for more than 48 h. After the infection is established, weather is not critical to the production of the toxin. Mold growth will continue throughout the season, and even afterward if not properly dried or if storage conditions are poor (too damp, too warm, poor air circulation). Mold growth can even occur during feed preparation and in poorly cleaned feed troughs. Late harvest may allow the growth of another fungus, *F. sporotrichioides*, which produces the toxins T-2, HT-2 and diacetoxyscirpenol. Contaminated grains thus will contain complex mixtures of toxins and metabolites.

Fusaria are capable of producing a number of tricothecene mycotoxins that are of economic and health importance. These are discussed subsequently. Tricothecenes, however, are produced by several mycotic organisms including *Tricothecium, Memnoniella,* and some species of *Trichoderma* of special importance is *Stachybotrys chartarum*. This is the agent that forms black mold in dwellings that have become waterlogged. Black mold is associated with several health problems including respiratory ones and is a suspected causative agent of idiopathic pulmonary hemosiderosis in infants. *S. chartarum* produces two groups of volatile, cytotoxic mycotoxins that most likely contribute to their respiratory toxicity. These are sesquiterpenes and the C_7–C_8 oxygenated compounds. One sesquiterpene, trichodiene, is an intermediate in the synthesis of tricothecenes. It is of potential use as a marker to identify the presence of trichothecene mycotoxins. In addition, *S. chartarum* mold produces a potpourri of highly toxic, macrocyclic tricothecenes. Some of these tricothecenes are discussed subsequently. They are produced mostly by the *Fusaria* (see life cycle discussed earlier) and are of particular importance to agriculture. They also have potential consequences for human health.

Zearalenone

This nonsteroidal, estrogen-like toxin causes (in swine) swollen, red vulva, vaginal and rectal prolapse, vulval enlargement in piglets, and fertility problems. Developmental defects and lethality have been shown in some laboratory species. While direct evidence of toxicity to humans is lacking, zearalenone can be added to the growing list of endocrine disrupters that pose a potential health threat to people. The hormonal growth promotant zerenol used in beef cattle is chemically related to zearalenone and may be a metabolite of it. Cases of precocious puberty associated with both naturally occurring and semisynthetic hormones are discussed under Growth Promotants in Chapter 8.

Vomitoxin (Deoxynivalenol or DON)

This trichothecene causes decreased feed intake and reduced weight gain in pigs at about 2 mg/kg of feed, vomiting and refusal of feed at very high concentrations (>20 mg/kg feed). DON will be used as a "prototype" mycotoxin to illustrate agricultural problems associated with these agents (see in the following).

Species Differences in DON Toxicokinetics: Swine appear to be much more sensitive to the anorexic and weight loss effects of DON than ruminants (cattle, sheep) or poultry, which are very tolerant. In one study, laying hens actually preferred a diet containing 5 ppm of DON in preference to clean feed. These differences are due in part to differences in absorption and in part to differences in biotransformation and elimination. Studies with radiolabeled DON indicated that sheep absorbed 9% or less of an orally administered dose, in turkeys 20% or less was absorbed, whereas pigs absorbed up to 85% of a single oral dose.

Intravenous administration of radiolabeled DON in sheep revealed an initial distribution phase (T1/2 = 18 min) followed by an elimination phase (T1/2 = 66 min). A glucuronide conjugate was formed and comprised 15%–20% of plasma levels. In turkeys, there was an extremely rapid distribution phase (T1/2 = 3.6 min), a rapid elimination phase (T1/2 = 46 min), and the formation of a conjugate (probably glucuronide) comprising up to 10% of the total dose. Again, swine showed a much different picture. There was a very rapid distribution phase (T1/2 = 5.8 min), a secondary slower distribution phase (T1/2 = 96.7 min), and a very prolonged terminal elimination phase (up to 510 min). There was no evidence of significant biotransformation in swine. Thus, it took seven to ten times longer for swine to clear the toxin than for the other two species. Toxicity is a function of many factors:

- The concentration of toxin reaching the target organ (which in turn is affected by the rate of absorption at the portal of entry)
- The extent of distribution to nontarget sites (i.e., where no toxic effects occur)
- The rate and extent of biotransformation to nontoxic metabolites (or to toxic ones as the case may be)
- The rate of elimination in urine and feces

The effect of species differences in some of these factors was introduced in Chapter 2. Volume of distribution (Vd) and clearance data provide some information regarding the fate of the absorbed toxin. The apparent Vd is a mathematical calculation of the volume of diluent required to dilute an administered dose of a substance (usually intravenously) to the observed concentration. Vd = M/C where M = mass (amount of substance) and C = concentration of substance. Calculations of Vd for DON yielded values of 0.167 L/kg for sheep versus 1.3 L/kg for swine, suggesting that in the former, DON was confined mainly to the extracellular fluid, whereas in the latter it was taken up by tissues. Initial systemic clearances were not all that different, being 1.37 mL/min/kg for sheep and 1.81 mL/min/kg for swine. An interpretation of this data suggests that DON is initially rapidly distributed to tissues and then slowly released back into the plasma, yielding the slow, terminal elimination phase. Turkeys also had a very large Vd (2.33 L/kg), but they also had an extremely rapid clearance (35.0 mL/min/kg), indicating that DON was rapidly distributed to tissue compartments but not held there.

Thus, the extreme sensitivity of swine to DON is the result of

- High oral bioavailability
- Wide distribution to tissues
- Slow elimination from the body
- Minimal detoxification through biotransformation

There is concern that DON, like other mycotoxins, may be a factor in growth faltering in human infants in highly contaminated areas.

Other Tricothecenes

T-2 and HT-2 toxins and diacetoxyscirpenol are more toxic than DON and cause reduced feed intake, vomiting, irritation of the skin and gastrointestinal tract, neurotoxicity, teratogenicity, impaired immune function, and hemorrhage. Adverse effects seen in farm animals are generally caused by mixtures of these toxins rather than by single ones. Blending of several grains in the preparation of feed may further contribute to the toxic diversity of the mixture. Potentiation of effects may occur. Thus, DON at the subthreshold level of 1 mg/kg plus low (ppb) concentrations of T-2 and other unidentified toxins may cause severe toxic manifestations in a sensitive species like swine. The chemical structures of some of these toxins are shown in Figure 10.2.

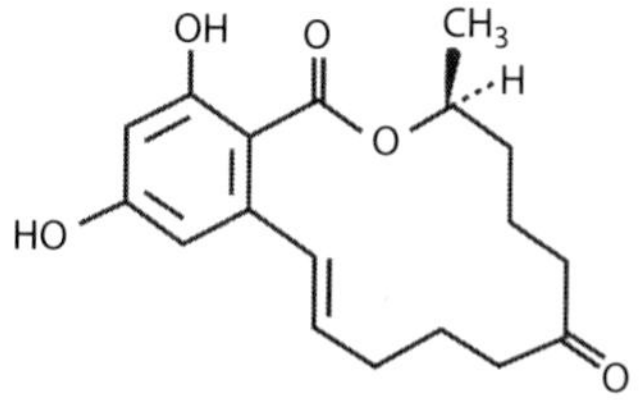

Zearalenone

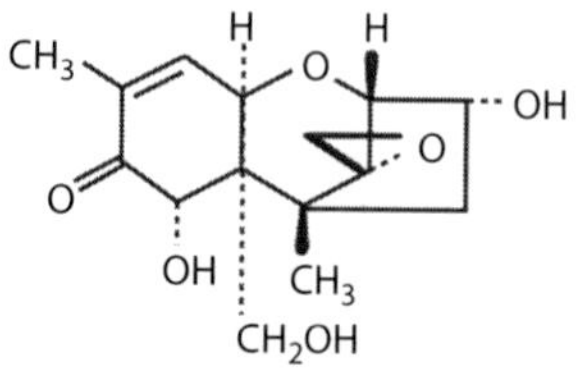

Deoxynivalenol (DON, vomitoxin)

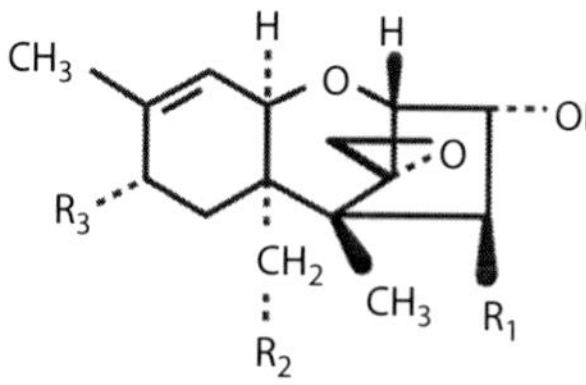

T-2: CH_3COO at R_1 and R_2, $(CH_3)_2CHCH_2COO$ at R_3

HT-2: OH at R_1, CH_3COO at R_2, $(CH_3)_2CHCH_2COO$ at R_3

Diacetoxyscirpenol (DAS): CH_3COO at R_1 and R_2, H at R_3

FIGURE 10.2
Chemical structures of several mycotoxins.

Economic Impact of Mycotoxins

In addition to their direct effects on human health, mycotoxins have a tremendous impact on agriculture. Through spoilage, field crops are rendered useless for animal or human consumption. (The loss of 20% of the corn crop noted earlier is one example.) Poor weight gain and outright illness occur in livestock that consume contaminated feeds. Losses are difficult to estimate but they undoubtedly run to many millions and possibly billions of dollars in North America. The presence of trace quantities in meat, dairy products, and eggs constitutes a further, if largely unconfirmed, health hazard to people. In the Great Lakes basin, various species of *Fusarium* are the most common offenders, especially in eastern Canada. They may produce a host of toxins with potent pharmacological actions.

There is a condition of rice called rice blast disease. It is caused by the fungus *Magnaporthe oryzae*. It forms a dome-shaped cell called an appressorium that invades the plant tissue, destroying the plant or disrupting grain formation. It is estimated that 30% of the rice crop worldwide is lost annually due to this infection. That is, enough rice to feed 60 million people. Food crop destruction by mycotic organisms is a major threat to the world food supply.

Detoxification of Grains

Because of the diverse chemical properties of the mycotoxins, physical and chemical procedures that are effective against one toxin may have little or no influence on the toxicity of others. Thus, there is no single process that can be utilized. The most important control factor must be the avoidance of conditions favoring fungal growth at all stages of food production.

Harvesting and Milling

Infected kernels may represent less than 5% of all the grain. They may be broken or shriveled and in wheat may take on a "tombstone" appearance. In corn, the tips of cobs may have shriveled, highly infected kernels containing up to 3000 mg/kg of DON. Grain dust may be very contaminated. Screening and blowing will remove much of the dust, particles, and withered kernels.

Wet milling of corn has been shown to remove about 2/3 of T-2 toxin. But milling had little effect on the DON content of flour from hard wheat, nor did baking the flour into bread. Some milling procedures may actually increase the DON content of the finished product. In mild infections, washing and roasting may significantly reduce toxin levels.

Chemical Treatments

Laboratory tests have shown that moist ozone, ammonia, microwaving, and convection heating reduce DON concentrations in moldy grain. Aqueous sodium bisulfite plus heat effected a complete detoxification. Studies have shown that this technique resulted in normal feed intake and weight gains when contaminated corn was treated and fed to swine.

Binding Agents

The addition of binding agents such as bentonite, anionic and cationic resins, and vermiculite–hydrobiotite were tested on the toxicity of T-2 in rats. Bentonite prevented T-2 toxicosis by blocking intestinal absorption. Polyvinylpyrrolidone or ammonium carbonate had no effect on DON toxicity in swine. Alfalfa fiber has been shown to partially overcome the growth-depressing effect of zearalenone in rats but not the estrogenic effects in swine.

Other Techniques

Dilution of contaminated feed with clean feed will improve palatability and feed consumption. More concentrated diets with respect to calories, protein, etc. may overcome the effects of a moderate reduction in feed intake. Experimentally, antibodies against zearalenone have been raised in swine and shown promise in protecting against its toxic effects.

Regarding species differences in capacity to biotransform xenobiotics, evolutionary factors have played a role. Given the plethora of toxic substances in the plant kingdom, pure herbivores would have been exposed to the greatest risk of poisoning as compared to carnivores or omnivores and would therefore have needed to evolve nonspecific detoxifying enzymes. The toxin most likely encountered by pure carnivores, however, would have been botulinum toxin, and their need would have been to evolve a method of eliminating it quickly before this potent neurotoxin could paralyze movement and respiration. Thus, canines and felines are very poor metabolizers of drugs (e.g., aspirin can be extremely toxic to cats and dogs) but they have exquisitely sensitive vomiting mechanisms to eliminate bad meat before the toxin is absorbed. Conversely, herbivores are efficient metabolizers but generally lack good vomiting reflexes. Emesis does not occur in equines, ruminants, or rodents. Omnivores, like humans and swine, fall in between. The importance of diet in the evolution of detoxifying systems applies even to primates. Despite their closeness to us on the phylogenetic tree, fruit- and plant-eating primates are generally more efficient metabolizers of xenobiotics than we are. In further support of this theory, it can be pointed out that the $t_{1/2}$ for amphetamine is about 86 min in both the rabbit and the horse but is 390 min for the cat and 300 min for humans.

The question of these mycotoxin residues in human food sources remains largely unanswered, but studies indicate that DON residues are not a problem. Feeding very high levels of DON to dairy cattle resulted in only trace quantities in milk, and when fed to poultry, no appreciable tissue residues were measured. Again, this is the consequence of the species pharmacokinetic characteristics.

Other Toxins in Unicellular Members of the Plant Kingdom

Many soil organisms produce substances that are toxic to others, and a continual state of chemical warfare exists in the microenvironment. These organisms provide the source of all of our antibacterial and antifungal antibiotics, many of which have significant mammalian toxicity. Many others were tested and discarded because of their high toxicity. Some antibiotics are teratogenic and are used in the treatment of cancer because of their effects on cell reproduction (e.g., actinomycin D, doxorubicin, adriamycin). Some organisms may be responsible directly for poisonings in humans.

The blue-green algae called *Cyanobacteria* produce pentapeptides that are hepatotoxic and that have caused numerous deaths when they contaminate drinking water. Some strains of that ubiquitous organism *Staphylococcus aureus* are capable of producing a protein enterotoxin, 1 µg of which can induce vomiting, severe colic, and profuse diarrhea. The bacteria are usually introduced to foods from infected handlers and they proliferate in warmth and especially in creamy foods (cream pies, salad dressings, etc.).

The most potent toxin known is the protein toxin from *Clostridium botulinum*, 1 µg of which may be fatal to a human. Lab animals may show symptoms at 10^{-6} µg/kg. Botulinum toxin blocks the release of acetylcholine from peripheral nerve endings.

Review Questions

For Questions 1–10 use the following code:

    Answer A if statements a, b, and c are correct.
    Answer B if statements a and c are correct.
    Answer C if statements b and d are correct.
    Answer D if statement d only is correct.
    Answer E if all statements (a, b, c, d) are correct.

1. Regarding mycotoxins, which of the following statements is/are true?
   a. *Fusarium* species produce numerous mycotoxins that are hazardous for many species.
   b. Deoxynivalenol (DON) causes fertility problems in swine.
   c. DON causes vomiting and poor weight gain in swine.
   d. Other "trichothecenes" are less toxic than DON.
2. Which of the following statements is/are true?
   a. All varieties of wheat and corn are susceptible to fungal infections in varying degrees.
   b. Rapid drying of grain minimizes the risk of postharvest mold growth.
   c. Mycotoxin toxicity in livestock usually results from mixed fungal contamination.
   d. There is no practical way of detoxifying corn and wheat contaminated with mycotoxins.
3. Which of the following statements is/are true?
   a. There is no evidence that fumonisins are carcinogenic for humans.
   b. Zearalenone has estrogen-like activity.
   c. Aflatoxins are not toxic for poultry.
   d. Ochratoxin causes renal damage in swine and poultry.
4. The symptoms of ergot poisoning
   a. Are similar for humans and animals
   b. Include nausea, dizziness, burning pain in the extremities, and abortion
   c. Arise largely from consuming rye contaminated with ergot alkaloids
   d. Are caused by ergot alkaloids produced by *Fusarium moniliforme*
5. Which of the following statements is true regarding Fumonisin B_2 (FB_2)?
   a. It is strongly suspected of being a carcinogen for humans.
   b. It produces brain damage in horses.
   c. Horses are very susceptible to FB2 toxicity.
   d. In swine, the signs of toxicity are similar to those of DON.
6. Regarding ergot alkaloids:
   a. They are derivatives of lysergic acid.
   b. Chemically they resemble catecholamine neurotransmitters.
   c. They are chemically related to lysergic acid diethylamide (LSD).
   d. They may produce hallucinations.

7. Which of the following statements is/are true regarding ergot alkaloids?
   a. Ergometrine (ergonovine) causes contractions of the gravid uterus.
   b. Ergometrine has never had any medical application.
   c. Ergotamine in low doses has been used to treat migraine headaches.
   d. Ergotamine is nontoxic except at very high doses.

8. Regarding aflatoxin:
   a. The mold that produces them is very hardy and survives in cold climates.
   b. They are oxygenated heterocyclic compounds.
   c. They cause hepatic damage but are not carcinogenic.
   d. Epidemiological studies have shown a high degree of correlation between liver cancer and the consumption of foods contaminated with aflatoxin B_1.

9. Regarding detoxification of grains contaminated with mycotoxins:
   a. A technique that works on one mycotoxin may not work on others.
   b. DON concentration may be reduced by heating.
   c. Binding agents may reduce toxicity of T-2.
   d. Moist ozone and ammonia have reduced DON concentrations in lab tests.

10. Regarding DON toxicity:
    a. Swine are most susceptible than poultry or ruminants.
    b. Turkeys form a conjugate and rapidly eliminate DON.
    c. Sheep form a glucuronide conjugate.
    d. A high volume of distribution (Vd) is generally associated with higher toxicity in that species.

11. Match the statement with the correct toxin or effect.
    a. Cyanobacteria cause this
    b. Blocks acetylcholine release from peripheral nerve endings
    c. Adriamycin
    d. Sta*phylococcus aureus*
       i. Produces a protein enterotoxin
       ii. Produces hepatotoxic pentapeptides
       iii. A teratogen used in cancer chemotherapy
       iv. Botulinum toxin

12. Answer the following: True or false:
    a. Aflatoxin B1 8,9-epoxide is genotoxic.
    b. Aflatoxin B1 epoxides are formed exclusively after ingestion of AFB1.

c. Conjugation with glutathione is an important means of detoxifying AFB1 epoxide.

d. The 8,9-epoxide is formed exclusively by Cyp 1A2.

e. Cyp 3A4 is the main source of 8,9-endo-epoxide.

f. Species differences in the rate and extent of glutathione conjugation of 8,9-epoxide may account for differences in toxicity.

g. Several mycotoxins are implicated in growth faltering in human infants.

h. Ochratoxin causes renal disease but not kidney cancer.

i. After a mold infection is established in a crop, weather is not a critical factor in its growth.

Answers

1. B
2. A
3. C
4. A
5. A
6. E
7. B
8. C
9. E
10. A.
11. (i) d, (ii) a, (iii) c, (iv) b.
12. a. True
    b. False
    c. True
    d. False
    e. True
    f. True
    g. True
    h. False
    i. True

Further Reading

Dagas, Y.F., Yoshino, K., Dagas, G., Ryder, L.S., Bielska, E., Steinberg, G., and Talbot, N.J., Septin-mediated plant cell invasion by the rice blast fungus, *Magnaporthe oryzae*, *Science*, 336, 1590–1595, 2012.

Duarte, S.G., Lino, C.M., and Pena, A., Food safety implications of ochratoxin A in animal-derived food products, *Vet. J.*, 192, 266–292, 2012.

Eaton, L. and Gallagher, E.P., Mechanisms of aflatoxin carcinogenesis. *Ann. Rev. Pharmacol. Toxicol.*, 34, 135–172, 1994.

Frizzell, C., Ndossi, D., Verhaehen, S., Dahl, E., Eriksen, G., Sorlie, M., Ropstad, E., Muller, M., Elliott, C.T., and Connolly, L., Endocrin disrupting effects of zeara-lenone, alpha- and beta-zearalenol at the level of nuclear receptor binding and steroidogenesis, *Toxicol. Lett.*, 206, 210–217, 2012.

Glasser, N. and Stopper, H., Patulin: Mechanism of genotoxicity, *Food Chem. Toxicol.*, 50, 1796–2801, 2012.

Haighton, L.A., Lynch, B.S., Magnuson, B.A., and Nestmann, E.R., A reassessment of risk associated with dietary intake of ochratoxin A based on a lifetime exposure model, *Crit. Rev. Toxicol.*, 42, 147–168, 2012.

Hope, J.H. and Hope, B., A review of the diagnosis and treatment of ochratoxin A inhalation exposure associated with human illness and kidney disease includ-ing focal segmental glomerulosclerosis, *J. Environ. Public Health*, 2012, Article ID: 835095.

Marasas, W.F., Fumonisins: Their implications for human and animal health, *Nat. Toxins*, 3, 193–198, 1995.

Nair, M.G., Fumonisins and human health, *Ann. Trop. Paediatr.*, 18 (Suppl), S47–S52, 1998.

Pitt, J.I., Toxigenic fungi and mycotoxins, *Br. Med. Bull.* 56, 184–192, 2000.

Prouillac, C., Koraichi, F., Videmann, B., Mazallon, M., Rodriguez, F., Baltas, M., and Lecoeur, S., In vitro toxicological effects of estrogenic mycotoxins on human placental cells: Structure activity relationships, *Toxicol. Appl. Pharmacol.*, 259, 366–375, 2012.

Raney, K.D., Meyer, D.J., Ketterer, B., Harris, T.M., and Guengerich, F.P., Glutathione conjugation of aflatoxin B1 exo- and endo-epoxides by rat and human glutathione-S-transferases, *Chem. Res. Toxicol.*, 5, 470–478, 1992.

Raney, K.D., Shimada, T., Kim, D.H., Groopman, J.D., Harris, T.M., and Guengerich, F.P., Oxidation of aflatoxins and sterigmatocystin by human liver microsomes: Significance of aflatoxin Q1 as a detoxication product of aflatoxin B1, *Chem. Res. Toxicol.*, 5, 202–210, 1992.

Schroeder, J.J., Crane, H.M., Xia, J., Liotta, D.C., and Merrill, A.M., Disruption of sphingolipid metabolism and stimulation of DNA synthesis by fumonisin B_1, A molecular mechanism for carcinogenesis associated with *Fusarium moniliforme*, *J. Biol. Chem.*, 269, 3475–3481, 1994.

Shepphard, G.S., Theil, P.G., Stockenstrom, S., and Sydenham E.W., World-wide sur-vey of fumonisin contamination of corn and corn-based products, *J. AOAC Int.*, 79, 671–687, 1996.

Shier, W.T., Sphingosine analogs: An emerging new class of toxins that includes the fumonisins, *J. Toxicol.*, 11, 241–257, 1992.

Turner, P.C., Flannery, B., Isitt, C., Ali, M., and Pestka, J., The role of biomarkers in evaluating human health concerns from fungal contaminants in food, *Nutr. Res. Rev.*, 25(1), 162–179, 2012.

Ueng, Y.F., Shimada, T., Yamazaki, H., and Guengrich, F.P., aflatoxin B1 by bacterial recombinant human cytochrome P450 enzymes, *Chem. Res. Toxicol.*, 8, 218–2235, 1995.

WHO, Patulin. In *Evaluation of Certain Food Additives and Contaminants. 37th Report of the Joint FAO/WHO Expert Committee on Food Additives*. WHO, Geneva, Switzerland, 1990, pp. 29–30.

WHO, Ochratoxin A. In *Evaluation of Certain Food Additives and Contaminants. 37th Report of the Joint FAO/WHO Expert Committee on Food Additives*. WHO, Geneva, Switzerland, 1991, pp. 29–31.

Wilkins, K., Nielsen, K.F., and Din, S.U., Patterns of volatile metabolites and nonvolatile trichothecenes: Produced by isolates of *Stachybotrys, Fusarium, Trichoderma, Trichothecium* and *Memnoniella, Environ. Sci. Pollut. Res.*, 10, 162–166, 2003.

11

Animal and Plant Poisons

He was a bold man that first ate an oyster.

Jonathan Swift

Massasauga rattlesnake eat brown bread.
Massasauga rattlesnake fall down dead.
If you catch a caterpillar give it apple juice,
But if you catch a rattlesnake, TURN IT LOOSE!

Old Ontario skipping rhyme

Introduction

Chemical warfare is widely practiced in the animal and plant kingdoms. Just as microorganisms produce antibiotics that inhibit the growth and reproduction of competing organisms, more complex plants synthesize chemicals that render them unpalatable to potential predators or that are truly toxic, thus selecting for individuals that avoid them, or producing aversive reactions that limit consumption to once or twice only. Occasionally, however, these plants accidentally enter the food chain of livestock or humans, directly or indirectly, leading to a toxic reaction.

Similarly, toxins and venoms are used in the animal kingdom for defense and prey capture. Toxins are consumed and operate much like those of plants. Toxins can also be secreted by special cells or glands in the skin, so that toxicity may occur simply by taking the intended victim into the mouth. Toxic reactions have occurred among students indulging in the fad of "toad-licking." These toxins tend to be low molecular weight peptides with neurotoxicity. Some fish actually swim in a "cloud" of toxin secreted by these skin glands. Their toxins are usually steroid glycosides and choline esters. A neurotoxin secreted in this way by a species of flounder of the Red Sea is so powerful that a shark attempting to bite it is incapable of closing its jaws and instantly convulses. It has been proposed as a shark repellant for downed fliers and divers. These skin toxins also serve as antibacterial agents to prevent infection, as the slime secreted by the skin of amphibians is an ideal culture medium for bacteria.

Venoms are injected in some way, and may be employed both for defense and for prey capture. Humans and animals sometimes accidentally become the victims of envenomation or intoxication by poisons of animal origin. The term toxin is used also to refer to the individual components of venoms. Numerous texts have been written on these subjects, and this chapter can do no more than skim the field and discuss some of the more important, or more interesting, examples. Emphasis will be placed on agents that have become important to the biological sciences, either as drugs or as research tools.

Toxic and Venomous Animals

Toxic and Venomous Marine Animals

Venoms and toxins are distinguished by the manner in which they are inflicted upon the victim. A venom is a substance kept in a special poison gland and administered by a complex injection apparatus or by lacerating spines. This type of system may be employed in defense or in prey capture. A toxin is a poison usually ingested as an accidental component of tissues or organs. Toxins likely evolved as a species (rather than an individual) protection. Predators with a preference for that prey would be selected out if the toxin is fatal. Conditioned avoidance could occur in survivors. The term for poisoning by the muscle tissue of scale fish, as opposed to shellfish, is ichthyosarcotoxism. It includes toxins that are accumulated up the food chain from plankton as well as toxins that are synthesized by the fish itself. In the former situation, poisoning is usually by a mixture of toxins.

Scale Fish Toxins

Ciguatoxin

The condition ciguatera, or ciguatera fish poisoning (CFP) is caused by several related toxins that concentrate up the food chain from a photosynthetic dinoflagellate, *Gambierdiscus toxicus*, to reach toxic levels in large marine fish such as grouper, snapper, amberjack, barracuda, parrot fish, etc. These species are coral reef predators and browsers. CTX neurotoxins are far more potent than the brevetoxins associated with red tides. The LD_{50} for Ciguatoxin (CTX) is 0.25–4.0 µg/kg versus >100 µg/kg for brevetoxin. Toxic symptoms are gastrointestinal (nausea, vomiting, cramps, diarrhea), neurological (numbness, tingling of lips, throat and tongue, dizziness, headache), and cardiovascular (arrhythmias, cardiac arrest). CTX is thought to increase membrane permeability to sodium and it is responsible for the neurological signs and symptoms. There may be at least 20 different toxins involved in ciguatera fish poisoning. CTX is very toxic to mice that are the test species for detecting its presence. CTXs and brevetoxin are large, colorless, heat-stable molecules classified as ladder-like polyethers, the structure of which has now been elucidated. CTX is 3 nm long. Other unidentified toxins are likely involved in ciguatera poisoning. Ciguatera is by far the most common cause of scale-fish poisoning. An estimated 20,000 people are affected annually in subtropical and tropical climes. Any large, warm water species is a potential source of the toxin.

CTX and brevetoxin bind to a common site on voltage-sensitive sodium channels causing persistent activation and/or prolonged opening of the channel. A component of CTX, maitotoxin (MTX), has recently been purified. MTX is a water-soluble polyether that is a potent inducer of an increase in cytosolic free calcium ion $[Ca^{2+}]i$ in a wide variety of organisms including mammalian cells. MTX appears to activate a voltage-independent, nonselective cation channel that may require an extracellular source of Ca^{2+} for activation. MTX has become a useful research tool for studying ion channels. MTX is a large, complex, multiringed compound. The LD_{50} in mice is less than 0.2 µg/kg. This is at least five times more toxic than tetrodotoxin (see also http://www.rehablink.com.ciguatera/poison.htm). Barracuda also sometimes contain a related neurotoxin that is very toxic to cats. In some parts of the West Indies, it is customary to feed some of the meat from a large barracuda to a local cat before humans eat it.

Tetrodotoxin

The term comes from *Tetraodontidae*, meaning four teeth, the name of the genus that carries the toxin. At least six different neurotoxin receptor sites have been identified on the channel protein for voltage-gated ion channels using the various dinoflagellate neurotoxins. Tetrodotoxin (TTX) toxin is present in puffer fish, which are called "fugu" in Japan. It is a very potent blocker of fast sodium channels, and therefore inhibits nerve conduction in a manner like local anesthetics. It causes paresthesia, paralysis, anesthesia, and loss of speech, but consciousness is retained. Prognosis is improved if vomiting occurs early after ingestion. In Japan there are about 150 cases of poisoning annually with 50% mortality. Fugu is considered a delicacy in Japan and special chefs are licensed to prepare it. The toxin is concentrated in the liver and gonads of the fish. Connoisseurs of fugu prefer that just enough toxin remains to cause a slight tingling of the lips when it is consumed. In Japan, this custom has spawned a somewhat macabre poetry form: "Last night he and I ate fugu; today I helped carry his coffin."

TTX is also found in some newts and frogs. The blue-ringed octopus, a small octopus that inhabits tidal pools and shallow waters around Australia and other central Indo-Pacific waters, produces a toxin, administered by a bite that is believed to be identical to TTX. The bite of a larger specimen can be fatal in minutes. In recent years, the geographic and species distribution of

TTX has widened considerably. Species reported to carry TTX include gobies, gastropods, frogs, starfish, crabs, flatworms, and ribbon worms. Poisoning has occurred not only in Japan but in China, Taiwan, New Zealand, and even Europe. Whether this spread is related to climate change or to alteration of the marine ecology through human activity or is a natural phenomenon is unknown at this time. The toxin is not synthesized by the fish and other marine species but is concentrated up the food chain from marine bacteria. The spread of these bacteria, for whatever reason, could account for the increased number of species affected.

There is a fascinating theory that the zombie tales associated with Haiti arose from poisonings with TTX or from the deliberate of it by voodoo practitioners. Someone who has been poisoned with TTX can appear to be dead but later recover, but in a much impaired state.

TTX is of interest as a research tool because of its potent sodium channel blocking activity. Its occurrence in such diverse species can be explained by the fact that it is not synthesized by the animal itself, but rather by certain species of *Vibrio* bacteria that exist in a symbiotic relationship with the host.

Scombroid Toxins

The name "scombroid" comes from "*Scombridae*," referring to dark-muscled fish like tuna, albacore, and mackerel. The poisoning results from eating fish rich in histidine. However, according to information released by the (U.S.) Centers for Disease control (CDC), the most common vectors are not the *Scombridae*. Other fish like mahi-mahi and amberjack are frequently responsible. Improper refrigeration results in the decarboxylation of histidine to histamine by surface bacteria. The signs and symptoms resemble a histamine reaction. They may onset in minutes to hours and include dizziness, headache, diarrhea, facial flushing, tachycardia, pruritis, and wheezing. Fish contaminated in this way are often described as having a spicy or peppery taste. Levels of histamine in contaminated fish may exceed 100 mg/100 g. The FDA has set 50 mg/100 g as the hazard level. Histamine is not destroyed by cooking. There is some evidence that another product of decomposition, saurine, may contribute to the symptomatology. Treatment with histamine blockers (H1 and H2) is nearly always successful.

Ichthyotoxin

Ichthyotoxin is a term that is applied to a number of toxins produced by algae and concentrated by scale fish. Freshwater and ocean species are both affected. The chemical nature of many of these toxins remains largely unknown. Karlotoxin-2 is an ichthyotoxin from the dinoflagellate *Karlodinium veneficum*. It is believed to be responsible for numerous fish kills in an estuarine aquaculture facility in Maryland. During a 1996 fish kill of about 15,000 hybrid striped bass associated with an algal bloom, human health effects were also observed. These included skin lesions, respiratory difficulty, and neurological symptoms primarily consisting of short-term memory loss. Similar past events were generally attributed to the dinoflagellate *Pfiesteria piscicida*. A highly labile, copper-containing metallo-organic compound has been proposed as the putative *P. piscicida* toxin but despite countless fish kills in the region a definite link has not been established. The complex chemical nature of karlotoxin-2 has been elucidated. It kills fish in a dose-dependent manner. Karlotoxins were also shown to be cytotoxic and hemolytic.

Red Tide Dinoflagellate Toxicity for Higher Species

Numerous species of dinoflagellate algae can cause the red tides that periodically plague the shores of the Gulf of Mexico and elsewhere. The aerosol created by wave action can cause respiratory problems in people onshore. This can be serious in people with preexisting respiratory problems. Skin contact with the algae can cause an irritating rash. Other animals also can be affected, especially by the aerosol. In the northern Gulf of Mexico, there were 757 dolphin and whale strandings between February 2010 and June 17, 2012. An additional 123 bottle-nosed dolphins washed ashore along the Texas Gulf coast from November 2011 to March 2012. These strandings coincided with a red tide caused by the dinoflagellate *Karenia brevis*. Although cause and effect has yet to be proven, this temporal association makes the red tide highly suspect as the culprit.

Shellfish Toxins

Dinoflagellates produce a host of toxic substances with a wide array of toxic effects. Their chemical nature is often unknown. The subject of dinoflagellate toxicity was introduced earlier under CTX and Ichthyotoxins. They are also responsible for a variety of shellfish poisonings and the major ones will be reviewed here. An excellent review article by Wang identifies the dinoflagellate responsible for the various toxins. The majority act as neurotoxins.

Saxitoxins

Saxitoxin is the best known of a family of neurotoxins that cause paralytic shellfish poisoning and it concentrates in bivalves such as oysters, clams, and mussels. There have been at least 24 structurally related imidazoline guanidinium derivatives identified that are paralytic shellfish poisons. The mechanism and symptomatology are the same as for tetrodotoxin; they block voltage-gated ion channels (voltage-gated sodium channel 1) for sodium, calcium, and potassium. The LD_{50} for mice is 3–10 µg/kg intraperitoneally. The lethal oral dose for humans is 1–4 mg. "Never eat oysters in a month without an R" is an old adage that still is good advice in the Northern Hemisphere as dinoflagellates bloom in the summer months.

Brevetoxins

Brevetoxins (there are two main ones) have both gastrointestinal and neurological toxicity and has been responsible for numerous fish kills. These toxins open voltage-gated sodium channels (channel 5) in cell membranes. Signs of poisoning in humans include gastroenteritis, nausea, tingling and numbness around the mouth, loss of motor control, and severe muscle pain.

Domoic Acid

In 1987 there was an outbreak of mussel poisoning in Canada caused by mussels from the waters around Prince Edward Island. The toxin was domoic acid, which concentrates from a seaweed called chondria. It is also found in a diatom and it is rare in North Atlantic waters, being more common in Japan. The toxin was identified by the Canadian National Research Council Atlantic Laboratory. There were three fatalities in elderly nursing home residents in Quebec, and over 100 individuals were affected. Several have been left with short-term memory deficit, a condition called *amnesic shellfish poisoning*. It has been suggested that marine pollution might have changed the environment to favor the growth of the offending diatom. Domoic acid binds to a specific subset of glutamic acid receptors, known as kainate receptors, in the brain.

Okadaic Acid

Also produced by a dinoflagellate, okadaic acid is responsible for a condition known as diarrhetic shellfish poisoning. There is some evidence that there has been a dramatic increase in toxic algal blooms, the so-called red tides. The first confirmed outbreak of diarrhetic shellfish poisoning in North America occurred in 1990 and it was traced to dinoflagellates in Canadian waters. Brown pelicans eating anchovies off California were dying of domoic acid poisoning in 1991, saxitoxin has been found in Alaska crabs, and in 1987–1988 shell fishing off North Carolina was shut down because of a red tide of a dinoflagellate that produces a neurotoxin, brevetoxin (see also under CTX in the preceding text).

Okadaic acid is a selective inhibitor of phosphatases 1 and 2A with interference to protein activation/inactivation. It can induce contraction of smooth muscle and cardiac muscle. It has been shown experimentally to be a tumor promoter but, unlike phorbol esters, it does not activate protein kinase C.

Azaspiracid Toxin

Shellfish poisoning with Azaspiracid toxin (AZP) toxin was first reported in the Netherlands from eating Irish mussels in the 1990s but has since become a widespread problem throughout Europe. It is produced by the dinoflagellate *Protoperidinium crassipes*. There have been over a dozen derivatives of this unique lipophilic, polyether toxin. Symptoms of poisoning are primarily gastrointestinal (nausea, vomiting, severe diarrhea, stomach cramps) but neurological symptoms may also occur. Experiments in mice have shown that repeated injections may cause the formation of lung tumors. The mechanism of action is presently unknown.

Yessotoxin

Yessotoxin (YTX) and its analogues are disulfated polyethers and are becoming more commonly associated with shellfish poisoning, having been identified in Japan, New Zealand, Europe, and South America. They were originally thought to be responsible for the diarrheic shellfish poisoning (DSP) because they were isolated along with other DSP toxins. Animal studies did not confirm this effect but indicated that the heart was the target organ. Neurotoxicity has also been shown experimentally.

Palytoxin

Palytoxin (PTX) is a very large, highly complex polycyclic compound that has biological activity at extremely low concentrations. It is believed to have the longest continuous chain of carbon atoms of any natural substance. Originally isolated from a soft coral, it has since been identified in seaweeds, shellfish, and other marine organisms including a benthic dinoflagellate, *Ostreopsis siamensis*, which causes algal blooms along the coast of Europe. It is of increasing concern for both economic and public health reasons. It has caused excessive die-offs of edible mollusks and echinoderms, as well as illness in people. Deaths from PTX have been recorded from eating contaminated crabs in the Philippines, sea urchins in Brazil, and fish in Japan. The mechanism of action is unknown but symptoms include weakness, ataxia, drowsiness, fever, and death. The intravenous LD_{50} in experimental laboratory animals ranges from 0.25 to 0.9 µg/kg.

Stinging Fish Venoms

In venomous fish such as the stonefish, lionfish, scorpionfish and stingray, spines are located ahead of the dorsal fin, on the tail, or around the mouth. A heat-labile protein causes intense pain and cardiac shock. Heat above 50°C may afford some relief. Stonefish and stingrays are most often stepped upon because they conceal themselves in the sand bottom. Fatalities have occurred from envenomation by these fish, generally as a result of cardiovascular shock, with AV block and bradycardia. Other symptoms include numbness, inflammation and edema at the site of injury, severe pain in surrounding tissues, delirium, nausea, vomiting, and sweating. Stingray venom is a large, heat-labile protein (molecular weight >100,000) with neurotoxic as well as cardiotoxic properties. There is an antivenin for stonefish venom. Figure 11.1 shows a scorpion fish, one of the less hazardous members of this group.

FIGURE 11.1
(See color insert.) A staff member of the Gulf Specimen Marine Laboratory holds a scorpion fish in the palm of her hand. The dorsal spines (folded back here) are the means of conveying the venom into tissues.

Mollusk Venoms

Conotoxins

These are found in marine cone snails. All are strongly basic peptides highly cross-linked by disulfide bonds. Alpha-conotoxins are nondepolarizing, neuromuscular blocking agents like curare, and they therefore cause paralysis. They are 13–15 amino acid peptides. Mu-conotoxins are 22 amino acid peptides that block sodium channels, thus acting like tetrodotoxin and saxitoxin. These sodium channel blocking conopeptides are under investigation for use in treating chronic pain. Omega-conotoxins block presynaptic, voltage-dependent calcium channels. Recent research has identified a number of subtypes with specificity for various channel subtypes, making them useful research tools. Cone snails are univalve gastropods with a complex envenomation apparatus they use for prey capture (see http://grimwade.biochem.unimelb.edu.au/cone/newlog.html).

Cone shells such as the cloth of gold cone shell are highly prized by collectors. The envenomation apparatus can reach any exterior point on the shell and collectors have been seriously envenomated and even killed by handling live cone shells of the most toxic types.

Coelenterate Toxins

Many coelenterates (anemones, sea urchins, jellyfish, and corals) produce venoms that can cause pain on contact or even systemic envenomation. Fire coral produces a protein venom that causes intense local burning when touched. It feels like a cigarette burn (as this author can attest personally).

Physalia physalis (Portuguese man-o'-war, bluebottle, mauve stinger) is not a true jellyfish but rather belongs to the siphonophores, which are not single organisms but colonial ones composed of many single-celled units called zooids. These are linked together and physiologically interdependent. The bluebottle drifts on the wind by virtue of its sail which protrudes above the surface of the water. The bluebottle causes signs and symptoms like fire coral. Red streaks, called straps, occur where tentacles touch skin. Allergic reactions can occur, also generalized symptoms such as fever, nausea, and cardiac and respiratory distress. Bluebottles washed up on beaches remain toxic until completely dried out. First aid consists of washing the affected area with salt water followed by hot water (45°C, 115°F) both of which denature the venom. Even urine may help relieve the pain. This species is common around the world. Vinegar is not recommended as it promotes further release of venom from the nematocysts.

True jellyfish belong to the phylum *Cnidaria*. They are free-swimming animals with an umbrella-like bell that pulses to provide a form of jet propulsion. Although lacking a brain or true nervous system, jellyfish do have a loose nerve net and appear able to sense some environmental conditions. The sea wasp (*Chironex fleckeri*, box jellyfish) is the most venomous marine animal known. It inhabits the Indo-Pacific region and accounts for many deaths annually. In Australia, a registry has been kept since the mid-twentieth century and about 70 deaths have been attributed to the sea wasp. Contact with tentacles causes intense, agonizing pain, coma, and cardiac shock. Mortality is 25% and children, the elderly, and heart patients are most vulnerable. First aid is denaturation with vinegar, removal of tentacles using forceps, gloves, a knife and fork, or any means to avoid touching the tentacles, and cardiopulmonary resuscitation (CPR) if required. CPR should be continued for as long as possible as late recovery has been reported. A specific antivenom for the sea wasp is available. Local anesthetic spray such as is used for sunburn may be helpful.

The "Irukandji" (*Carukia barnesi* and *Malo kingi*) is a tiny, four-tentacled, Indo-Pacific jellyfish that causes similar signs and symptoms as the sea wasp plus massive sympathetic discharge. Irukandji syndrome is characterized by a massive release of adrenaline (epinephrine) and other catecholamines and histamine. This jellyfish can actually inject barbs into the skin and cause systemic symptoms including hypertension. First aid consists of vinegar washing, which denatures the venom on the surface but not that already injected. Antihistamines may be helpful.

Echinoderm Venoms

Sea urchins and sea anemones such as the long-spined or black sea urchin, and crown-of-thorns sea urchin possess toxins. Injury usually occurs to the feet and lower limbs when a diver or swimmer steps on these bottom dwellers. Spines are driven into the flesh and break off. The toxin produces local pain and burning similar to that of the bluebottle. It is heat labile, so immersing the affected part in water as hot as the person can stand may help. Unlike the usual first aid for envenomations, movement and trauma may actually help by breaking up the spines so they can be absorbed more quickly. Vinegar or even urine may also help.

Sea anemones also produce very potent toxins that act when they are ingested. Best characterized of these are the equinatoxins (EqTs I, II, and III) from *Actinia equina*. In rats they cause coronary vasospasm, cardiac arrest, and other cardiorespiratory toxicity. Hemolysis also occurs, and degranulation of blood platelets and white cells.

For more information on marine venoms and toxins see http://www.pmeh. uiowa.edu/fuortes/63260/MARINEAN/index.htm and http://www.merck.com/ pubs/mmanual/

Freshwater Algae

Cyanobacteria comprise a group of prokaryotes that share characteristics of both bacteria and algae, leading to a plethora of often confusing names. Here they shall be referred to by their most familiar name, the blue-green algae. Like our own species, these organisms are capable of being highly beneficial as well as causing harm. They are important primary producers with high nutritive value. Nitrogen-fixing species contribute to both soil and water fertility. Future use in food production and solar energy conversion has been proposed to be an important, future application.

On the opposite side of the coin, they can contaminate water reservoirs and pose a threat to drinking water supplies and human health. Records from the Han dynasty in China (c. 950) noted the death of troops who drank from a green river. Reports of poisoning of domestic animals are numerous. Eutrophication of water bodies usually related to high phosphate levels results in algal blooms and scum formation. Discharges from municipal wastewater discharges and agricultural runoff accelerate the process. At least 85 toxins from this group have been identified. Hepatotoxins are dominant (microcystins and nodularins). They are protein phosphatases and tumor promoters. Neurotoxins have also been identified. These include mimickers of acetylcholine, an anticholinesterase and even saxitoxin, the toxin of paralytic shellfish poisoning (see later).

Fish kills from cyanobacteria are common. A recent local kill occurred in a small stream killing hundreds of fish of several common freshwater species including rock bass, chub, and white suckers.

Prymnesium parvum (golden algae) has spread extensively to inland waterways in southern North America and has been associated with fish kill events. It is migrating northward and has been responsible for ecological devastation that threatens the economic and recreational value of freshwater systems. Studies have shown that the array of toxins produced by cultured *P. parvum* is markedly different from that produced by wild organisms. A highly labile Ichthyotoxin was present in both, however, and was proposed by the authors as the probable cause of fish kills.

As noted earlier, algae can be highly beneficial as well as harmful. A unique example of this came to light recently. There is a very large cement plant in the area. It is now directing its smokestack emissions to a 20,000 L tank on site. Algae, whose growth is stimulated with a specific wavelength of light, utilize the carbon dioxide, nitrogen oxide, and sulfur dioxide for growth. The biomass is compressed to extrude an oil that can be used as biodiesel and the residue is dried and made into bricks that burn like coal. Cement plants are believed to cause about 5% of global carbon dioxide emissions. When the facility is at full production it will produce about 225,000 ton of biomass annually.

Toxic and Venomous Land Animals

Venomous Snakes

Fear of snakes (ophidiophobia) is a very common phobia. Some psychologists feel it has its origins in the biblical story of the Garden of Eden, but the fear may not be all that irrational. It is estimated that 50–100 thousand people die annually from snakebite and thousands more endure permanent disability. Tropical and subtropical areas record the most bites and rural dwellers are most often affected. Farmers working in their fields, often barefoot, are especially vulnerable. Many snakes are venomous, but most have rear fangs that are designed to paralyze prey after it has been taken into the oral cavity. These are incapable of inflicting a venomous bite (Figure 11.2). The four genuses of poisonous snakes that are a danger to humans are *Viperidae*, *Crotalidae*, *Elapidae*, and *Hydrophiidae*.

1. *Viperidae.* These Old World vipers have hollow, needle-like fangs that are set in short, movable maxilla that rotate to bring the fangs into the biting position as the mouth is opened. The head is large and triangular. Signs and symptoms of the bite are as for the *Crotalidae*. Viper bites are not uncommon in Europe, occurring most often in children. One French paper reported on 58 children bitten by adders (vipers) between 2001 and 2009. Moderate to severe envenomation occurred

FIGURE 11.2
(See color insert.) Top: An Australian snake handler is holding three, red-bellied black snakes (*Pseudoechis porphyriacus*). While venomous, the bite of this snake is rarely fatal and it is not too aggressive. Middle: Two tiger snakes (*Notechis* sp.) have been just released from their bag by the handler. This snake is highly venomous and the mortality if untreated with anti-venom is 40 to 60%. Bottom: The handler restrains this eastern brown snake (*Pseudonaja* sp.), considered to be the second most venomous land snake. It can grow to two meters and is very fast. Mortality is similar to the tiger snake. All of these snakes are elapids.

in 17% of the patients. Bites to the upper extremities were associated with more severe envenomation. A polyvalent antivenin, Viperfav (*Vipera aspis, Vipera berus, Vipera ammodytes*), was used in severe cases.

2. *Crotalidae (or Crotalinae)*. The Peterson Field Guide on North American Venomous Animals and Poisonous Plants lists 31 species and subspecies of rattlesnakes, three of cottonmouths, and five of copperheads. These "pit vipers" are similar to the Old World vipers with folding fangs but they also have a deep, infrared-sensitive pit between the eye and the nostril that is used for tracking prey by body heat. This genus includes all rattlesnakes, the water moccasin (or cottonmouth) and copperhead. All are found in North America. Canada has the Western Diamondback and, in Ontario, the Massasauga rattlesnake. The timber rattlesnake is thought to be extinct but some feel it may have survived in areas of the Niagara Escarpment. The United States has about 25 species of venomous snakes and most are crotalids. About 8000 venomous snakebites occur annually with several deaths.

Signs and symptoms of envenomation include tissue swelling at the site, pain, ecchymosis (purplish, streaky hemorrhages into the skin), altered mental status, tachycardia, respiratory distress, and hypotension. Laboratory tests show evidence of coagulopathy (hypofibrinogenemia, low platelet count, prolonged clotting tests, e.g., activated partial thromboplastin time). A polyvalent antivenin now is available, FabAV or Crofab. It is composed of the Fab fragment of antibodies derived from ovine sources immunized with venoms from the Western Diamondback rattlesnake (*Crotalus atrox*), the Eastern Diamondback (*Crotalus adamanteus*), the Mojave rattlesnake (*Crotalus scutulatus*), and the Cottonmouth (*Agkistrodon piscivorus*). FabAV has much lower allergenic potential than older, polyvalent crotalid antivenin. Coagulopathy following pit viper envenomation may persist or recur for up to 2 weeks and require periodic treatment with Fab-based antivenins.

Pit vipers can be found worldwide. Asian species include the Chinese pit viper (*Protobothrops mangshanensis*) and the Green pit viper (*Trimeresurus albolabris, Trimeresurus trigonocephalus*). In Mali, a study in two national hospitals recorded 832 snakebites from 1993 to 2002. The offending species were not identified but the symptomatology was characteristic of viper venom. Through the use of antivenom the mortality rate was reduced from 12% in 1995 to 3% in 2002.

3. *Elapidae*. They are distributed worldwide and account for the most venomous and feared snakes of the tropics and subtropics. This group includes the cobras, the boomslang, and many Australian species (taipan, tiger snake, brown snake, etc.). There are three species in the United States: the eastern or Florida coral snake (*Micrurus fulvius*), the Arizona coral snake (*Micruroides euryxanthus*),

and the Texas coral snake (*Micrurus tener*) that apparently is found only in Harris County. The rhyme "red-on-black, friend of Jack, red-on-yellow, kills a fellow" helps to distinguish the coral snake from harmless look-alikes, such as the scarlet king snake and other king snakes and the Louisiana milk snake. This rhyme, however, applies only to North American coral snakes. The markings of the South American ones do not always have the typical pattern. Coral snakes have hollow, short, rear fangs and tend to hang on rather than strike and release like vipers. Bites are rare.

4. *Hydrophiidae (or Hydrophiinae)*. There are some 50 species of sea snakes distributed worldwide. They are restricted to the warmer waters of the Pacific and Indian Oceans. Most are marine dwellers but some may enter river estuaries. All but one genus (*Laticauda*, which lays eggs onshore) give birth to live young in the water (ovoviviparous). Although they are highly venomous, they are not usually aggressive except during mating season. Most bites occur to commercial fishermen because the snakes become trapped in the nets or when the fishers are wading in muddy waters. For some years, sporadic discussions have occurred concerning the feasibility of digging a sea-level canal across the Isthmus of Panama to connect the Atlantic and Pacific Oceans. One environmental concern that has been raised is that such a canal could introduce the yellow-bellied sea snake (*Pelamis platurus*) to the Gulf of Mexico and the Caribbean Ocean where conditions are favorable for their proliferation. This sea snake may occur inshore from southern California through Central and northern South America. Claims of sea snake sightings around Jamaica are thought to be due to mistaking the spotted eel for a snake. The clinical picture of sea snake envenomation includes rhabdomyolysis (dissolution of muscle tissue) with resulting myoglobinemia and myoglobinuria, muscle pain and tenderness, flaccid paralysis (neurotoxicity), and renal failure. Antivenom is available.

5. *Trimorpohodon* spp. In addition to the coral snakes there are other, venomous, rear-fanged snakes that rarely cause envenomation as they have difficulty reaching human limbs with their fangs. These so-called lyre snakes, so named from a lyre-shaped marking on the back of the head, include the Texas lyre snake, the Sonoran lyre snake, and the California lyre snake. Small children could conceivably be bitten on the finger due to their small size but this author is unaware of any recorded incidents.

6. *Heloderma* spp. Gila monsters are native to the desert areas of the American southwest and Mexico. There are two subspecies, *Heloderma suspectum suspectum* (reticulate) and *Heloderma suspectum cinctum* (banded). Venom glands are located at the base of the grooved teeth and Gilas have a tendency to hang on and "chew in"

the venom. Gilas are not generally as dangerous as venomous snakes. The venom lacks neurotoxins but contains coagulants and enzymes as well as serotonin. The reaction tends to be more local. Lethal doses in animals lead to cardiorespiratory collapse.

Snake Venoms

Over the last decade a massive literature has developed concerning research into the chemical nature and biological activities of snake venoms. Given the very large numbers of these, it is only possible to skim the surface of the field here. Snake venoms are complex mixtures of proteins and polypeptides, many of which are proteolytic enzymes. In general, venoms of the *Elapidae* and the *Hydrophiidae* tend to be neurotoxic with myonecrosis (breakdown of muscle tissue) occurring at the bite wound, whereas venoms of the vipers and crotalids are generalized coagulants with local anticoagulant activity to spread the venom, causing much local damage (pain, necrosis, bleeding). Neurotoxicity is less prominent.

Signs and symptoms of neurotoxic venoms include progressive paralysis, muscle spasm, respiratory distress or failure, muscle ache (myalgia), kidney failure with myoglobinuria (the appearance of myoglobin in the urine), and cardiac failure. These symptoms apply to coral snakebites in North America. The venom of the Banded Krait (*Bungarus fasciatus*), native to the Indian subcontinent, contains alpha-bungarotoxin, which is an irreversible blocker of acetylcholine receptors, and beta-bungarotoxin that causes a massive release of neurotransmitter vesicles. Alpha-bungarotoxin is used as a research tool in biomedical research. The Black Mamba (*Dendroaspis polylepis*) produces dendrotoxin-I, a K^+ channel blocker.

Signs and symptoms of viper and crotalid bites (including the Massasauga rattlesnake) include immediate, intense burning at the site of envenomation (like a bee sting), followed by numbness, swelling, shock, and hematuria. Necrosis and possibly gangrene may occur later at the bite wound. Subcutaneous hemorrhages may be present and severe cases may show signs of neurotoxicity. Table 11.1 lists some of the major components that have been identified in snake venoms. The venom of Russell's viper contains an activator of clotting Factor X and it is used in coagulation research.

Phospholipase A2 complexes in rattlesnake venoms are neurotoxic as well as induce tissue damage. Myotoxic components have exhibited mitogenic multiple effects on growing cultured myocytes. Two potent beta-neurotoxins, crotoxin (from *Crotalus durissus terrificus*) and ammodytoxin (from *V. ammodytes*) were recently characterized by x-ray crystallography and their three-dimensional structure determined. The active molecular sites for both neurotoxic and anticoagulant activities were identified.

Elapidae and *Viperidae* both possess α-neurotoxins that act postsynaptically to prevent acetylcholine from binding to its receptor, as well as neurotoxins that affect transmitter release presynaptically; beta-neurotoxins have phospholipase A2 activity. The mambas and other African snakes have

TABLE 11.1

Some Components of Snake Venoms

Component	Elapidae/ Hydrophiidae	Viperidae/ Crotalinae
Nicotine blocker (neuromuscular blockade)	+	+/−
Cholinesterase (neurotoxic)	+	−
Coagulant protease (increases clotting)	−	+
Anticoagulant protease (decreases local clotting)	+/−	+
Adenosine triphosphate (ATP) (shock, hemolysis)	+	+
Phospholipase (shock, hemolysis)	+	+
Phospholipase A (release of histamine and leukotrienes)	+	−
Bradykininogen (forms bradykinin, causes pain)	−	+
Hyaluronidase (breaks down interstitial glue, spreads venom)	+	+

voltage-dependent K$^+$ channel blockers (dendrotoxins), noncompetitive inhibitors of acetylcholine (fasciculins), muscarinic toxins, and L-type Ca^{++} channel blockers (caliseptins). All are small proteins containing about 60 amino acids and three or four sulfides.

Neurotoxins from venom of the sea snake *Laticauda semifasciata*, alpha- and beta-erabutoxin have been shown to belong to a superfamily of long-chain and short-chain amino acid neurotoxins and cytotoxins found in both sea snakes and terrestrial snakes. The erbutoxins are short amino acid chain ones that block postsynaptic, nicotinic acetylcholine receptors like curare.

As the chemical structures of snake venoms are slowly unraveled, it is hoped that this may lead to further advances in treatment and possibly even prevention. New tools for pharmacological research may also emerge and possibly new treatments for clotting disorders.

Table 11.1 lists some major components of snake venoms.

First Aid

Regardless of the type of snakebite, the most important first aid measure is a tension bandage applied to the entire affected limb with the same tension one would use to bandage a sprain. Splint immobilization may also be helpful if practical. Rest and reassurance are important. Transportation to a medical center possessing antivenin should occur as soon as possible. Modern hospitals should have the antivenin or polyvalent antivenin, appropriate to their area, in stock. Forced exercise and alcoholic beverages are definitely contraindicated, as is incision and suction at the site of envenomation.

Venomous Arthropods

Members of the order *Hymenoptera* (bees, wasps, ants) of the class *Insecta*, and of the class *Arachnida* (spiders, scorpions) have venoms that contain substances commonly involved in the mammalian pain response. The Old World scorpion *Leiurus quinquestriatus* produces charybdotoxin that affects calcium-activated K^+ channels as does the Israeli yellow scorpion (*Leiurus q. hebraeus)*, and the bee venom apamin. The Mexican scorpion *Centroides noxius* produces noxiustoxin, which blocks voltage-dependent K^+ channels (see Table 11.2). Mast cell degranulating peptide in bee venom also blocks these channels. Scorpion toxins that can target voltage-gated sodium channels consist of 60–76 amino acids cross-linked by four disulfide bridges. They constitute two groups classed as alpha and beta toxins. A very venomous Brazilian scorpion (*Tityus serrulatus*) contains a beta toxin, Ts2.

Allergic reactions may occur to any of these insect venoms and may be life threatening if severe or if multiple stings occur or if the sting is in the throat. Direct toxic effects account for less than 5% of all deaths from hymenopteran stings. The African honeybee is not dangerous because it is more venomous but because it is extremely aggressive so that multiple stings are common. In North America about 3.3% of adults and 0.8% of children are allergic to honeybee and wasp venom. Signs and symptoms include flushing, tachycardia, abdominal

TABLE 11.2

Components of *Hymenoptera* Venoms

Hymenoptera	Histamine Releaser	Bradykininogen/ Bradykinin	Serotonin	K^+ Channel Blocker	Melittin, Hyaluronidase, PLA2
Bees				Apamin (Ca$^+$ activated) Mast cell degranulating peptide (voltage-activated)	+
Ants	+	+	+		
Wasps	+	+	+		
Old world scorpions	+	+	+	Charybdotoxin (Ca$^+$ activated) R-agitoxin-2 (voltage-activated)	**Voltage-activated Na$^+$ Channels** +
Mexican scorpions				Noxiustoxin (voltage-activated)	+

colic, diarrhea, and, in severe cases, progressing to hypotension and coma. Severe cases require injection of epinephrine (adrenaline) with an Epi-pen, antihistamines, and corticosteroids. There have been about 20 identified enzymes, peptides, and active amines identified in bee venom. These include phospholipase A2 and hyaluronidase, which breaks down hyaluronic acid, the intercellular glue that maintains conformational integrity. Hyaluronidase acts as a spreading agent. Phospholipase A2 disrupts cell membranes. Mass bee attacks result in the release of large amounts of cytokines like interleukins 1, 6, and 8, and tumor necrosis factor (TNF).

All spiders are venomous but most lack jaws powerful enough to penetrate human skin or do not inject enough venom to cause anything more than some local irritation. Neurotoxins tend to predominate in spider venoms. So-called widow spiders (*Latrodectus* spp.) have a worldwide distribution in temperate and tropical climes. The black widow, also known as the death spider, hourglass spider, and by many other names, inhabits all of Ontario south of Sudbury but it is rarely encountered since the demise of the outdoor privy. The venom is extremely toxic but very little is injected so fatalities are rare. The toxin is complex. At least seven "latrotoxins" have been identified; five are specific for insects, one for crustaceans, and one for vertebrates. The latter is alpha-latrotoxin (α-LTX). It is a medium-sized protein now available in pure form (see later) that induces the release of most if not all neurotransmitters. It bears similarity to beta-bungarotoxin (found in the Banded Krait). Both cause a massive release of cholinergic vesicles.

Fasciculations and board-like rigidity of the muscles of the trunk occur rapidly followed by muscle cramps, pain in the muscles, and respiratory distress. The presence of a specific receptor on vertebrate neurons for α-LTX has been identified and this is the likely explanation for the species specificity of the toxin. Recovery from black widow spider bite takes about 12 h.

The brown recluse (*Loxosceles* spp.) known also as the violin spider, has been working its way north from Mexico and Florida and has reached New York State and probably southern Ontario aided, no doubt, by climate change. The bite is painless. Local necrosis develops and expands over the next week due to local blood clotting and microthrombosis. Permanent scarring may result. Occasional deaths have occurred from hemolytic anemia and kidney failure. The venom is complex and includes coagulants, enzymes, and a complement inhibitor.

Funnel web spiders (*Atrax* spp. and *Hadronyche* spp.) from southeast Australia give a bite that resembles a scorpion sting and is characterized by massive catecholamine release. Banana spiders (*Phoneutria nigriventer*) can be shipped in banana bunches and cause serious bites, especially in countries where no immunity is likely. Funnel web and red-backed spiders are claimed to be especially venomous. Two funnel web toxins have been identified, a polyamine (FTX) and omega-agitoxin. Both block high-voltage-dependent P Ca^{2+} channels, and they are now used as research tools for this purpose. R-agitoxin-2, from the Israeli yellow scorpion, blocks some voltage-activated K^+ channels.

Toxic Plants and Mushrooms

Introduction

Folk knowledge, passed orally from generation to generation, usually determines what we can eat and what we cannot eat, and even what parts of a plant are edible. Thus, we make pies from rhubarb stalks but we never make salad from the leaves, which are toxic. Nor do we eat the bulbs or stalks of the many ornamental flowers and shrubs in our gardens. Seldom do we consider the fact that these decisions are toxicologically based. The number of plants that are potentially harmful is too voluminous for extensive coverage. The subject has formed the basis of many texts. The Peterson field guide, *Venomous Animals and Poisonous Plants* by Foster and Caras lists over 250 species of plants and mushrooms with potential toxicity and concedes that it is by no means exhaustive. The agricultural costs from livestock consuming toxic plants can be extreme. One study reported that in the 17 southwestern states of the United States, direct losses were estimated at $340 million annually using 1989 data. This figure did not include indirect losses due to such things as poor growth rate, prolonged gestation period, infertility, etc. nor costs of preventing poisoning through fencing, destruction of weeds, and other measures.

The following list contains examples of major groups of toxicants chosen because they are common or because of their pharmacological significance.

Vesicants

Many plants contain oxalates that can cause corrosive burns to the mouth, esophagus, and stomach. Symptoms also include vomiting and diarrhea. Since oxalates are anticoagulant, bleeding may also occur. *Dieffenbachia* contains calcium oxalate, which is a vesicant. Rhubarb leaves contain a variety of oxalates and may be fatal if consumed in quantity due to their renal toxicity (see also ethylene glycol). May apple, buttercup, and philodendron contain vesicants. Philodendron sometimes causes poisoning in cats if they chew the leaves. Poison ivy, poison oak, and poison sumac all contain urushiol, which is a phenolic vesicant.

Cardiac Glycosides

White and purple foxgloves (*Digitalis lanata* and *D. purpura*) are the commercial source of medicinal digitalis. Many garden plants possess similar active components, including lily of the valley, star of Bethlehem, and oleander. Symptoms are those of digitalis overdose; nausea, vomiting, visual disturbances (a green halo seen around objects), and cardiac arrhythmias.

Astringents and Gastrointestinal Irritants (Pyrogallol Tannins)

Acorns, geraniums, sumac berries, hemlock bark, rhubarb leaves, horse chestnut, all contain these. North American natives learned to remove tannins from acorns by steeping the crushed nuts repeatedly in freshwater. This will not work for horse chestnuts, however.

Autonomic Agents

Deadly and wooded nightshades contain atropine (hyoscyamine) and scopolamine (hyoscine). The signs and symptoms are those of blockade of muscarinic, cholinergic receptors plus CNS symptoms (disorientation, hallucinations). Poisoning from consuming the berries continues to occur, usually in children. A group from Turkey reported on 49 poisoned children, six of whom were considered to be severely affected. Signs and symptoms included meaningless speech, tachycardia, mydriasis, flushing, and in severe cases, coma. Artificial ventilation was not necessary for any patient and all recovered fully. As physostigmine was not available, neostigmine was used successfully. Practitioners of folk medicine sometimes use belladonna with potentially harmful consequences. A report from Africa (Morocco) deals with an 11-year-old girl under treatment for tuberculosis (rifampicin and isoniazid). She was given a preparation of *Atropa belladonna* by a herbalist. She presented at the local hospital 24 h later with dry mouth, incoherent speech, confusion, inability to recognize members of her family, uncontrollable vomiting, visual hallucinations, visual and hearing disturbances, and intermittent coma. She was treated successfully with symptomatic treatment that included oxygen, diazepam (tranquilizer/sedative), antiemetics, and i.v. electrolytes.

If ingestion has been recent, activated charcoal is often given orally to neutralize any toxin still in the gastrointestinal tract.

Amanita muscaria is the common poisonous mushroom. Signs of poisoning are those of massive cholinergic discharge due to the toxin muscarine, plus CNS effects like those of nightshade due to the presence of anticholinergic agents. CNS depression and cardiac failure may follow. Symptoms may be delayed 12–24 h.

Solanine and chaconine are glycogenic alkaloids that occur in potatoes and tomatoes (members of the *Solanaceae* family) when they are exposed to excessive sunlight, blight, sprouting, and prolonged storage. The compounds have cholinergic activity and have been shown to be teratogenic. Symptoms of poisoning include vomiting, diarrhea, cramps, dizziness, visual disturbances, and other CNS manifestations. In 1820, a man named Johnson defied conventional wisdom by publicly eating a tomato in Salem, New Jersey. To the surprise of onlookers, he did not die a horrible death. At the time it was widely believed that tomatoes, because of their relationship to deadly nightshade, were highly toxic. Johnson single-handedly launched the North American tomato industry.

Another member of the *Solanaceae* is *Datura stramonium*, commonly known as jimsonweed or thorn apple. Other names are devil's trumpet and angel's trumpet. This is a common weed throughout North and Central America, although it was first introduced from Europe. It typically grows in rough ground such as vacant lots. It has a large, white or mauve trumpet flower and a spiny seedpod. The seeds can contain significant levels of atropine and scopolamine that can induce hallucinations. For this reason it is sometimes used as a recreational drug, usually by adolescents. Due to its anticholinergic activity, signs and symptoms of jimsonweed poisoning can include dilated pupils, dry mouth, rapid and irregular heartbeat, stasis of the bowel, agitation and disorientation, as well as hallucinations. Several deaths have occurred.

Dissolvers of Microtubles

Colchicine from the autumn crocus is used in the treatment of gouty arthritis and as an experimental tool. It dissolves microtubules to prevent mitosis and also phagocytosis. Overdose causes severe diarrhea. Vincristine and vinblastine from the periwinkle plant share this property and are used to treat childhood leukemias. Podophyllotoxin from the May apple also dissolves microtubules.

Phorbol Esters (for Example, Phorbol Myristate Acetate, PMA)

These are components of croton oil from spurge plants. These substances are cancer promoters. They directly activate protein kinase C (substituting for diacylglycerol) independently of extracellular calcium and are they are used as experimental tools for this reason. They act as drastic purgatives. The site of action of PMA is shown in Figure 11.3. It affects an important control mechanism for intracellular regulation.

Many other carcinogens, co-carcinogens, promoters, and anticarcinogens exist in plants. Safrol is a liver carcinogen found in some spices (nutmeg, cinnamon) and in oil of anise (licorice flavoring). The use of anise oil and of oil of sassafras has been banned. Some tannins are liver carcinogens, and the polyaromatic hydrocarbon (PAH) benzo-[a]-pyrene is a potent carcinogen that occurs in green vegetables, unrefined vegetable oils, coconut oil, and chicory. Benzanthracenes are other PAHs that occur in vegetables. Many others exist.

Thapsigargin is a plant-derived sesquiterpene lactone capable of inhibiting Ca^{2+}-ATPase and causing discharge of internal Ca^{2+} stores. It is widely used as a research tool because of this action. It also has tumor-promoting properties.

Cyanogenic Glycosides

Substances such as amygdalin in almonds, dhurrin in sorghum, linamarin and lotaustralin in cassava and lima beans, and prunasin in stone fruit (cherries, peaches, and chokecherries) are cyanogenic glycosides. They are capable of forming hydrogen cyanide (HCN) with the beta-glucuronidases from the plants when cells break down or from the microflora of the gastrointestinal tract. Cyanide poisoning can occur in ruminant animals from eating vegetation

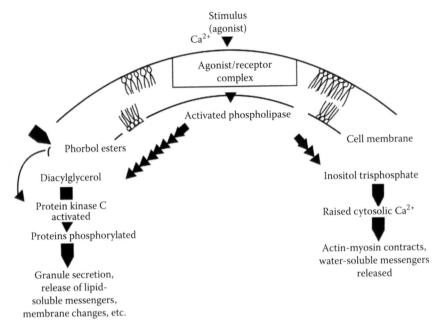

FIGURE 11.3
Site of action of phorbol esters as activators of protein kinase C.

high in cyanogenic glycosides or in humans who have consumed improperly stored or prepared foods such as lima beans or cassava. In humans, CN can be formed from organic nitriles by Cytochrome P450-dependent monooxygenases, and from organic thiocyanates by glutathione S-transferases.

Detoxification of Hydrogen Cyanide

Hydrogen cyanide (HCN) is detoxified by conversion to thiocyanate, which requires sulfur-containing amino acids and vitamin B_{12}. Deficiencies of these increase the risk of toxicity. The metabolic detoxification system is overwhelmed and hydrogen cyanide interferes with electron transport in the cytochrome a–a_3 complex, with resulting in tissue hypoxia. This leads to rapid failure of the CNS and death. Treatment is the administration of intravenous nitrites, which form methemoglobin. Methemoglobin has a high affinity for HCN and binds it to protect the cytochrome and allow time for biotransformation to occur. These events are summarized in Figure 11.4.

Convulsants

Water hemlock, the poison of Socrates, typifies this group. The toxin is cicutoxin (from the plant's Latin name *Cicuta maculata*). This plant resembles parsnips, smells like turnips, tastes sweet, and it is the most toxic indigenous plant in North America. The toxin is present in all parts of the plant

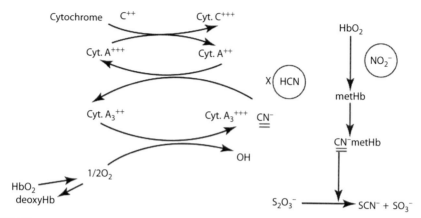

FIGURE 11.4
Site of action of HCN and of detoxification by nitrites.

Used in Research and Treatment

Many chemicals of animal and plant origin are useful as research tools in physiology and pharmacology. A partial list follows.

1. Tetrodotoxin from puffer fish and saxitoxin from shellfish block fast sodium channels and are used to study nerve conduction.

2. Alpha-conotoxin from cone snails is a nondepolarizing neuromuscular blocking agent.

3. Mu-conotoxin from cone snails acts like tetrodotoxin.

4. Omega-conotoxin from cone snails is a specific inhibitor of presynaptic, voltage-dependent Ca^{2+} channels.

5. Russell's viper venom activates Factor X in the clotting system and is used in certain clotting tests and coagulation research.

6. Alpha-bungarotoxin from the banded krait is an irreversible blocker of acetylcholine receptors.

7. Beta-bungarotoxin from the banded krait and alpha-latrotoxin from the black widow spider cause massive release of peripheral neurotransmitter vesicles.

8. Apamin from bees is a blocker of K^+ channels.

9. Charybdotoxin from the scorpion is also a potent K^+ channel blocker.

10. Digoxin from foxglove blocks Na^+/K^+ ATPase and is used to treat congestive heart failure.

11. Atropine from nightshade is a muscarinic blocker. It has many uses.

12. Tannins from a variety of plants are used in astringent lotions.

13. Cytochalasins from fungi fix cell membranes and microtubules *in vivo*.

14. Phorbol myristate from croton oil activates of protein kinase C. It is used to study calcium intracellularly.

15. Colchicine from autumn crocus dissolves microtubules and arrests mitosis. It is also used to treat acute attacks of gouty arthritis.

16. Capsaicin from chili peppers is used as a counterirritant in lineaments and ointments to provide heat by causing vasodilation.

17. Vincristine and vinblastine from periwinkle arrest mitosis, and are used as anticancer drugs.

but is concentrated in the root. It is most toxic in springtime. Mild intoxication produces nausea, abdominal pain, epigastric distress, and vomiting in 15–90 min. Early vomiting may be protective. Severe poisoning produces profuse salivation, sweating, bronchial secretion, respiratory distress, and cyanosis. Convulsions occur and *status epilepticus* precedes death. Mortality rates are of the order of 30%. There is no known antidote. Fatal poisonings in children have occurred from using toy whistles made from the stem.

In the period 1978–1989, 58 persons in the United States are known to have died from ingesting toxic plants mistaken for edible wild fruit or vegetables. Water hemlock was responsible for at least five of these.

Recent research into the nature and chemical composition of polypeptide venoms has led to their availability in pure form as research tools, mainly from Alomone Labs in Jerusalem. Some examples are as follows:

1. From the eastern green mamba (*Dendroaspis angusticeps*), alpha-dendrotoxin blocks certain voltage-gated K^+ channels. Beta-dendrotoxin blocks certain voltage-gated K^+ channels in synaptosomes and smooth muscle cells.

2. From the Australian taipan (*Oxyuranus scutellatus*), taicatoxin selectively blocks high-threshold voltage-gated Ca^{2+} channels in heart cells.

3. From the black widow spider (*Latrodectus tredecimguttatus*), α-latrotoxin: A 130,000 Da protein, it is the principal toxic component of the venom, causing massive exocytotic secretion of neurotransmitter vesicles both centrally and peripherally.

Case Study 20

In 1988, several patrons of a restaurant experienced signs and symptoms of illness including nausea, headache, dizziness, facial flushing, and diarrhea. The symptoms onset about 5–60 min after the meal (median 38 min) and persisted for about 9 h. Only these six patrons (four males, two females) experienced problems even though an estimated 50–60 had partaken of the same buffet lunch.

Q. What questions would you wish to ask of the affected and the unaffected patrons?

Q. What possible causes of this reaction could there be?

Several of the affected individuals noted upon questioning that a particular fish dish had a "Cajun" or peppery flavor.

Q. Does this help to identify the problem?

Case Study 21

In June of 1990, six fishermen aboard a private fishing boat off the Nantucket coast of Massachusetts developed symptoms that included numbness and tingling of the mouth, tongue, throat, and face, vomiting, loss of sensation in the extremities, periorbital edema, and 24 h later, low back pain (in all six). The initial symptoms persisted for about 14 h, the back pain for 2–3 days.

Q. What organ system is primarily affected?

Q. What information would you want to obtain from the victims?

It emerged that all six men had consumed blue mussels at the same meal. The blue mussels had been harvested in deep water about 115 miles off-shore. The mussels had been boiled for about 90 min, and were consumed with boiled rice, baked fish, and a salad. There appeared to be a correlation between the severity of the symptoms and the number of mussels consumed.

Q. What is the likely cause of the poisoning?

Q. What fish could have been responsible for the same array of signs and symptoms?

Q. What other marine toxin would produce the same signs and symptoms?

Case Study 22

Over a period of 72 h in August, eight seasonal tobacco workers were admitted to a regional hospital with a variety of signs and symptoms that included weakness, nausea, vomiting, dizziness, abdominal cramps, headache, and difficulty in breathing. They had all been working in the fields in the morning following an evening of steady rain. The average time of onset of the symptoms was 10 h after commencing work. All patients were males, 18–32 years of age. All required hospitalization for 1 or 2 days.

Q. Is this likely an occupational disease, a food poisoning from something in the breakfast meal, or an infection?

Q. What occupational hazards might these workers encounter?

Q. What lab tests might help in the differential diagnosis?

Case Study 23

On August 9, eight persons were admitted to the emergency department of a Florida hospital with one or more of these symptoms: cramps, nausea, vomiting, diarrhea, chills, and sweats. All reported having eaten amberjack, a predatory scale fish, at a local restaurant within the preceding 9 h (mean time to symptoms 5 h). Three of the victims required hospitalization. These symptoms persisted for 12–24 h. Within 48–72 h, most of these patients developed pruritis and parathesias of the hands and feet and muscle weakness.

Subsequent investigation uncovered 14 similar cases, all of which had eaten amberjack at one of several local restaurants. These received the fish from the same supplier in Key West.

Q. What organ systems are involved in this intoxication?

Q. Does the evidence point to restaurant kitchens as the source of the toxin?

Q. What potential causes of this problem must be considered?

Q. Which is your choice?

Case Study 24

In the fall of 1992, two young men were foraging in the Maine woods for wild ginseng. Several plants were collected. The younger man, aged 23, took three bites from the root and his 39-year-old brother took one bite from the same root. Within 30 min, the younger man vomited and began to convulse. They walked out of the woods and received emergency rescue within 45 min of the onset of symptoms. At this point the man was unresponsive, cyanotic, and had tachycardia, dilated pupils, and perfuse salivation. He had several clonic–tonic convulsions, developed ventricular fibrillation and was dead on arrival at the local hospital despite resuscitative attempts. The older brother was not showing symptoms at this time and was given gastric lavage and activated charcoal. Sometime later, he developed delirium and seizures. He recovered with symptomatic treatment.

Q. What is the likely source of this problem?

Q. What organ systems are involved?

Q. Which plant and toxin would cause this array of symptoms?

Case Study 25

During the summer of 1999, in London, Ontario, several teenagers were admitted to the emergency department of a local hospital, one in critical condition. Signs and symptoms included stomach cramps, irregular heartbeat, hallucinations, and dilated pupils. One boy was found unconscious in the basement of his home. The teens admitted to eating the seeds from the seedpods of a wild plant with large, white trumpet flowers and spiny seedpods.

Q. What is the probable identity of this plant?

Q. What are the active ingredients that impart its toxicity?

Q. Where does this plant grow?

Case Study 26

In late spring, a 10-year-old boy was playing around the edge of a marshy area in southwestern Ontario near the city of Windsor. When reaching into some undergrowth to retrieve a ball he felt a sharp sting on his hand. He thought he had been stung by a bee or wasp and decided to run home which was about 20 min away. By the time he reached home the hand was beginning to swell and some purple streaks were visible as were two small puncture wounds. He was feeling a bit faint so he was taken to the emergency department of the closest hospital.

Q. What is the most likely cause of the boy's symptoms?

Q. What first aid measures might have been instituted at home before the trip to the hospital?

Q. What would be the likely treatment given at the hospital?

Q. Was the boy's life likely in any danger?

Review Questions

For Questions 1–15 answer true or false:

1. Crotalid venoms are predominantly anticoagulant.
2. Hyaluronidase in snake venoms helps to disseminate the poison at the site of the bite.

3. Omega-conotoxin blocks presynaptic, voltage-gated calcium channels.

4. Ciguatoxin does not biomagnify up the food chain.

5. Beta-bungarotoxin is found in the venom of the banded krait.

6. Potassium channel blockers are found in the venom of the brown recluse spider.

7. Urushiol is an astringent found in the horse chestnut.

8. Hyoscyamine is the same as scopolamine.

9. Alpha-bungarotoxin is an irreversible blocker of acetylcholinesterase.

10. Cytochalasin fixes microtubles *in situ*.

11. The venom of vipers is predominantly neurotoxic.

12. Tetrodotoxin is a potassium channel blocker.

13. Saxitoxin is synthesized by dinoflagellates.

14. Lily of the valley contains cardiac glycosides.

15. The bite of the brown recluse spider is extremely painful.

For Questions 16–23 match the statements with appropriate response from the following:

a. Vinegar

b. Traumatizing the area

c. A tension bandage over the entire affected limb

d. Phorbol myristate acetate

e. Tetrodotoxin

f. Domoic acid

g. Ciguatoxin

h. Okadaic acid

16. Causes generalized paralysis due to fast sodium channel blockade

17. The cause of amnesic shellfish poisoning

18. Directly activates phosphokinase C

19. General first aid for any snakebite

20. First aid for the sting of a bluebottle (*Physallis*)

21. May reduce the pain of an imbedded sea urchin spine

22. The cause of diarrhetic shellfish poisoning

23. May cause gastrointestinal and neurological symptoms when large marine scale fish are eaten

Answers

1. True
2. True
3. True
4. False
5. True
6. False
7. False
8. False
9. False
10. True
11. False
12. False
13. True
14. True
15. False
16. e
17. f
18. d
19. c
20. a
21. b
22. h
23. g

Further Reading

Anderson, D.M., Red tides, *Sci. Am.*, 271, 62–68, 1994.

Ashton, J., Baker, S.N., and Weant, K.A., When snakes bite: The management of North American *Crotalinae* snake envenomation, *Adv. Emerg. Nurs. J.*, 33, 15–22, 2011.

Berdai, M.A., Labib, S., Chetouani, K., and Harandou, M., *Atropa belladonna* intoxication: A case report, *Pan. Afr. Med. J.*, 11, 72, 2012.

Caksen, H., Odabas, D., Akbayram, S., Cesur, Y., Arslan, S., Uner, A., and Oner, A.F., Deadly nightshade (*Atropa belladonna*) intoxication: an analysis of 49 children, *Hum. Exp. Toxicol.*, 12, 665–668, 2003.

Cataldi, M., Secondo, A., d'Alessio, A., Taglialatela, M., Hoffmann, F., Klugbauer, N., Di Renzo, D., and Annunziato, L., Studies on maitotoxin-induced intracellular Ca^{2+} elevation in Chinese hamster ovary cells stably transfected with cDNAs encoding for L-type Ca^{2+} channel subunits, *J. Pharmacol. Exp. Ther.*, 290, 725–730, 1999.

Chorus, I. and Bartram, J. (eds.), *Toxic Cyanobacteria in Water: A Guide to their Public Health Consequences, Monitoring and Management*, World Health Organization, London, U.K., 1999.

Claudet, I., Gurrera, E., Maréchal, C., Cordier, L., Honorat, R., and Grouteau, E., Pediatric adder bites (Fr.), *Arch. Pediatr.*, 18, 1278–1283, 2011.

Codd, G.A., Cyanobacterial toxins: Occurrence, properties and biological significance, *Water Sci. Technol.*, 32, 149–156, 1995.

Culotta, E., Red menace in the world's oceans, *Science*, 257, 1476–1477, 1992.

Dare, R.K., Conner, K.B., Tan, P.C., and Hopkins, R.H. Jr., Brown recluse spider bite to the upper lip, *J. Ark. Med. Soc.*, 108, 208–210, 2012.

Dept. of Surgical Education, Orlando regional medical center, Snakebite/crotalid antivenoms, http://www.surgicalcriticalcare.net/Guidelines/envenomation%20 2010.pdf 2010 (accessed on November 13, 2011).

Dickey, R.W., Fryxell, G.A., Granade, H., and Roelke, D., Detection of the marine toxins okadaic and domoic acid in shellfish and phytoplankton in the Gulf of Mexico, *Toxicon*, 30, 355–359, 1992.

Dramé, B.S., Diarra, A., Diani, N., and Dabo, A., Epidemiological, clinical and therapeutics aspects of snakebites in the Gabriel-Touréand Kati national hospitals of Mali: A ten-year retrospective study (Fr.), *Bull. Soc. Pathol. Exot.*, 105, 184–188, 2012.

Edmonds, C., Venomous marine animals, Chapter 32 in Diving and Subaquatic Medicine, Edmonds, C., Lowry, C., Pennefather, J. and Walker, R. (eds), 4th Edn., Hodder Arnold, London, U.K., 335–352, 2005.

Escobar, L.I., Salvador, C., Martinez, M., and Vaca, L., Maitotoxin, a cationic channel activator. *Neurobiology (Budapest)*, 6, 59–74, 1998.

Faure, G. and Saul, F., Crystallographic characterization of functional sites of crotoxin and ammodytoxin, potent β-neurotoxins from *Viperidae* venom, *Toxicon*, 60, 531–538, 2012.

Ferreira, R.S., Almeida, R.A., Barraviera, S.R., and Barraviera, B., Historical perspective and human consequences of Africanized bee stings in the Americas, *J. Toxicol. Environ. Health B Crit. Rev.*, 15, 97–108, 2012.

Foster, S. and Caras, R.A., Venomous animals and poisonous plants. A *Roger Tory Peterson Field Guides*, Peterson, R.T. (ed.), Easton Press, Norwalk, CT, 1994.

Hashimoto, Y., *Marine Toxins and Other Bioactive Marine Metabolite*, Japan Science Society Press, Tokyo, Japan, 1979.

Henrikson, J.C., Gharfeh, M.S., Easton, A.C., Easton, J.D., Glenn, K.L., Shadfan, M., Mooberry, S.L., Hambright, K.D., and Cichwicz, R.H., Reassessing the icthytoxin profile of cultured *Prymesium parvum* (golden algae) and comparing it to samples collected from recent freshwater bloom and fish kill events in North America, *Toxicon*, 55, 1396–1404, 2010.

FIGURE 3.3
Specimen of *Microciona prolifera*, the "red-bearded sponge of Moses," collected from St. Joseph Bay, Florida. (Photo courtesy of Richard B. Philp.)

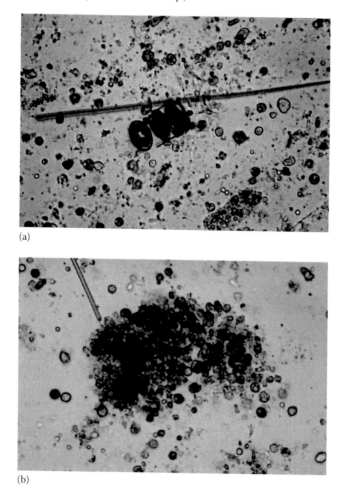

(a)

(b)

FIGURE 3.4
(a) Sponge cells dissociated in a suspension of calcium/magnesium-free artificial seawater. (b) A large mass of aggregated sponge cells formed after the addition of $CdCl_2$ (final concentration 24 µM) (microscope power × 100). (Reprinted from *Comp. Biochem. Physiol.*, 118C, Philp, R.B., 347–351, Copyright 1997, with permission from Elsevier.)

FIGURE 11.1
A staff member of the Gulf Specimen Marine Laboratory holds a scorpion fish in the palm of her hand. The dorsal spines (folded back here) are the means of conveying the venom into tissues.

FIGURE 11.2
Top: An Australian snake handler is holding three, red-bellied black snakes (*Pseudoechis por-phyriacus*). While venomous, the bite of this snake is rarely fatal and it is not too aggressive. Middle: Two tiger snakes (*Notechis* sp.) have been just released from their bag by the handler. This snake is highly venomous and the mortality if untreated with anti-venom is 40 to 60%. Bottom: The handler restrains this eastern brown snake (*Pseudonaja* sp.), considered to be the second most venomous land snake. It can grow to two meters and is very fast. Mortality is similar to the tiger snake. All of these snakes are elapids.

Hirama, M., Oishi, T., Uehara, H., Inoue, M., Maruyama, M., Oguri, H., and Sataki, M., Total synthesis of ciguatoxin CTX3C, *Science*, 294, 1904–1907, 2001.

James, L.F., Kip, E., Panter, E., Darwin, B., and Molyneux, R.J., The effect of natural toxins on reproduction in livestock, *J. Anim. Sci.*, 70, 1573–1579, 1992.

Junghanss, T. and Bodio, M., Medically important venomous animals: Biology, prevention, first aid, and clinical management, *Clin. Infect. Dis.*, 43, 1309–1317, 2006.

Kini, R.M., Anticoagulant proteins from snake venoms: Structure, function and mechanism, *Biochem. J.*, 397, 377–387, 2006.

Larm, J.A., Beart, PM., and Cheung, N.S., Neurotoxin domoic acid produces cytotoxicity via kainate- and AMPA-sensitive receptors in cultured cortical neurons, *Neurochem. Int.*, 31, 677–682, 1997.

Larréché, S., Mion, G., Mornand, P., and Imbert, P., Adder bites in France (Fr.), *Arch. Pediatr.*, 19, 660–662, 2012.

Malins, D.C. and Ostrander, G.K. (eds.), *Aquatic Toxicology: Molecular, Biochemical and Cellular Perspectives*, Lewis Publishing, Boca Raton, FL, 1994.

Manners, G.D., Plant toxins: The essences of diversity and a challenge to research. *Adv. Exp. Biol. Med.*, 391, 9–35, 1996.

Noguchi, T., Onuki, K., and Arakawa, O., Tetrodotoxin poisoning due to puffer fish and gastropods, and their intoxication mechanism, *Int. School Res. Net. Toxicol.*, 2011, doi:10.5402/2011/276939, Article ID 276939, 2011.

Ostrander, G.K. (ed.), *Techniques in Aquatic Toxicology*, Lewis, Boca Raton, FL, 1996.

Patrick, J.D., Scombroid toxicity, Medscape Ref., drugs, diseases and procedures, http://emedicine.medscape.com/article/818338-overview (accessed on November 17, 2011).

Peng, J., Place, A.R., Yoshida, W., Anklin, C., and Hamann, M.T., Structure and absolute configuration of karlotoxin-2, an ichthyotoxin from the marine dinoflagellate *Karlodinium veneficum*, *J. Am. Chem. Soc.*, 132, 3277–3279, 2010.

Scheuer, P.J., Marine natural products: Diversity in molecular structure and biodiversity. *Adv. Exp. Biol. Med.*, 391, 1–8, 1996.

Scombroid Fish Poisoning, *Morbid. Mortal. Wk. Rep.*, 38, 140–147, 1989.

Sivonen, K., Cyanobactrial toxins. In *Encyclopedia of Microbiology*, Schaechter, M. (ed.), 3rd Edn, Elsevier, Waltham, MA, 290–307, 2012.

Tamiya, N. and Yagi, T., Studies on sea snake venom, *Proc. Jpn. Acad. Ser. B Biol. Sci.*, 87, 41–52, 2011.

Tu, A.T., Overview of snake venom chemistry. *Adv. Exp. Biol. Med.*, 391, 37–62, 1996.

Valenta, J., Stach, Z., and Otahal, M., Protobothrops mangshanensis bite: First clinical report of envenoming and its treatment, *Biomed. Pap. Med. Fac. Univ. Palacky Olomouc. Czech Repub.*, 156, 183–185, 2012.

Warrell, D.A., Venomous bites, stings and poisoning, *Infect. Dis. Clin. North Am.*, 26, 207–223, 2012.

Water hemlock poisoning-Maine, 1992, *Morbid. Mortal. Wk. Rep.*, 43, 229–231, 1994.

12

Environmental Hormone Disrupters

Every man in this room is half the man his grandfather was.

**Louis J. Guilette Jr., zoologist, to Congressional Committee,
on endocrine disrupters**

Introduction

Burlington and Lindeman in 1950 made one of the earliest observations that synthetic chemicals could seriously impair normal reproductive function. They showed that leghorn cockerels exposed to DDT had impaired testicular growth and diminished secondary sexual characteristics. A decade or so later, reports began to emerge that women whose mothers had received diethylstilbestrol (DES) during pregnancy experienced difficulties in conceiving and an increased incidence of cervical deformities and cancer (see also Chapter 8). The first reported case of clear cell carcinoma in a DES daughter was in 1971. More recently, the men exposed to DES *in utero* have also been shown to have an increased incidence of reproductive dysfunction. It is now well known but poorly publicized that the source of infertility problems can be traced to the male 50% of the time. About the same time that the DES problem was emerging, Glen Fox, a scientist with the Canadian Wildlife Service, discovered that gulls in Lake Ontario were showing numerous signs of disrupted reproductive function. Females were sitting on eggs that refused to hatch. Males were losing interest in sex, forcing females to pair up to brood over sterile eggs. The eggs themselves frequently were misshapen and fragile. Fox speculated that pollutants such as DDT and PCBs, persistent organic pollutants (POPs) that are now banned, could be responsible. In 1963, Rachel Carson, formerly an aquatic biologist with the U.S. Fish and Wildlife Service, published her book *Silent Spring* in which she warned that the indiscriminate use of pesticides could have catastrophic environmental effects. By the mid-1970s her words were appearing highly prophetic. Since then, numerous classes of chemicals have been shown to possess the ability to modulate or severely disrupt hormonal function, either experimentally or in wildlife.

Lake Apopka Incident

In 1980, a small, chemical mixing company spilled massive amounts of sulfuric acid dicofol, a pesticide for mite control containing DDT, into Lake Apopka, the fourth largest body of freshwater in Florida. Although the size of the spill was undetermined, 2 weeks after it, the pH of the lake water was still as low as 1.7, about that of stomach acid. Six years later, a study by the University of Florida at Gainesville found high numbers of unhatched eggs in alligator nests and by 1988 only 4% of the eggs were hatching. In the hatchlings following the spill, both testes and ovaries showed anatomical abnormalities. In both sexes, estrogens had come to dominate. The male/female ratio of hatchlings is controlled by the temperature of the nest. This species may be very sensitive to outside disturbances of hormonal balance. In females, the ratio of estradiol to testosterone was twice normal. In males, levels of testosterone were depressed and penises were abnormally small. This episode clearly indicated that exposure to high concentrations of synthetic chemicals could severely affect reproduction. The question of the effects of long-term exposure to very low levels of such chemicals is not as clear.

Brief Review of the Physiology of Estrogens and Androgens

Estrogens and androgens are, respectively, the male and female steroid hormones that regulate reproductive function and behavior and the development of secondary sex characteristics. The most potent are 17β-estradiol (E2) (female) and testosterone (male). Estrogen synthesis primarily by ovarian follicles is controlled by Follicle-Stimulating Hormone (FSH) from the anterior pituitary gland (adenohypophyseal gland). Release of FSH is regulated by a negative feedback system. Androgen synthesis by interstitial Leydig cells of the testes is stimulated by luteinizing hormone (LH). This also is regulated by a negative feedback loop. FSH and LH, along with thyroid-stimulating hormone (TSH) constitute the glycoprotein group of hormones.

Following synthesis, both sex hormones are transported in blood 99% bound to plasma proteins, primarily to Sex Hormone Binding Globulin (SHBG). In females of reproductive age, total E2 serum levels are 50–300 pg/mL, depending on the stage of the estrous cycle. Males also have circulating E2 levels (10–60 pg/mL). At the target cell, cytosolic receptors complex with their respective hormones and act as transcription factors, transporting the hormone to the nucleus where they bind to specific response elements and induce the transcription-regulated genes. When E2 binds to its receptor (ER), it induces a conformational change, forming a dimer with another ER complex. Testosterone is acted on intracellularly by 5-α-reductase to form 5-α-dihydrotestosterone. In some tissues, bone marrow, and skeletal muscle, for example, testosterone itself and other metabolites are the active forms.

Disruption of Endocrine Function

Mechanisms

Xenobiotic estrogens (xenoestrogens) belong to a group of chemicals known as persistent organic poisons (POPs) that are characterized by long biological half-lives and high lipid solubility. These characteristics contribute to their tendency to biomagnify in the environment and to accumulate and persist in lipid stores. There are several mechanisms by which xenoestrogens could, at least in theory, disrupt or modulate normal hormone function. With regard to estrogen (17β-estradiol, hereafter referred to as E2)

1. They could mimic estrogen by binding directly with its receptor.
2. By this same mechanism they could also prevent E2 from interacting normally with its receptor.
3. They could react directly or indirectly with E2 carrier proteins.
4. They could react directly with free or bound E2 to alter plasma E2 levels.
5. They could interfere with E2 synthesis.
6. They could up or down regulate the number of E2 receptors available.
7. They could antagonize E2 by virtue of inherent androgenic activity.
8. Although seldom included in a discussion of environmental hormone disruption, any toxic agent that is present in sufficient concentration to cause overt toxicity, either acute or chronic, can have a negative effect on reproductive function.

The same mechanisms could also apply to interference with normal androgenic function. Most research, however, has focused on interference with E2 activity.

Methods of Testing for Hormone Disruption

Two main types of laboratory studies have been used to assess the potential of xenoestrogens for hormone-modulating activity. The first involves receptor-binding studies using E2 receptor proteins from a variety of species including human. These E2 receptor proteins etc. may be in their natural cell membrane, as in cultured, E-sensitive breast cancer cells such as the MCF-7 cell line, or they may be transfected into another cell type such bacteria or cultures human cell lines. The second involves *in vivo* studies in rodents, usually rats. Since much of the theoretical concern about environmental hormone disrupters centers on their potential to cause damage to offspring *in utero*, studies often involve administration of the test substance to pregnant rats, with examination of the pups for abnormalities. These include delayed vaginal opening, anatomical differences such as weights of the uterus or testes, and anatomical defects.

More recently, the use of an ovoviviparous species of rice-fish, the medaka (*Oryzias latipes*) has been employed. This fish is popular for developmental studies. A version is transfected with the green fluorescent protein (GFP) placed under the control of a gene-regulating gonadotropin-releasing hormone (GnRH). Medaka embryos are transparent so that the green fluorescence can be readily observed *in vivo*. Exposure to xenoestrogens such as bisphenol A (BPA), nonylphenol, as well as 17-β-estradiol suppressed the fluorescence as well as lowered heart rate and lengthened the time to hatch. Transfected cell lines have also been used as test models for estrogenic activity.

Field studies include the examination of wildlife species that have been exposed to high concentrations of xenobiotics, notably marine and aquatic ones, for evidence of impaired reproductive function. In humans, populations exposed accidentally to such agents (e.g., the women in Seveso exposed to high levels of TCDD) are monitored for reproductive problems. Epidemiological studies attempt to examine trends over a long time course during which it is assumed that exposure levels to hormone disrupters have increased. The sperm count studies are of this category. Populations of workers or others known to have unusual exposures are also useful, especially if effects can be correlated to serum levels of the suspect agent.

Some Examples of Xenoestrogen Interactions with E2 Receptors *In Vitro* or Effects in *In Vivo* Tests

BPA, a chemical used in the manufacture of plastics, is an environmental estrogen. Its interaction with the human E2 receptor Er-alpha has been investigated using human hepG2 hepatoma cells with the transfected receptor. Compared to E it was 26 times less potent and acted as a partial agonist. By using ER-alpha mutants in which the AF1 or AF2 regions were inactivated, differing patterns of activity were demonstrated for BPA, weak agonists, estrol and estriol, partial agonists, antagonists. BPA had no effect on uterine weight when given to immature female rats but shared some effects with E2 on peroxidase activity.

Of the hundreds of PCBs examined, only a handful has shown estrogenic activity. When the effects on age of vaginal opening were compared in rats, their potency as compared to diethylstilbestrol (DES) was 1/80,000–1/1,000,000. Even the most potent required four to five times the dose of E2 to affect the pituitary secretion of LH and FSH. Estrogenic activity seems to require substitution in the ortho position and non-ortho-substituted PCBs may actually possess antiestrogenic activity.

In studies using MCF-7 breast cancer cell cultures, technical-grade dichlorodiphenyltrichloroethane (DDT) and its metabolite *o,p'*-DDT had estrogenic activity about 1/1,000,000 that of E2s. When 5 mg/kg/day was given for 27 days to female rat pups, vaginal opening was delayed and uterine and ovary weights were increased. Lower doses had no effects. Mature animals required a dose of 25 mg/kg/day for 7 days to develop an increase in uterine weight.

Several environmental chemicals were tested in a yeast-based, human E2 receptor assay. Most potent were the DDT metabolites *o,p'*-DDT, *o,p'*-DDD, and *o,p'*-DDE. Their potencies were, respectively, $1/8$, $1/15$, and $1/24 \times 10^{-6}$ less potent than E2. Other organochlorines like dieldrin were of similar potencies. In the *in vivo* assay, dieldrin, aldrin, and mirex had no effect on time of vaginal opening, estrous activity, or ovulation. Kepone, however, increased uterine weight and had E2 receptor binding characteristics that were 0.01%–0.04% that of E2.

Several compounds have been shown to reduce E2 binding to its receptor. Most potent of these were butyl benzyl phthalate (BPP) and di-*n*-butyl phthalate (DBP) from plastics and the antioxidant food additive, butylated hydroxyanisole (BHA). Their potencies were also about 1/1,000,000 that of E2. The plasticizer bisphenol-A was considerably more potent, being 1/1000–5000 times less potent than E2 in the MCF-7 cell culture assay.

Some Effects of Xenoestrogens on the Male Reproductive System

Male mice exposed to DES *in utero* had, as adults, decreased sperm production and a high frequency of abnormal morphology. It is axiomatic that exposure to high levels of estrogenic substances will have a deleterious effect on the male reproductive system, whether such exposure occurs *in utero* or not. A variety of xenobiotics have been examined for their ability to cause DNA breaks in rat and human cultured testicular cells. Methoxychlor, benomyl, thiotepa, acrylonitrile, and Cd^{++} had little effect on either. Styrene oxide, 1,2,-dibromoethane and chlordecone induced a significant number of strand breaks in both types of cell although not always at the same concentration. Several others, including aflatoxin B1, induced strand breaks only in rat testicular cells. Octylphenol has been shown to impair spermatogenesis in male rats and to decrease the percentage of viable cultured spermatogenic and Sertoli cells.

One of the most potent environmental antiandrogens yet identified is the dicarboximide antifungal agent vinclozoline. Administration of 100 or 200 mg/kg/day to pregnant, female rats did not produce any signs of toxicity nor did it affect the viability of pups. All male pups were misclassified as females at birth, had delayed puberty, and impaired copulation. Numerous anatomical anomalies were seen at necropsy including vaginal pouches and ectopic testes. These effects appear to be due to antiandrogenic activity and not to estrogenic activity. There was no evidence of estrogenic effects in female pups. No anatomical anomalies were noted and fecundity was normal. Primary metabolites of vinclozoline, but not vinclozoline itself, were shown to bind to the androgen receptor. 2,2,-bis(p-chlorophenyl)-1,1 dichloroethane (p,p'-DDE) also has antiandrogenic activity and tributyltin is an environmental androgen.

Modulation of Hormone Activity through Effects on the Ah Receptor

Estrogen/Androgen Effects

A great number of compounds are capable of modulating both estrogenic and androgenic function either by acting as ligands for the aryl hydrocarbon (Ah) receptor (AhR) or by acting as Ah inducers. In the former category are dibenzo-p-dioxins, dibenzofurans, biphenyls, diphenyl ethers, substituted PAHs, hydroxylated benzo[a]pyrenes, and many others. These may induce genes in a tissue- and species-dependent manner, including those responsible for the synthesis of estrogens and androgens. Thus, both estrogenic and androgenic effects can occur.

Effects on Thyroid Function

A number of studies have shown hypothyroid effects of PAHs. Rat pups exposed perinatally to PCBs had T_4 levels that were depressed whereas T_3 levels were not. Studies with TCDD required doses much higher than the no effect level to depress T_4 and again, T_3 levels were not altered. Studies in adult rats have shown a depression of plasma T_4 following the administration of a number of PAHs and organohalogens. Both transport and metabolism of T_4 were felt to play a role. The results generally indicated that PCBs and related compounds affected thyroid function to a greater extent than TCDD.

Other compounds bind poorly to the AhR but are capable of inducing it. These include imidazoles, pyridenes, oxidized carotenoids heterocyclic amines, fumonisin B1, and others. TCDD has been shown to induce endometriosis in female monkeys and goiter, hypothyroidism, and hyperthyroidism in rats. See Chapter 5 for more on the reproductive toxicity of TCDD and other AhR ligands.

A host of synthetic hormone disrupters have entered the waterways draining into the Great Lakes. They pose a threat to both humans and animals. PCBs have been shown to lower circulating thyroid levels in humans. Fetuses and infants may be especially vulnerable to this effect. Vietnam veterans who were exposed to high levels of TCDD had elevated levels of TSH.

The Michigan accident involving the fire retardant containing PBBs revealed depressed circulating levels of thyroid hormone. This accident, including its adverse effects on human health, is discussed in Chapter 5.

To date, there has been no epidemiological study that demonstrated a link between environmental xenoestrogens and either depressed sperm counts or reproduction function in men, or breast cancer in women. The problems with *in utero* exposure to DES are discussed in Chapter 8.

Plastic-Associated Chemicals

BPA, as noted earlier, is used in the production of plastics as are phthalates. Both have been implicated in disrupting other hormonal functions besides estrogen- and androgen-related ones. They have produced depressed thyroid levels when given to experimental animals. Thus, there is a wide variety of synthetic chemicals capable of disrupting thyroid function and they are widely distributed in the environment.

Recently, concerns have been expressed that exposure to BPA and phthalates could be contributing to the epidemic of obesity pervading Western societies, especially in North America, and to the looming health crisis of skyrocketing cases of type 2 diabetes. BPA and phthalates as well as organotins are activators of the peroxisome proliferator-activated receptor (PPARγ), which activates adipocytes, and hence promotes obesity, as well as affecting glucose and lipid regulation and insulin responsiveness, creating a condition favorable to the development of type 2 diabetes. The association of these chemicals with the increased risk of type 2 diabetes is most prominent when combined with high calorie, high carbohydrate, and high fat diets later in life. An association with increased risk of cardiovascular disease has also begun to emerge.

Phytoestrogens

Phytoestrogens are natural estrogenic substances found in cereal grains, vegetables, and fruit. Estrogenic activity in plant extracts was first identified in 1926 and by 1975 several hundred plants were shown to contain substances with estrogenic activity. Three principal classes of phytoestrogens occur in plants and their seeds: isoflavones, coumestans, and lignans. Different parts or stages of the same plant may contain different phytoestrogens. Thus, soy sprouts are rich in coumestrol, whereas the bean has high levels of isoflavones. Mycoestrogens such as zearalenone also exist and have caused reproductive problems in livestock (see Chapter 10).

Some relative potencies, taking that of E2 as 100 in the MCF-7 bioassay, are

Coumestrol (the principal coumestan), found in sunflower, alfalfa, legumes, soybean products, 0.03–0.2.

Genistein and daidzein, found in barley, oats, rye, rice, wheat, and soy products, 0.001–0.01.

By comparison, the mycoestrogens zearalenone and its metabolite zearalenol have relative potencies 0.001–0.1.

Coumestrol is, thus, the most potent of the phytoestrogens.

Results of Human Studies on Xenoestrogens

Males

Golden et al. published an extensive review of the literature in 1998. Concern over the effects of environmental xenobiotic disruption of male reproductive function stems mainly from a report by Carlsen et al. in 1992 presenting evidence that sperm counts in men had declined steadily from 1938 to 1990. It was assumed that, during this period, exposure to xenobiotics would have been high and that this could have been responsible for the decline. Recent findings that men exposed *in utero* to DES had impaired sperm quality lent weight to this supposition. The methodology of the Carlsen study was criticized for failure to control for such confounders as smoking, marijuana use, and environmental temperature, all known to affect sperm count. When their data were reexamined, it was found that all of the low sperm counts occurred before 1970. In a more recent study of men who banked sperm before vasectomy there was a slight but significant increase in sperm count from 1970 to 1994 and no change in motility or volume. The Airforce Health Study of Vietnam veterans exposed to Agent Orange during Operation Ranch Hand (the application of herbicides to defoliate the jungle) found no correlation between serum dioxin levels and either sperm quality or the number of conceptions. As noted earlier, there have been no conclusive epidemiological studies linking low sperm count to exposure to xenoestrogens.

Experimental evidence that xenoestrogens increased the incidence of undescended testes (cryptorchidism) prompted epidemiological studies. A study of 6935 live births at Mount Sinai Hospital in New York did not indicate any change in the incidence of cryptorchidism between 1950 and 1993. The incidence at 3 months of age stayed at about 1%. In Great Britain there was a slight increase in the incidence but no conclusions could be drawn as to the cause.

The incidence of prostate cancer has been increasing for several years. Much of this increase is attributed to earlier and better diagnosis and to an aging population. Epidemiological studies, however, have shown that agricultural workers, golf course managers, and pesticide applicators have a slightly higher incidence of prostate cancer. The applicators had a relative risk factor of 2.38. The 95% confidence limit was 1.38–3.04.

There has been a dramatic increase in the incidence of testicular cancer over the past few decades. Studies in men exposed perinatally to DES do not indicate that it is a risk factor for testicular cancer. One study suggested an association between pesticide use and a slight increase in the incidence of both prostate and testicular cancer in Hispanic and Black agricultural workers in California. No firm conclusions can be drawn at this time regarding exposure to pesticides or other chemicals as a risk factor for testicular cancer.

Since organochlorines have been shown experimentally to produce a number of thyroid-related endocrine disturbances (see the preceding text), there has naturally been great interest in the significance of this for humans. The Airforce Health Study, in 1987, examined TCDD serum levels in 866 Ranch Hand veterans. For 319 officers the median level was 7.8 parts/trillion (ppt, 0–42.6), for 148 enlisted air crew it was 18.1 ppt (0–195.5), and for 399 enlisted ground crew it was 24 ppt (0–617.8). Extrapolating back to the initial exposure levels 130 had 52–93 ppt, 260 had 93–292 ppt, and 131 had greater than 292 ppt. A decade later the highest level group had slightly elevated TSH levels compared to normal background levels. This is suggestive of a slight decrease in thyroid function. T_3 uptake by the thyroid gland was 29.99% compared to 30.65% for normal subjects with background levels of TCDD.

Females

Xenoestrogens have been implicated as risk factors for breast cancer and endometriosis. To date there has been no demonstrated association between risk for endometriosis and exposure to xenoestrogens other than the reported findings in DES daughters. Several studies have been conducted in which levels of PCBs and other xenoestrogens in breast fat, blood, or serum were measured in breast cancer patients. Cancer incidence was then correlated with levels of the xenobiotics. A study conducted at Mount Sinai hospital in New York and published in 1992 found that women with the highest levels of DDE in their blood had a breast cancer incidence that was four times higher than that of women with the lowest DDE levels. By 1994, four studies had been published showing an association between levels of DDE and PCBs in fluids or tissues and an increased incidence of breast cancer. (It is a commentary on the ubiquitous nature of POPs that no population totally free of contamination can be found as a control group.) Some of these studies were criticized for failure to control for known risk factors for breast cancer and for small numbers of participants. The New York study, for example, involved 58 women. In April of 1994 a study led by Nancy Krieger of the

Kaiser Foundation Research Institute examined blood that had been frozen and stored in the late 1960s from women who developed breast cancer, on average, 14 years later. There were 50 Caucasian, 50 African–American, and 50 Asian women. Blood levels of DDE in these samples were four to five times higher than those found in the New York samples (drawn 1984–1991) because the Kaiser samples were drawn prior to the 1972 ban on DDT. When the results were compared to matched controls, no association could be demonstrated between serum levels of DDE and risk of breast cancer. Similar findings were obtained for PCBs.

In 1999 the (U.S.) National Cancer Institute released results of a similar study. Women who donated blood over a 10 year period from 1977 were followed for up to 9.5 years. For each subject, two controls were selected matched for age and date of blood collection; 105 donors who matched the selection criteria were diagnosed with breast cancer. Serum was analyzed for 5 DDT analogs, 13 other organochlorine pesticides, and 27 PCBs. For hexachlorobenzene, the women in the upper three quartiles of serum levels had twice the risk of breast cancer as those in the lowest quartile. There was no evidence of a dose–response relationship and the finding was limited to women whose blood was collected close to the time of diagnosis. Positive but weak associations were suggested for PCBs 118 and 138. Again, this association was noted only in women whose blood was collected close to the date of diagnosis. There would normally be a significant latency period between exposure to the carcinogenic agent and the development of malignancy. No other positive associations were found.

Two studies on organochlorines and risk of breast cancer in South America were published recently. One from Brazil found no correlation between blood levels of DDE and breast cancer risk when 177 hospital patients with invasive breast cancer were compared to 350 controls (hospital visitors). The other, from Columbia, compared 153 cases of breast cancer with 153 age-matched controls. Odds ratios were adjusted for a number of factors such as first-child breast feeding. An odds ratio of 1.95 (confidence interval 1.10–3.52) suggested increased risk for the higher category of DDE exposure but the test for trend was not statistically significant ($p = 0.09$). Taken overall, there is no strong evidence for or against the premise that environmental exposure to xenoestrogens constitutes a significant risk factor for breast cancer. Further studies may clarify the situation. The massive exposures to PBBs that occurred in Yusho Japan in 1968 and in Yu-Cheng Taiwan in 1979 (see also Chapter 5) have revealed a number of developmental defects in children exposed in the womb. Total body burdens in the mothers were calculated to be 200 times higher than the average for North America. The Michigan study of the effects of pregnant women eating large amounts of lake fish on their offspring is discussed in Chapter 5.

The situation with regard to phytoestrogens appears to be quite different. Epidemiological studies suggest that consumption of a diet rich in phytoestrogens, such as those eaten in Asian societies, is associated with lower incidences of breast, prostate, and colon cancers, as well as cardiovascular disease. Similar findings have been reported for people on strict vegetarian diets. As noted earlier, coumestrol is the most potent of the phytoestrogens and soy protein, a staple in Asian diets, is rich in coumestrol. People in Japan, Taiwan, and Korea are estimated to consume 20–150 mg/day of isoflavones. Menopausal women in Japan are reported to have fewer hot flushes than their North American counterparts.

The phytoestrogen content of plants can vary widely with climate, time, and geography, making estimates of intake difficult. In one study, 12 oz of soy milk three times daily decreased serum E2 and progesterone levels in 22- to 29-year-old women. Other effects of phytoestrogens on hormonal balance have been observed experimentally. While the clinical use of phytoestrogens has not yet occurred widely in Western medicine, their use in herbal medicine is becoming widespread. In his review of the subject, Sheenan points out the problems of lack of standardization of content and of the likelihood that herbal preparations can contain many other active ingredients, some of which could be toxic.

Effects of Xenoestrogens and Phytoestrogens in Livestock and Wildlife

In 1946, a report appeared in the literature of sheep in Western Australia developing infertility as a result of eating subterranean clover. This plant is rich in the equol precursor formonestein. Equol is a known phytoestrogen. Reproductive problems in swine consuming feed grains with high zearalenol content (a mycotoxin) have been discussed. There are numerous reports of livestock experiencing fertility problems as a result of grazing on poor pastures containing phytoestrogen-rich weeds. Wild quail in California also have been reported to have similar problems.

The reproductive problems in gulls and other fish-eating birds associated with DDT, and the Apopka chemical spill and its effects on alligators, have been discussed. Vertebrate and invertebrate aquatic and marine animals also are vulnerable to the effects of POPs. Estrogenic activity has been detected in treated sewage outfalls and industrial effluents. Vertebrates such as fish, downstream from such effluents, have become feminized. Vitellogenin induction in males, morphological changes, and complete sex reversal have been reported. Xenoestrogens, natural sterols, organochlorines, and alkylphenols have been identified as suspected agents.

Invertebrates also have been shown to be vulnerable to endocrine disruption by environmental xenobiotics. Intersex in crustaceans exposed to sewage discharges has been noted and endocrine disruptions in gastropod mollusks. Barnacle larvae displayed reduced resettlement activity following exposure to the xenoestrogen 4-nonyl phenol. Endocrine disrupters have been shown experimentally to have a negative effect on growth and reproduction of amphipods and marine worms, both of which inhabit and feed in bottom sediment.

Reproductive problems in livestock have resulted from the consumption of other toxic plants. *Veratrum californicum*, known as false hellebore or California corn lily, is rich in toxic alkaloids that affect the heart and nervous system. These can cross the placental membrane and interfere with fetal development in sheep. Some species of lupines, notably elegant lupine (*Lupinus lepidus*), can also cause fetal abnormalities when consumed by cattle. Locoweed (*Astragalus lentiginosus*) has estrogenic properties and has been shown to interfere with spermatogenesis and oogenesis in cattle and sheep. There are several *Astragalus* species and all are toxic. Bizarre behavior is typical of poisoned animals and gives rise to the common name. There is even a suggestion that animals may become addicted to locoweed, foregoing other pasture plants in favor of it. Horses are especially vulnerable to poisoning. For excellent reviews of this extensive subject (see James et al. 1992; Safe et al. 2001).

Dozens of toxic plants may be consumed accidently by milk cows. Their alkaloid toxins are secreted in the milk and can pose a threat to human health. The deaths of Abraham Lincoln's mother and of several other people were attributed to drinking milk containing the toxin of white snake root (*Eupatorium rugosum*, now known as *Ageratina altissima*). It and rayless goldenrod (*Isocoma pluriflora*) contain tremetol and trematone, perhaps the best-known plant toxins secreted in milk. They cause a condition known as milk sickness. It is characterized by muscle tremors, loss of appetite, abdominal cramps, vomiting, severe thirst, delirium, and coma, which may lead to fatality. See the excellent reviews by Knight and Walter for more information. Before the discovery of the cause of milk sickness, thousands died in the early 1800s from it. White snakeroot can be found throughout North America east of the Rocky Mountains. It can grow to over 1 m in height, has heart-shaped leaves on opposing stalks, and clusters of small, white flowers branching from the end of a single, otherwise bare stalk.

Problems in Interpreting and Extrapolating Results to the Human Setting

There is difficulty in extrapolating from the laboratory findings to the whole organism, whether it is mammal, fish, or reptile. Species differences abound with respect to all sites of potential interference, and differing sensitivities of E2 receptors have been recorded. One study compared the binding of 44 PCBs, 9 hydroxylated PCBs, and 8 arochlors to recombinant human, rainbow trout, and green anole (lizard) E2 receptors (fusion proteins) linked to the glutathione-*S*-transferase protein and expressed bacterially. Only PCBs 104, 184, and 188 competed effectively with E2 for the rainbow trout protein and only displaced E2 by 30% from the others. Other congeners showed varying affinities for the rainbow trout protein but none for the others.

A further confounding factor is the fact that many agents have multiple effects that can all impact on reproduction and some of these may be mutually opposed. Thus, the isoflavone genistein is a potent inhibitor of both E receptor-positive and E receptor-negative breast cancer cells of the MCF-7 strain. It is also a potent inhibitor of the tyrosine protein kinase activity of several growth factor receptors and oncogenes that might be involved in the growth of tumor cells. Other effects of genistein have been recorded.

Lack of potency is another factor that raises questions about the health impact of environmental hormone modulators. Much of the present concern stems from the DES experience where exposures *in utero* were much higher than what could be encountered environmentally. Coumestrol is the most potent of the phytoestrogens and the one most associated with mild estrogenic effects in women. Taking E2 potency as 100 in MCF-7 cell culture tests, coumestrol had a potency of 0.3–0.02. It is several orders of magnitude more potent than the xenoestrogens. In areas where the dietary intake of phytoestrogens is high, their metabolites are present in urine in easily measured amounts. This is seldom true of xenoestrogens except in populations with abnormally high exposures.

Since we are exposed to a veritable potpourri of environmental hormone modulators, many of them natural agents with potential benefit, the net effect is likely to be the result of interactions among many agents, some with mutually opposed activity. At present, we can only say that the evidence for environmental hormone modulators having a significant impact on human health is inconclusive and that more evidence is required. There is, however, a good case to be made for invoking the precautionary principle in calling for the eventual elimination of POPs. This is especially true with regard to the environmental damage POPs can cause.

Case Study 27

A family, which belongs to a religious sect, operates a farm in southwestern Ontario. The sect eschews modern technology of any kind. They till their fields with horse-drawn implements, grow their own food, raise their own crops, and sell the excess at a local farmer's market. In mid-July one day, the family had dinner around 5:30 pm. It consisted of cold, smoked ham, homemade potato salad, homemade apple sauce, fresh green peas, and cold milk. Dessert was peaches and cream with brown sugar. About 30 min after the meal, families began to complain of stomach cramps. This was followed by vomiting, severe thirst, and muscle tremors. One of the family members became delirious, so a neighbor with a motor vehicle was contacted, and he drove them to the nearest emergency department. Gastric lavage was performed and i.v. fluids and sedatives were administered. The family all recovered over the course of several days.

The neighbor returned to their farm to feed the livestock and discovered three calves staggering around and trembling. The local veterinarian was called and identified the source of the problem.

Q. Given that the time course would suggest the meal as the source of the family's problem, what could they have ingested that would explain the stomach cramps and vomiting? Ignore the other symptoms for now.

Q. Taking into account the whole array of symptoms, what is your initial differential diagnosis?

Q. Does the presence of the sick calves help with the diagnosis?

Q. Why was the veterinarian able to pinpoint the problem so quickly?

Review Questions

For Questions 1–20 answer true or false.

1. Impairment of reproductive function and testicular growth by DDT was first demonstrated in chickens in 1950.
2. Some men exposed to DES *in utero* appear to have impaired reproductive function.
3. High levels of estrogens in male alligators lead to increased penis size.
4. Follicle-Stimulating Hormone (FSH) primarily regulates the synthesis of testosterone.
5. Synthesis of sex hormones is regulated by negative feedback loops.
6. Males have circulating levels of estrogens (E2).
7. Testosterone is the active form of androgen in all tissues.
8. The estrogen (E2) receptor is located on the cell membrane.
9. Xenoestrogens have long biological half-lives.
10. The affinity of xenoestrogens for the E2 receptor is about the same as that of E2 itself.
11. Only a few PCBs have significant estrogenic activity.
12. Non-*ortho*-substituted PCBs may possess antiestrogenic activity.

13. DDT and its metabolite *o,p'*-DDT are equipotent with E2 in the breast cancer cell culture assay.

14. The insecticide kepone, one of the most potent xenoestrogens, has E2 receptor binding characteristics about 0.01%–0.4% of that of E2.

15. Aflatoxin B1 induces DNA strand breaks in rat testicular cells.

16. Vinclozoline is probably the most potent xenobiotic antiandrogen yet identified.

17. TCDD has no adverse effects on reproductive function.

18. Phytoestrogens tend to be less potent than xenoestrogens.

19. Pleiotropic effects modulated by the Ah receptor can include disruption of reproduction.

20. Fish and other wildlife are more vulnerable to xenoestrogen toxicity than mammals.

Answers

1. True
2. True
3. False
4. False
5. True
6. True
7. False
8. False
9. True
10. False
11. True
12. True
13. False
14. True
15. True
16. True
17. False
18. False
19. True
20. True

Further Reading

Burlington, H. and Lindeman, V.F., Effect of DDT on testes and secondary sex characteristics of white leghorn cockerels, *Proc. Soc. Exp. Biol. Med.*, 74, 48–51, 1950.

Bjorge, C., Bruneborg, G., Wiger, R., Holme, J.A., Scholz, T., Dybing, E., and Soderlund, E.J., A comparative study of chemically induced DNA damage in isolated human and rat testicular cells, *Reprod. Toxicol.*, 10, 509–519, 1996.

Boas, M., Main, K., and Feldt-Rasmussen, U., Environmental chemicals and thyroid function: An update, *Curr. Opin. Endocrinal Diabetes Obes.*, 16, 385–391, 2009.

Carlsen, E., Giwercman, A., Keiding, N., and Skakkebaek, N.E., Evidence for decreasing quality of sperm during the last 50 years, *Br. Med. J.*, 305, 1228–1229, 1992.

Denison, M.S. and Helferich, W.G. (eds.), *Toxicant-Receptor Interactions,* Taylor & Francis Group, Philadelphia, PA, 1998.

Depledg, M.H. and Billinghurst, Z., Ecological significance of endocrine disruption in marine invertebrates, *Mar. Pollut. Bull.*, 39, 32–38, 1999.

Dorgan, J.F., Brock, J.W., Rothman, N., Needham, L.L., Miller, R., Stephenson, H.E., Schussler, N., and Taylor, P.R., Serum organochlorine pesticides and PCBs and breast cancer risk: Results from a prospective analysis (USA), *Cancer Causes Control*, 10, 1–11, 1999.

Fisch, H., Goluboff, E.T., Olson, J.H., Feldshuh, J., Broder, S.J., and Barad, D.H., Semen analysis in 1,283 men from the United States over a 25-year period: No decline in semen quality, *Fertil. Steril.*, 65, 909–911, 1996.

Fleming, L.E., Bean, J.A., Rudolph, M., and Hamilton, K., Mortality in a cohort of licensed pesticide applicators in Florida, *Occup. Environ. Med.*, 56, 14–21, 1999.

Foster, S. and Caras, R., *Venomous Animals and Poisonous Plants, Roger Tory Peterson Field Guides*, Easton Press, Norwalk, CT, 1994.

Golden, R.J., Noller, L., Titus-Ernstoff, L., Kaufman, R.H., Mittendorf, R., Stillman, R., and Resse, E.A., Environmental endocrine modulators and human health: An assessment of the biological evidence, *Crit. Rev. Toxicol.*, 28, 109–227, 1998.

Gould, J.C., Leonard, L.S., Maness, S.C., Wagner, B.L., Conner, K., Zacharewski, T., Safe, F., McDonnell, D.P., and Gaido, K.W., Bisphenol A interacts with the estrogen receptor alpha in a distinct manner from estradiol, *Mol. Cell. Endocrinol.*, 142, 203–214, 1998.

Gray, L.E. Jr., Ostby, J.S., and Kelce, W.R., Developmental effects of an environmental antiandrogen: The fungicide vinclozolin alters sex differentiation of the male rat, *Toxicol. Appl. Pharmacol.*, 129, 46–52, 1994.

Hutchinson, T.H., Reproductive and development effects of endocrine disrupters in invertebrates: In vitro and in vivo approaches, *Toxicol. Lett.*, 131, 75–81, 2002.

James, L.F., Panter, K.E., Nielsen, D.B., and Molyneux, J., The effect of natural toxins on reproduction in livestock, *J. Anim. Sci.*, 70, 1573–1579, 1992.

Knight, A.P. and Walter, R.G., Plants affecting the mammary gland. In *A Guide to Plant Poisoning in Animals in North America*, Knight, A.P. and Walter, R.G. (eds.), Teton New Media, Jackson, WY, 2004.

Knight, A.P. and Walter, R.G., Plants associated with congenital defects and reproductive failure. In *A Guide to Plant Poisoning in Animals in North America*, Knight, A.P. and Walter, R.G. (eds.), Teton New Media, Jackson, WY, 2004. http://www.ivis.org/special_books/knight/chap8/ivis.pdf

Krieger, N., Wolff, M.S., Hiatt, R.A., Rivera, M., Vogelman, J., and Orentreich, N., Breast cancer and serum organochlorines: A prospective study among white, black and Asian women, *J. Natl. Cancer Inst.*, 86, 589–599, 1994.

Kurzer, M.S. and Xu, X., Dietary phytoestrogens, *Annu. Rev. Nutr.*, 17, 353–381, 1997.

Lee, W., Kang, C.W., Su, C.K., Okubo, K., and Nagahama, Y., Screening estrogenic activity of environmental contaminants and water samples using a transgenic medaka embryo bioassay, *Chemosphere*, 88, 945–952, 2012.

Lind, L. and Lind, P.M., Can persistent organic pollutants and plastic-associated chemicals cause cardiovascular disease? *J. Intern. Med.*, 271, 537–553, 2012.

Lind, P.M., Zethelius, B., and Lind, L., Circulating levels of phthalate metabolites are associated with prevalent diabetes in the elderly, *Diabetes Care*, 35, 1519–1524, 2012.

Matthews, J. and Zacharewski, T., Differential binding affinities of PCBs, HO-PCBs, and arochlors with recombinant human, rainbow trout (*Onchorhynkiss mykiss*) and green anole (*Anolis carolinensis*) estrogen receptors, using a semi-high, throughput competitive binding assay, *Toxicol. Sci.*, 53, 326–339, 2000.

Mendonca, G.A., Eluf-Neto, J., Andrada-Serpa, M.J., Carmo, P.A., Barreto, H.H., Inomata, O.N., and Kusumi, T.A., Organochlorines and breast cancer: A case-control study in Brazil, *Int. J. Cancer*, 83, 596–600, 1999.

Mills, P.K., Correlation analysis of pesticide use data and cancer incidence rates in California counties, *Arch. Environ. Health*, 53, 410–413, 1998.

Murkies, A.L., Wilcox, G., and Davis, S.R., Phytoestrogens, *J. Clin. Endocrinol. Metab.*, 83, 297–303, 1998.

Olaya-Contreras, P., Rodriguez-Villamil, J., Posso-Valencia, H.J., and Cortez. J.E., Organochlorine exposure and breast cancer risk in Columbian women, *Cad. Saude Publica.*, 14 (Suppl 2), 125–132, 1998.

Raychaudhury, S.S., Blake, C.A., and Millette, C.F., Toxic effects of octylphenol on cultured rat spermatogenic cells and Sertoli cells, *Toxicol. Appl. Pharmacol.*, 157, 192–202, 1999.

Riet-Correa, F., Medieros, R.M., and Schild, A.L., A review of poisonous plants that cause reproductive failure and malformations in the ruminants of Brazil, *J. Appl. Toxicol.*, 32, 245–254, 2012.

Safe, S., Pallaroni, L., Yoon, K., Gaido, K., Ross, S., Saville, B., and McDonnell, D., Toxicology of environmental estrogens, *Reprod. Fertil. Dev.*, 13, 307–315, 2001.

Sheehan, D.M., Herbal medicines, phytoestrogens and toxicity: Benefit considerations, *Proc. Soc. Exp. Biol. Med.*, 217, 379–385, 1998.

Thayer, K.A., Heindel, J.J., Bucher, J.R., and Gallo, M.A., Role of environmental chemicals in diabetes and obesity: A national toxicology program workshop review, *Environ. Health Persp.*, 120, 779–789, 2012.

Van Der Gulden, J.W. and Vogelzang, P.F., Farmers at risk for prostate cancer, *Br. J. Urol.*, 77, 6–14, 1996.

13

Radiation Hazards

Introduction

The electromagnetic spectrum encompasses all forms of radiant energy. The following table (Table 13.1) lists the various components and their wavelengths in decreasing order.

Ionizing radiation, that portion of the spectrum that can cause serious cell damage, includes all wavelengths of 1000 Å or less (Table 13.1). Ionizing radiation, by stripping electrons from molecules as it passes through tissues, produces ionized species of everything from H_2O to macromolecules like DNA. These ionized species are unstable and reactive, and can produce dramatic disruptions in cell function including mutation.

There is still controversy regarding the degree of risk. Radiophobia, an illogical fear of radiation hazards, has led to considerable controversy over the extent of the environmental risks of ionizing radiation. Consider the following conflicting statements taken from Cobb (1989):

- Dr. K. Z. Morgan, a pioneer in health physics—"It is incontestable that radiation risks are greater than published."

- C. Rasmussen, nuclear engineer at MIT—"There is a lot of evidence that low doses of radiation not only don't cause harm but may in fact do some good. After all, humankind evolved in a world of natural low-level radiation." (About 82% of our total radiation exposure comes from natural sources). This is another example of hormesis.

- R. Guimond, EPA—"We can't avoid living in a sea of radiation."

In some cultures, deliberate exposure to natural-source radiation is done in the belief that it has curative powers. In Japan, exposure to radon is courted in "radon spas" where natural hot springs occur.

TABLE 13.1

Components of the Electromagnetic Spectrum

Type of Radiation	Wavelength
Radio waves	30 km–3 cm
Microwaves	3 cm–10 mm
Thermal (heat)	0.078–0.001 mm
Infrared (includes thermal portion)	0.5 mm–1,000 Å
Visible	7,800–4,000 Å
Ultraviolet	4,000–1,850 Å
Extreme ultraviolet	1,850–150 Å
Soft x-rays	1,000–5 Å
X-rays	5–0.06 Å
Gamma rays	1.4–0.01 Å
Cosmic rays (protons 85%, alpha-particles 12%, electrons, gamma rays, etc.	To 1/10,000 Å

Sources and Types of Radiation

Sources

Natural Sources of Radiation

These include cosmic rays from space, solar rays that intensify during solar storms (sun spots), radiation emanating from rocks and ground water, and radiation coming from within our own bodies, mainly from decay of radioactive potassium in muscle. Of considerable concern at present is the exposure to radon gas, a radioactive decay product of radium, a common radioactive element in soil and rock. This will be discussed in more detail later.

Man-Made Sources of Radiation

These include medical x-rays and radioisotopes, ion sensors in smoke detectors, uranium used to provide the gleam in dentures, mantles in camping lanterns, radioactive wastes, nuclear accidents (e.g., Chernobyl), and careless handling of nuclear materials. Cesium 137 was discarded in a dump in Brazil and ultimately killed 4 people and contaminated 249. Until fairly recently, radium was used to hand-paint luminous watch dials. Workers used to "point" their brushes by running them between their lips, a practice that led to cases of radiation sickness.

Cause of Radiation

Elements that exist in an unstable form are continually decaying to more stable ones. In the process, they give off energy in several ways. Ionizing radiation arises when an unstable nucleus gives off energy. An unstable nucleus is called a radionuclide. In contrast, x-irradiation is a form of cosmic ray and also occurs when a suitable target such as tungsten is bombarded with electrons. It does not arise from nuclear decay.

Types of Radioactive Energy Resulting from Nuclear Decay

1. A nucleus can eject two protons and two neutrons to lose mass and convert itself into another element. The ejected components constitute an alpha-particle. An alpha-particle is slow moving and will not penetrate skin, but it can cause dangerous ionization if ingested. An alpha-particle is actually identical to the nucleus of helium.

2. The neutron of a nucleus can lose an electron to become a positron. The lost, negatively charged particle is called a beta-particle.

3. Even after emitting alpha- or beta-particles, a nucleus may remain in an agitated state. It may rid itself of excess energy by giving off gamma rays. These are short, intense bursts of electromagnetic energy with no electrical charge. They can penetrate lead and concrete and can cause extensive tissue damage by ionization.

4. Neutrons are ejected from the nucleus during nuclear chain reactions. They collide and combine with the nuclei of other atoms and induce radioactivity of the aforementioned types. This is the primary source of radioactivity following a nuclear explosion.

Measurement of Radiation

There are two different types of measurements of ionizing radiation. One is concerned with the level of energy actually emitted from the source, and the other is concerned with the amount of tissue damage that can be produced by a particular form of radiation. The field is unfortunately further confused by a more recent shift to the international (SI) system. The equivalent values are shown in Table 13.2. They are not easily interchangeable.

Measures of Energy

1. As the nucleus of an atom decays, it gives off a burst of energy (the ionizing radiation) called a "disintegration." The number of disintegrations per unit time varies with the nature of the source. Various counting instruments (scintillation counters, etc.) measure

TABLE 13.2

Units for Measuring Radiation: Old vs. New (SI) Systems

Old System	New System (SI Units)
1 curie (Ci)	37 gigabecquerels (Gbq)
1 rem (rem)	10 millisieverts (mSv)
100 rem	1 Sievert (Sv)
1 rad (rad)	10 milligrays (mGy)
100 rad	1 Gray (Gy)
1 roentgen (R)	258 millicoulombs/kilogram (mC/kg)

disintegrations per minute (dpm). The basic unit of measuring radiation energy is the curie (Ci) or 37 Gbq and it is the number of disintegrations (3.7×10^{10}) occurring in 1 s in 1 g of radium-226. Radioisotopes used in science are usually provided in milli Ci (mCi) amounts. The new unit is the Becquerel (Bq). A Bq represents one disintegration per second.

2. The first unit of radiation was the roentgen (abbr. r). It has a complicated definition based on the amount of x-rays or gamma rays required to cause a standard degree of ionization in air.

Measures of Damage

1. The earlier measure of radiation damage is the rem, which stands for roentgen–equivalent–man. It is that amount of ionizing radiation of any type that produces in humans, the same biological effect as 1 r. The new international unit is the sievert (Sv). See Table 13.2.

2. The rad is the amount of radiation absorbed by 1 g of tissue. The new international unit is the gray (Gy). See Table 13.2.

A rough scale of toxicity is as follows: 10,000 rem (100 Sv) is rapidly fatal because of damage to the central nervous system (CNS). Whole body exposure to 300 rem (3Sv) is about the LD_{50}. Between these values, radiation damage occurs. The assumption is made that the risk associated with radiation is linear all the way to zero. When it comes to assessing carcinogenic potential however, accuracy is extremely difficult. Below 10 rem (0.1 Sv), effects are unclear due to confounding factors like smoking, pollution, and diet. Over 300 agents have been shown to be carcinogenic in animal tests. Residents of Denver have lower cancer death rates than those of New Orleans despite higher radiation exposures because of increased levels of cosmic radiation at their high altitude.

Some Major Nuclear Disasters of Historic and Current Importance

Hiroshima

The group of people from whom the most reliable data have been gathered concerning radiation hazards are the survivors of the atom bomb dropped on Hiroshima at 8:15 a.m., August 6, 1945. In 1947, the Radiation Effects Research Foundation was established. Exhaustive studies have shown that the heavily exposed people, called the "hibakusha," had a 29% greater chance of dying from cancer than normal. Excess numbers of leukemia cases began appearing in the late 1940s and peaked in the early 1950s, but by the early 1970s they had dropped to levels near those of unexposed Japanese. Now the surviving hibakusha have longer life expectancies than the overall population, perhaps because of closer medical supervision. One of the most feared hazards of radiation is that of congenitally deformed infants because of radiation-induced genetic defects in the mother. While such defects have been demonstrated experimentally, the Hiroshima study compared 8000 children of hibakusha with 8000 of unexposed Japanese and found an incidence of chromosome damage in 5/1000 of the former and 6/1000 of the latter. Protective mechanisms may be functioning in humans (e.g., spontaneous early abortion). Fetuses exposed *in utero* are a different story. Dozens of mentally retarded infants were born in the areas around Hiroshima and Nagasaki (target of the second atomic bomb) in the months following the blasts. Abortions were also numerous. The fetus appears to be most vulnerable between 8 and 15 weeks. One more recent development, however, is that a special panel formed by the U.S. National Research Council released a report following a reassessment of the Hiroshima and Nagasaki data and concluded that the levels of exposure were much lower than previously calculated. Original estimates were based on tests at the Nevada nuclear test site using much flimsier buildings than were actually present in those cities. As a consequence, the established safe limits have had to be revised downward. This is mainly a concern for persons exposed to radiation on the job, but it has revived the controversy about whether there is any safe level of exposure. Ironically, evidence is now surfacing that scientists and technicians who worked on the atomic bomb project during the war are showing up with elevated incidences of cancer which, when adjusted for exposure level, may be even greater than those of the hibakusha. The new safe exposure limit is 20 mSv/year averaged over 5 years with no more than 50 in any 1 year. The old level was 50 mSv.

Hiroshima Update

In 2011, Douple et al. published a review of the data compiled by the Radiation Effects Research Foundation in the 63 years (to 2008) following the Hiroshima blast. Of the 200,000 people who were identified and followed,

40% were still alive in 2008. Few people who were within 1 km of the center of the blast survived past 1950. A Life Span Study (LSS) was established in 1955 with a fixed cohort of 120,000 people from both Hiroshima and Nagasaki (A-bomb survivors). Excess leukemia deaths became the first radiation-associated, long-term health effect observed in the LSS, first noted in 1952. A linear exposure–response affect was noted. In 2002, there were 315 leukemia deaths in the cohort of which 98 (45%) were estimated to be excess deaths due to radiation. There has been a continuing increase in cancer deaths each year in the LSS. There was evidence that there was a latency period (the increase began several years after exposure) and that cancer risk continues through-out life. Numerous body sites demonstrated increased cancer incidence, a reflection of the whole-body nature of the exposures.

Excess cancer risks were highly dependent on age at exposure and attained age. Risk with exposure at age 10 was about double that associated with exposure at age 40. Conversely, risk continued to increase with age attained.

Significant exposure-related increases in late-onset diseases such as car-diovascular disease, gastrointestinal diseases, and respiratory diseases were also observed. People who were in the city at the time of the bombing were given an in-depth interview 17–20 years later and, compared to a control group not in the city, reported higher frequencies of anxiety and somatiza-tion symptoms, disorders frequently associated with posttraumatic stress disorder. See "Multiple Chemical Sensitivity" (in Chapter 4) for more on these psychological conditions.

Life expectancy was reduced by about 2.6 years for those with doses greater than 1 Gy. About 60% of this reduction was attributed to solid cancers, 10% from leukemia, and 30% from other diseases. Numerous other health effects were recorded.

Three Mile Island

The partial core meltdown of the Unit 2 reactor at TMI in March 1979 was largely responsible for bringing the nuclear power program in the United States to a halt. Of the "defense in depth" safety features, all but the outer water shield failed. Some authorities claim that even a complete meltdown would not have breached this defense. Despite concerns of nearby residents, only 15 Ci (555 Gbq) of radioactivity were actually released. The news media exploited the event with sensational reports of "deadly clouds of radioac-tive gas" and made much of the potential for explosion of a large bubble of hydrogen in the reactor. In fact, there was none because no oxygen was pres-ent. The people at greatest risk from radiation were the workers who were involved in the cleanup. U.S. federal regulations limit the maximum, annual exposure of workers in the nuclear industry to 12 rem (120 mSv). Workers in the "hot" areas receive about 1 millirem/h, the equivalent of one chest x-ray. In 1987, totals for such workers averaged about 710 millirem (7.1 mSv). See Chapter 2 for more on Three Mile Island.

Hanford Release

In contrast, massive amounts of ^{131}I were deliberately released (for purposes still classified as top secret) from the military nuclear facility at Hanford, Washington State, in the 1940s and 1950s. Some residents may have received as much as 2295 rem! Again, the greatest source of exposure may have been the consumption of contaminated meat and vegetables. Multimillion-dollar studies have recently been commissioned to seek answers to the degree of risk and to assign responsibility. Obviously, the potential for civil action is considerable.

Chernobyl

The more recent and highly publicized nuclear disaster was Chernobyl in April of 1986 (a much worse but largely concealed disaster occurred in Russia in 1958). About 20% of the plant's radioactive iodine escaped along with 10%–20% of its radioactive cesium and other isotopes; 135,000 people lived in a 30 km radius of the power plant. There were 30 deaths and 237 cases of severe radiation injury; 2000 children have been born to women who were living in the accident zone at the time of the disaster. Early studies, 10 years after the event, did not find any abnormalities in this group. An examination of about 700,000 people over a wider area did not reveal any physical problems. Russian scientists estimate an increase in the cancer rate of 0.04% over the next 20 years. In Western Europe, exposed to the drift of radioactive dust, it is estimated that, over the next 50 years, 1000 additional cancer deaths will occur. Normally, there would be 30,000,000 cancer deaths in this period, so the increase is 0.003%. Aside from those people directly exposed to the effects of the explosion, the greatest risk of exposure seems to come from eating contaminated food. Thousands of reindeer had to be destroyed in northern Scandinavia because they had grazed on contaminated pasture.

Dr. Marvin Goldwin, chief of the Joint-U.S.-Soviet Medical Team, made the following points in a 10 year follow-up report.

1. Everyone in the Northern hemisphere received a small dose of radiation. The degree of exposure of those people at highest risk cannot be accurately identified.

2. Radioactive iodine posed an early risk to the thyroid glands of exposed people.

3. As of 1991 there was no detectable increase in the incidence of cancer but leukemia may yet show up and solid tumors may not show up for 10 years.

Chernobyl provides another example of how psychological damage can often exceed physical damage in the first years following an environmental disaster. Thousands of people received a radiation exposure, which exceeded the maximum recommended lifetime allowance. Since radiation levels in their locales have fallen to low levels similar to those of surrounding areas, it made no medical or scientific sense to relocate them. Stress and fear, however, create an understandable desire in these people to be moved out of the area, and their wishes were acted upon.

In 2011, a 25 year follow-up report on Chernobyl health issues was released. It shed considerable light on the risks of cancer and other health issues and on the accuracy of earlier predictions. Studies of the cleanup workers provided evidence of increased incidences of leukemia and other hematologic cancers and of cataracts as well as cardiovascular diseases following low doses and low dose rates of radiation. Thyroid cancer, first linked with radiation exposure following the atom bomb explosions in Hiroshima and Nagasaki, was significantly increased in pediatric populations in the first years after exposure in the most exposed areas, notably the Ukraine and Belarus. It was especially pronounced in those who were 4 years old or less at the time of exposure and was detectable within 4 years later. Thyroid cancer in adults was found to be significantly increased from 1986 to 1991 in the more contaminated areas of the Ukraine and Russia and in Belarus from 1991 to 2002. No increases, however, were observed in adults in a study of the exposed population in the Russian region of Bryansk, which used individual doses from a population registry. Iodine-131 (I-131) was a chief component of fallout and it is rapidly taken up by the thyroid gland. There was no conclusive evidence of thyroid cancer associated with exposure *in utero*. This area appears to require additional study.

Leukemia was the first malignancy to be linked to radiation exposure in the studies of exposed Japanese. It was detectable 2–5 years after exposure. The European Childhood Leukemia-Lymphoma Study (ECLIS) did not find evidence of an increase in these malignancies in the first 5 years after exposure. An association between radiation dose and incidence of leukemia was found in the Ukraine but it was not statistically significant. Leukemia in children exposed *in utero* was found to be increased in a small Greek study but nowhere else. This area probably needs further investigation. There was some evidence of increased dose-related risk of leukemia and chronic lymphocytic leukemia (CLL) in cleanup workers. The differences were not statistically significant for CLL however.

In summary, it is well established that thyroid cancer was significantly increased in children and adolescents with young children at greatest risk. The cleanup workers, the most heavily exposed group, had increased risk of leukemia and other hematologic malignancies and of cataracts and cardiovascular disease. Evidence of increased malignancy in the exposed general population was inconclusive and warrants further study.

Sensational accidents like Chernobyl do much to undermine the public's confidence in nuclear power. This is unfortunate at a time when the air-polluting and global-warming aspects of coal-fired electrical generators are becoming a great concern. From 2000 to 2009, 311 coal miners died in mining accidents in the United States. In 1980 alone there were 133 deaths. In China from 2000 to 2009, 50,152 coal miners died [298]. Since methane is commonly found in seams of coal, explosions are an ever-present danger. It is estimated that 6000 coal miners die annually in accidents around the world. This seems a reasonable figure given the statistics for China alone. In the 25 years since Chernobyl, conceivably 1,500,000 coal miners have died in accidents. This does not take into account the hundreds of thousands who likely succumbed to pneumoconiosis (black lung disease) during this period. According to a *Times* online article of September 5, 2005, which quotes a United Nations watchdog report, only 56 people had died to that date from the direct result of radiation. The report stated that the projected death toll would be much less than expected and predicted that some 4000 people might eventually succumb to radiation. Another report by the group representing those who worked in relief operations claimed that 15,000 were killed in the cleanup and 50,000 were injured. Even accepting these much larger figures, there is a huge difference between them and the 1.5 million coal miners who have died in mining accidents alone in the ensuing 25 years. The seriousness and tragedy of the Chernobyl accident notwithstanding, it is obvious that the death toll from coal mining is seldom taken into account when evaluating the risk–benefit equation of nuclear versus other sources of electrical power. It is probably the only source capable of fully replacing coal-fired electric generators.

An old proverb has it that "It's an ill wind that blows no good," and this appears to apply to Chernobyl. The huge area of the evacuation, devoid of human presence for over 25 years, has become a vast, accidental wildlife preserve. Species that have not been seen in centuries in the area have reappeared. These include Eurasian brown bear, wolves, lynx, wild boar, deer, and many bird species. The Ukrainian government has declared the area an official wildlife sanctuary.

Fukushima, Japan

On March 11, 2011, an earthquake and subsequent tsunami devastated the nuclear power plant at Fukushima, Japan. Reactors 1, 2, and 3 suffered full meltdown. Two workers were killed by the tsunami but none died from radiation exposure, although eventually six workers were deemed to have exceeded their lifetime limit. An area 20 km (12 mile) around the nuclear plant was evacuated.

The total radiation released was calculated by the Japanese government as one-tenth of that released at Chernobyl. However, cesium-137 levels were found to be high enough to cause concern up to 50 km from the plant.

Recently, evidence has emerged that not all of the problems were due to natural disasters. Charges have been made that collusion occurred among the government in Tokyo, the regulators, and the operators. This set up conditions that caused a preventable disaster. This was the finding of an expert panel that published its report in July, 2012. Regulators had been reluctant to accept global standards that could have helped prevent or limit the disaster.

The consistent element present in all of these disasters, including the discarding of cesium-137 in Brazil, has been human error, misjudgment, and negligence. Deliberate callousness may have been involved at Hanford, where charges of unsafe practices and antiquated, dangerous equipment are still being made. Poor construction at Chernobyl, corruption at Fukushima, and faulty design at Three Mile Island all contributed to these nuclear accidents.

Radon Gas: The Natural Radiation

As noted earlier, 82% of the ionizing radiation to which North Americans are exposed comes from natural sources. It has been estimated that there is enough natural radioactive material in the human body that, if it were a laboratory animal, it would have to be disposed of as hazardous waste! The average, annual, natural background exposure in Great Britain is about 1 mSv (0.1 rem or 100 mrem). In Canada, it is somewhat higher because the Canadian Shield (the band of granite rock that spans mid-northern Ontario, Quebec, and Manitoba) is rich in uranium deposits. Radon gas is by far the biggest potential health hazard from natural radiation. It is the decay product of uranium and it seeps up through faults in the substrata of soil and may leak into houses through cracks in basement walls, drains, etc. The advent of airtight houses has increased the risk by trapping radon gas in the house. In the British study, it was calculated that 1,000,000 people were exposed to radon at levels of 5–15 mSv annually. In contrast, only 5100 workers in their nuclear industry were exposed to levels as high or higher. This is the equivalent of 50–150 chest x-rays annually! Radon homes are distributed very randomly with highly contaminated homes located right beside radon-free ones. The federal government conducted a survey of Canadian homes and found that Winnipeg had the highest radon levels of any Canadian city, with high levels also found in parts of Saskatchewan and northern Ontario and Quebec (Table 13.3). Toronto has low levels.

A map of radon risk areas in the United States, published by *National Geographic*, shows a band running slightly east of north through the middle of Ohio, and another running east–west through New York State. A U.S. federal study surveyed 20,000 homes in 17 states and found that 25% had potentially hazardous levels of radon. Radon was described as the largest environmental radiation health hazard in America. Debate over the degree of risk plagues this area as it does others. The EPA study measured radon levels in basements where they would be highest. Calculations of risk have estimated the lifetime risk of dying from radon-induced cancer at 0.4% for exposed individuals, which is the same as your chances of dying in a fire or a fall. If radon were a man-made carcinogen it would unquestionably

TABLE 13.3

Radon Concentrations in Canadian Homes in pCi/L (37×10^{-3} Gbq) Air

City	No. Homes Tested	No. Homes >4.5 pCi/L (1.7 Gbq)	% Homes >4.5, etc.
Quebec Province			
Sherbrooke	905	64	7.1
Montreal	600	9	1.5
Quebec City	584	9	1.5
St. Lawrence	432	63	14.6
New Brunswick			
St. John	866	51	5.9
Fredericton	455	26	5.7
Prince Edward Island			
Charlottetown	813	35	4.3
Newfoundland and Labrador			
St. John's	585	17	2.9
Ontario			
Sudbury	722	29	3.8
Thunder Bay	627	29	4.6
Toronto	751	1	0.1
British Columbia			
Vancouver	823	0	0

Source: Compiled from data reported by McGregor, R.G. et al., *Health Phys.*, 39, 285, 1980.

be banned, and most certainly would be the target of antinuclear activists. Nevertheless, there is still controversy regarding the degree of risk, or perhaps more correctly, the degree of exposure. A British study calculated that 6%–12% of all cases of myeloid leukemia might be attributed to radon exposure, with levels rising to 23%–43% in Cornwall, where the highest exposures occur. Worldwide, their calculations suggested 13%–25% of all myeloid leukemia in all age groups could be due to radon.

In 1984, a construction worker at the Limerick nuclear generating station near Reading, Pennsylvania consistently set off radiation alarms despite the fact that he had never worked in a "hot" area. An examination of his home subsequently revealed radon levels of 2600 pCi (96 bq)/L, the highest ever recorded. North Dakota, with 63% of homes showing levels of 4 pCi (0.15 bq)/L or greater, leads the states in radon exposure, followed by Minnesota (46%), Colorado (39%), Pennsylvania (37%), and Wyoming and Indiana (26%). After cigarette smoking, radon is probably the most common cause of lung cancer. Radon-222 (^{222}Rn) and ^{220}Rn are the only gaseous decay products of uranium. Their half-life is 3.8 days and they decay to particles (not gases) including radioactive poloniums, which actually are the alpha-emitting toxins causing cell damage. Alpha-particles, unlike gamma rays, can only cause cell damage for a radius of about 70 μ, hence the risk of lung cancer. The increased incidence of leukemia is hard to explain on this basis.

Perrson and Holm published in 2011 the results of a study of the levels of ^{210}Po and ^{210}Pb in various soil and biological species. They reported that ground level air had 0.03–0.3, and 0.2–1.5, Bq/m^3, respectively. Mosses, lichens, and peat have a high efficiency in capturing these radioisotopes from atmospheric fallout, having levels of 0.5–5 kBq/m^2. They may concentrate to 250 Bq/kg. Reindeer (caribou) that graze on lichens have levels in meat of 1–15 Bq/kg. People with a high consumption of reindeer meat may have an annual dose of 260 (Po) and 132 (Pb) μSv. Seafood may also have fairly high levels of these isotopes.

Tissue Sensitivity to Radiation

In general, tissues with a high rate of turnover are more susceptible to the effects of ionizing radiation. Thus, thyroid, lung, breast, stomach, colon, and bone marrow have high sensitivity; brain, lymph tissue, esophagus, liver, pancreas, small intestine, and ovaries are intermediate; and skin, gall bladder, spleen, kidneys, and dense bone are low. This order of sensitivity roughly parallels the frequency of primary cancer in these tissues. If molecular disruption is sufficient, the cell will die. Since hair follicles and gastrointestinal mucosa have high turnovers, radiation sickness involves hair loss and severe diarrhea. Because bone marrow cells also have a high turnover, repair of DNA may not be complete before replication occurs and the daughter cells may be malignant. This is why leukemia is the commonest cancer associated with radiation injury.

Questions regarding the safety of the nuclear industry continue to emerge. In August 1989, two workers at the Pickering Nuclear Plant in Ontario were mistakenly given unshielded practice equipment to change a new type of fuel rod just recently introduced. They received what were widely reported as the highest levels of radiation ever encountered by workers in the Canadian nuclear industry—5.6 and 12.2 rem (56 and 120 mSv). The annual allowable limit set by the Atomic Energy Commission is 2 rem (20 mSv). Radiation exposure from an average chest x-ray is 15 mrem (150 mSv). This has been calculated to (theoretically) cause one additional cancer per 100,000,000 people. In other words, if the entire population of North America were to receive one chest x-ray, one could expect three additional cancer cases as a result. These workers received about 700 times this amount, which would cause an additional cancer per 143,000 people. This risk will be lessened if they are removed to areas where there is no possibility of additional exposure for at

TABLE 13.4

Annual Average U.S. Radiation Exposures

Source of Radiation	mrem (mSv)
Natural sources	<100 (<1000)
Medical and dental x-rays	78 (0.78)
Radioisotopes	14 (0.14)
Weapons testing	4–5 (0.04–0.05)
Nuclear industry	<1 (<0.01)
Building materials (radon from masonry and brick)	3–4 (0.03–0.04)
One chest x-ray	10–15 (0.1–0.15)
Total average annual exposure	<200 (<2)

least 1 year. The safety maxim that should apply in all situations involving exposure to radiation is ALARA (as little as reasonably achievable). It should be noted that, once again, human error was responsible for this accident.

In Great Britain, a disturbing report was released early in 1990 to the effect that offspring of nuclear plant workers had an increased incidence of birth defects. Since these children are not directly exposed to radioactive material, the conclusion, if the data are correct, is that exposure of the parents (mostly men) caused genetic damage. These results contradict earlier studies, and it has been pointed out that clusters of birth defects occur geographically in the absence of nuclear generators (or other identifiable causes) but the data will be of concern until they can be explained or disproved. Public pressure has resulted in the cancellation of 50 new nuclear power plants in the United States. As a result, there is greater reliance on coal-fired generators. About 200 coal miners die each year in mine accidents and an equal number from "black lung disease" (pneumoconiosis). Recent (1992) mine accidents include 26 killed in Nova Scotia, 400 in Turkey, and 38 in Russia. Such events rarely cause a ripple of concern amongst opponents of nuclear energy (see also this chapter under Chernobyl). There is also the problem of acid rain resulting from sulfur pollution of the atmosphere by coal-fired generators. Has the public traded a potential but high-profile risk for a real and greater one that is less visible? Table 13.4 compares various sources of radiation encountered by Americans.

Microwaves

Microwaves are the shortest waves in the radio portion of the electromagnetic spectrum (1 mm–30 cm). They are at very high frequencies (1,000–300,000 megacycles/s) and they are used in radar, for long-distance transmission of phone signals and TV signals, and, of course, in microwave ovens.

Because of its high dielectric constant, water dissipates energy as heat when exposed to microwaves. At high-enough energies, thermal damage may occur in living tissues.

Cell Phone Use and Brain Tumors

Recent concern has been expressed over possible carcinogenic effects of microwaves given off by cellular phones. The energy level is so low, however (<5 W), that no thermal effects can be detected. Little information is available concerning nonthermal effects of microwaves, but evidence for carcinogenicity is scanty and anecdotal. This is not to say that there is no concern regarding cell phones.

Cell phones emit nonionizing (300 MHz–300 GHz) radiofrequencies (RF). Tissues exposed to RF have a specific absorption rate (SAR) measured in watts (W)/kg tissue. Regarding cranial exposure, this is affected by the thickness of the bone (thinner in children) and the distance the phone is held from the ear. The law of inverse squares dictates that the level of any radiant energy falls as the square of the increase in distance from the source; thus, if the distance from the source is doubled, the energy levels fall to one quarter of its previous value. Even under maximum exposure conditions, including frequent cell phone use, it could take 10 years or more for effects to be manifested.

In the early 1990s, a woman in Florida claimed that her brain tumor was the result of extensive cell phone use and her husband sued the cell phone company. Media coverage and the burgeoning use of cell phones, now numbering one for every person in the United States, prompted the scientific community to undertake studies in an effort to resolve this controversy. The results, however, have generally failed to do so, and it may be some time before this question is resolved.

An extensive literature has emerged in the last few years that attempts to identify the risk of developing brain tumors with cell phone use. Studies have concentrated on children and adolescents as the most vulnerable segment of the population. Two distinct camps seem to have developed. Epidemiological studies compare changes in the incidence of brain tumors, notably gliomas, with the exponential growth of cell phone use in recent years. The other camp used case-control studies to compare the incidence of tumors in an exposed population with a matched population of unexposed individuals. A refined version attempts to identify the degree of exposure (frequency and duration of use) with tumor incidence.

Two major European studies were reported in 2011; the Hardell study (Sweden) and the Multicenter European Interphone Studies. These and other studies (13 in all) were reviewed by Corle et al. in 2012. They point out that, according to the U.S. Central Brain Tumor Registry, from 1995 to 2004 the total incidence of benign and malignant brain tumors rose from 13.4 to 18.2; a rise of 36%. This increase fed the concern about cell phone use. The gliomas, which include astrocytomas and oligodendrogliomas, are the most feared brain tumors in both children and adults. Glioblastomas comprise 33% of all brain tumors and 79% of malignant ones. Grade IV astrocytomas are also known as glioblastomas and have a mean, post-diagnosis survival rate of only 14 months despite aggressive treatment. It is thus not surprising that there is concern regarding the risk of cell phone use.

One meta-analysis of 22 case-control studies in 2009 concluded that there was a slight increase in risk of brain tumors for frequent cell phone users. The conclusions of Corle et al. were that there was no clear-cut evidence that cell phone use increased the incidence of brain tumors, although they recognized that some human studies suggested such an association in people with a 10 year or more latency period and a high number of cumulative call hours. They also recommended further studies using a standardized design, noting that several design weaknesses compromised existing studies. Among their suggestions for improvement were the following:

1. Cell phone energy levels need to be tabulated and matched between studies.
2. Study populations need to be subdivided in a standard fashion as to age, sex, ethnicity, health, etc.
3. The range of pathologies needs to be defined, e.g., brain tumors, parotid tumors, oral cancers.
4. Questionnaires need to be standardized and standard blinding procedures employed.
5. Actual cell phone use records should be employed where possible, instead of using subject recall.
6. Latency periods should be uniformly defined.
7. The statistical approach should be well defined in advance.

Little et al. (U.S. National Cancer Institute) took an epidemiological approach based on the premise that if cell phone use was a risk factor for the development of brain tumors there should be a correlation between increased phone usage and tumor incidence. The increased phone usage between 1997 and 2008 predicted that tumor incidence should have been 44% higher than the actual observed incidence. This does not support the contention that phone use is a risk factor. The authors identify some weaknesses in the study. Phone use was defined as phone ownership. No information regarding actual exposure level was available. The study focused on Caucasians because black and Hispanic people are known to have a lower incidence of brain tumors.

The CEFALO study (the meta-analysis of 22 studies noted earlier) authors concluded that there was a slight risk from phone usage but only for the longest and most frequent users. This conclusion was challenged by Morgan et al. who concluded that these data indicated increased risk for contralateral tumors as opposed to ipsilateral ones and pointed out what they felt were several design flaws. Debate over this study continues.

From this mass of complex and often conflicting data your author has reached the following conclusions:

1. A definitive answer to the question of whether cell phone use constitutes a risk for brain tumors remains elusive.

2. If there is such a risk, it is small and likely significant only for long-term and very frequent users.

3. Design flaws in existing studies will be difficult and very expensive to correct.

4. In consequence, the definitive answer will likely remain elusive.

5. Cell phone usage is increasingly shifting from talking to texting, especially by younger users, and hands-free devices in motor vehicles are becoming more common. These factors will reduce the degree of exposure but also make more difficult to quantify it.

In conclusion, there is one statement that can be made with absolute confidence. One is infinitely more likely to die from using a cell phone while driving than from a brain tumor caused by one.

Ultraviolet Radiation

Ultraviolet radiation occupies the electromagnetic spectrum between 400 and 4 nm: UVa—400–320 nm, UVb—320–280 nm, UVc—below 280 nm.

An increase of 1%–2% of UVb radiation is associated with an increase of 2%–4% in skin cancer. UVb is in the ionizing radiation range, and can therefore damage DNA, leading to mutations and cancer. The effect on melanocytes appears to be more complicated. Melanomas often appear first at sites not directly exposed to sunlight. Tropical and subtropical areas have much higher incidences of skin cancer than temperate zones. In North America, it has been claimed that the incidence of skin cancer has increased by 400% in recent years, presumably because of the destruction of the ozone layer.

Of concern is the use of tanning beds by young people. There is now overwhelming evidence that exposure to irradiation in tanning beds increases the risk of both forms of skin cancer, basal cell carcinoma (BCC) and melanoma. A U.S. study followed 73,494 female nurses for 20 years. They examined whether frequency of tanning bed use during high school and college (25–35 years of age) increased the risk of skin cancer. There were 5505 cases of BCC and 394 cases of melanoma. The hazard ratios comparing users of tanning beds to nonusers were 1.13 and 1.11, respectively. For use during high school the HR increased to 1.78 if usage was more than six times yearly. Thus, young girls who used tanning beds more than six times yearly almost doubled their risk of developing skin cancer. A claim by the industry that tanning beds can increase levels of vitamin D has been dismissed by the scientific community as insufficient justification for their use.

Medical Uses of UV Radiation

1. Photophoresis—blood is removed from a patient and exposed to UV light, then returned to the patient. The technique is useful for treating mycosis fungoides, a complication of skin cancer, and it is promising for some leukemias.

2. UVa is used in conjunction with a photosensitive drug called psoralen to treat the skin lesions of psoriasis. The technique is called puva (psoralen-UVa).

Extra-Low Frequency Electromagnetic Radiation

ELF waves are nonionizing radiation waves with extremely long wavelengths (several hundred km) and very low frequencies (<300 Hz). Exposure to artificial ELF fields occurs near high-tension electrical lines and much lower levels emanate from household appliances. Electric blankets especially are thought to be potent sources of ELF exposure because of the close proximity and prolonged contact they entail. Other types of low-frequency waves include microwaves, emissions from TVs, and radio-frequency fields. High-tension lines create both electrical and magnetic fields and, although insulation will shield the former, the latter have great penetrating powers and will penetrate the human body. The unit of measure of magnetic fields is the gauss, which has a very complicated definition. Electromagnetic fields (EMFs) are also measured in volts/meter (V/m). The field immediately below a 400,000 V transmission line is 10,000 V/m, dropping to 500–1,000 V/m at a distance of 100 m. The law of inverse squares is at work here.

Numerous studies have shown deleterious effects from low levels of ELF radiation including deformities of chick embryos, behavioral alterations, and physiological changes. Mice and snails have shown increased sensitivity to heat following exposure to ELF. Experimental findings have not always been consistent, however. Calcium uptake by cells is inhibited in some systems. Most of these studies employed exposures ranging from

30 mG (the level emitted by some electric blankets) to 1 gauss and exposure times ranging from 30 min to several days or weeks (in the case of the chick embryo studies). The field strength of the Earth is about 0.3–0.6 gauss.

The greatest controversy centers on the interpretation of epidemiological data. Several studies have been conducted that purport to show increased incidences of cancer, especially leukemia, recurrent headache and depression, congenital deformities, and other health problems in people living near high-voltage power lines or in high exposure working environments. These studies have been reviewed by several epidemiologists and have been faulted on several counts, including failure to account for other risk factors, failure to balance control and test groups and the use of inappropriate statistics. In particular, the use of the Proportional Mortality Rate (PMR) has been criticized. This is the practice of expressing mortality rates due to certain causes as a percentage of the total. With this method, a decrease in one cause (e.g., traffic accidents) leads to an apparent increase in the other causes even though none has actually occurred, or even when a decrease has occurred. Most reviewers concede that there is some evidence of a marginal increase in the risk of leukemia in electrical workers, but qualify this by stating that the question remains open. This has not discouraged Paul Brodeur, a staff writer for the "New Yorker," from publishing a book on the subject entitled "Currents of Death," to follow up on his previous success with "The Zapping of America" on the subject of the "hazards" of video display terminals.

Two studies in 1988 provided further conflicting evidence. One, conducted in Washington state, found no correlation between the incidence of acute lymphocytic leukemia and exposure to electromagnetic radiation from high-power lines located within 140 ft of homes whether compared to actual measures within the homes or to computed values based on the wiring configurations. The other, conducted in Colorado, looked at cancer incidences in children 14 years old and under and found no association between total cancer incidence and exposure levels but did find a modest association for lymphomas and sarcomas. The situation was not clarified at a 1992 meeting in San Diego at which a Swedish group, using more accurate data regarding exposure levels than has been previously available, showed a correlation between exposure and cancer incidence. Conversely, a literature study commissioned by the U.S. government found no convincing evidence of an increased cancer risk for exposure to ELFs. In 1999, a study in Toronto, conducted by the Hospital for Sick Children, reported that children exposed to high levels of electromagnetic radiation had an incidence of leukemia two to four times higher than did children not so exposed. In contrast, a similar study conducted in British Columbia did not find any association between exposure level and leukemia incidence.

The question of ELF exposure and risk of leukemia in children is still being investigated with continuing unclear results. Typical is a study by Draper et al. who found an association between the incidence of childhood leukemia and proximity at birth to high-voltage lines. Their case-control study looked at 28,081 Welsh and English children with cancer including 9700 with leukemia. The children were 14 years of age or less. Children who lived within 200 m had a relative risk of 1.69 whereas those living 200–600 m from a line had a relative risk of 1.23. This inverse relationship was significant ($p < 0.01$). The association did not apply to other forms of cancer. The relationship extended further out than reported in previous studies. The authors state that there is no known biological mechanism to explain these results and they may be due to chance or confounding factors. A study by Malagoli et al. looked at a corridor along a high-voltage power line with a calculated magnetic field of 0.4 µT and found a relative risk factor for ALL of 5.3, but conceded the numbers were too small to draw firm conclusions. A study in Iran found a significant association between power line proximity and the incidence of ALL and concluded that living close to power lines was a risk factor for ALL. Draper conducted another study in 2010 on 28,968 Welsh and English children under the age of 15 and diagnosed with cancer. Matched controls were selected from birth registries. Once again results were suggestive of an association. For each increase in exposure of 0.2 µT, the relative risk was 1.4 for leukemia, 0.8 for CNS/brain tumors, and 1.34 for other cancers. These findings were not statistically significant, however. In summary, these studies appear to parallel the situation of the cell phone–brain tumor association. They frequently suggest that there is one but definitive evidence remains elusive.

The results of a massive study of some 224,000 utility workers in Quebec, Ontario, and France were released in the spring of 1994. This was a case-control study in which 4151 new cancer cases identified during the study period were matched with 6106 cancer-free controls from the same population to eliminate confounding factors such as smoking, etc. Past exposure to EMFs was documented. There was no association observed between the overall cancer incidence and past EMF exposure. This was true for lymphoma, melanoma, lung, stomach, and colon cancer. There was a statistically significant association between cumulative exposure and a rare form of adult leukemia, acute nonlymphoid leukemia and its subtype, acute myeloid leukemia. Risk was increased by a factor of 2.41, but a causal relationship was not demonstrated. The level of significance, moreover, was at the 95% confidence limit, which is usually considered to be the minimum acceptable. There was a nonsignificant association between cumulative exposure to the highest levels of EMF and astrocytoma, a rare brain cancer. Given the size of this study it seems unlikely that any more conclusive evidence soon will be forthcoming.

One rather oddball theory attempts to associate geographic areas of high EMF with deep underground tension between tectonic plates of the Earth. One such area is in Canada, and the suggestion (largely unfounded) is that these areas are associated with luminous atmospheric phenomena, UFO reports, and increased incidence of cancer, notably brain tumors!

Irradiation of Foodstuffs

The use of ionizing radiation to preserve foodstuffs by killing spoilage micro-organisms was proposed as early as 1905 when British patent No. 1609 was issued to "J. Appleby, miller, and A.J. Banks, analytical chemist." They noted that sterilization might be accomplished in the complete absence of foreign chemicals. It was not until the postwar arrival of the "atomic age," however, that sources of ionizing radiation were sufficiently abundant to make this procedure practical and economical. In the immediate postwar era there was considerable public enthusiasm for the development of nuclear technology for peaceful purposes, and there was good acceptance of the notion of irradiated foods such as milk and vegetables. By 1965, the United States, Canada, and the Soviet Union had approved the marketing of certain foods treated with low-dose irradiation, but no manufacturers had taken advantage of this situation. By this time the health concern over Strontium 90 levels in milk was widely known, the antinuclear movement was in full swing, and radio-phobia was growing. Acceptance of irradiated foods was in rapid decline, and it became replaced with concerns over potential health hazards. These concerns tended to fall into three areas:

1. Foods would be made radioactive and those who consumed them might develop radiation sickness.
2. Foods might be altered in some way, rendering them toxic, even carcinogenic.
3. Microorganisms could mutate to new and horrific pathogenic forms.

The first objection is easily dismissed. Foods are not rendered radioactive by low-dose ionizing radiation. Extensive toxicological testing, including tests or carcinogenicity, over several decades led a United Nations Joint Expert Committee on Irradiated Foods to conclude, in 1980, that "… the irradiation of any food commodity up to an overall average dose of 10 kGy presents no toxicological hazard; hence, toxicological testing of foods so treated is no longer required." The second objection is therefore readily disposed of. The third concern stems from experimental evidence that pure, dilute solutions of glucose become mutagenic for *Salmonella typhimurium* after irradiation, and that polyunsaturated fats in an oxygen atmosphere form lipid peroxides when irradiated. Neither change has ever been noted in complex food materials. Nor is there any evidence of a direct mutagenic effect on microorganisms by ionizing radiation under the conditions in which it is employed. Nevertheless, resistance to irradiation varies greatly from bacterium to bacterium with spore formers like *Clostridium botulinum* being very resistant. Ionizing radiation is thus likely more useful to prevent spoilage than to eliminate potential pathogenic organisms. Public acceptance of irradiated foods remains a problem.

The pendulum has been swinging back toward acceptance of irradiation of foods. Several large outbreaks of so-called hamburger disease, resulting from contamination of ground beef by strain O157:H7 of *Escherichia coli*, have led to the use of irradiation of ground beef. In June of 2000, 77 metric tons of ground beef was recalled from a number of supermarket chains in Canada because of *E. coli* O157:H7 contamination. The public may be becoming aware that this is a far greater, and very real, health risk than any posed by irradiated food.

Reports confirming the safety of the process continue to emerge. Gamma (γ) radiation has been confirmed safe to treat chestnuts to prevent bacterial spoilage and insect damage during storage of vegetables, and fruits, and has been approved by the U.S. Food and Drug Administration (FDA) for use on spinach and lettuce up to 4.0 kilograys (kGy). Reviews of the process have declared it safe and note that it is highly regulated by the FDA.

Irradiation of Insect Pests

The principle of irradiating insect pests to control their population has been around for a long time. In the early 1950s, Dr. Raymond C. Bushland and Dr. Edward F. Knipling received the World Food Prize for demonstrating that male screwworm flies could be sterilized by γ irradiation. Since the female fly will only allow mating to occur once, releasing millions of sterilized males effectively interrupts the breeding cycle (a form of *coitus interruptus?*). The screwworm lays her eggs on the skins of cattle and sheep, causing them great misery and reducing weight gain, milk production, and damaging hides. The technique has virtually eliminated the problem from the southern United States and eradicated the pest from the island of Curacao and elsewhere. The technique has been used successfully on several species of fruit flies as well.

Case Study 28

A 72-year-old Navajo man resided on an American Native reservation in the four corners region of the American southwest. In 2000, he attended the Navajo Medical Center with respiratory symptoms. He was hospitalized and treated for pneumonia when a chest x-ray revealed an infiltrate in the lower right lung. He developed progressive respiratory failure and died 19 days after admission. Autopsy revealed a 2.5 cm squamous cell carcinoma of the right lung, a broncho-esophageal fistula, and a right lower lung abscess. This individual had worked in the local mining industry for 17 years, leaving it in 1978.

Q. What mining industry was predominant in the region in the 1970s?

Q. What risk factors might have existed that would predispose to his medical condition?

Q. What safety measures should have been instituted to minimize the risk?

Review Questions

For the Questions 1–8 use the following code:

Answer A if statements a, b, and c are correct.

Answer B if statements a and c are correct.

Answer C if statements b and d are correct.

Answer D if only statement d is correct.

Answer E if all statements (a, b, c, d) are correct.

1. a. Gamma rays have a longer wavelength than microwaves.
   b. Ionizing radiation includes all wavelengths of 1000 Å or less.
   c. Only a small fraction of radiation comes from natural sources.
   d. Ionizing radiation strips electrons from molecules as it passes through tissues.

2. a. Alpha-particles consist of two neutrons and two protons.
   b. An alpha-particle is identical to the helium nucleus.
   c. Gamma rays have no charge.
   d. Beta-particles have no charge.

3. a. A roentgen defines the amount of ionization in air caused by any radiation.
   b. rem stands for "reactive emission material."
   c. Predicting the carcinogenic potential for radiation is very difficult below 10 rem.
   d. Alara stands for "always let active radiation alone."

4. a. The main source of radon in the environment is nuclear power plants.
   b. Alpha-emitting daughters of radon bind to dust particles that lodge in the lung.
   c. Strontium-90 poses no threat to human health.
   d. Radon daughters are likely the second-most common cause of lung cancer after smoking.

5. a. In some geographic areas, the public may get higher radiation exposures than most nuclear power plant workers.
   b. The thyroid gland is very sensitive to radiation damage.
   c. Neutrons given off during radioactive decay may collide with other atoms to induce alpha, beta, and gamma radiation.
   d. Smoke detectors contain a radioactive component in the ion detector.

6. Which of the following statements is/are true?
   a. Alpha-particles cause tissue damage only over a very short distance (70 μ).
   b. Tissues with a rapid turnover are most vulnerable to radiation damage.
   c. Tissues sensitive to radiation damage include hair follicles, bone marrow, and gastrointestinal mucosa.
   d. Safe limits for radiation exposure have been reduced recently because of new evidence that low exposures cause cancer in mice.

7. a. Extra-low frequency (ELF) electromagnetic radiation is emitted by high-voltage power lines.
   b. Evidence that ELF radiation causes cancer is inconclusive.
   c. ELF radiation has been shown to affect many biological systems adversely in single-celled organisms and chick embryos.
   d. ELF radiation is given off by electric blankets.

8. With respect to irradiated foods:
   a. Fruit are sometimes irradiated to prevent spoilage.
   b. Low-level irradiation has been shown conclusively to induce mutation in several species of bacteria.
   c. Lipid peroxides have been shown to form from polyunsaturated fats *in vitro* but not *in vivo*.
   d. Irradiated food becomes radioactive.

Answers

1. C
2. A
3. B
4. C
5. E
6. A
7. E
8. B

Further Reading

Antonio, A.L., Carocho, M., Bento, A., Quintana, B., Luisa-Botelho, M., and Ferreira, I.C., Effects of gamma radiation on the biological, physico-chemical, nutritional and antioxidant parameters of chestnuts—A review, *Food Chem. Toxicol.*, 2012, Jun. 23 (Epub ahead of print).

Arvanitoyannis, I.S., Stratakos, A.Ch., and Tsarouhas, P., Irradiation applications in vegetables and fruits: A review, *Crit. Rev. Food Sci. Nutr.*, 49, 427–472, 2009.

Bithell, J.F., Dutton, S.J., Draper, G.J., and Neary, N.M., Distribution of childhood leukemias and non-Hodgkin's lymphomas near nuclear installations in England and Wales, *BMJ*, 309, 501–505, 1994.

Bowie, C. and Bowie, S.H., Radon and health, *Lancet*, 1, 409–413, 1991.

Cardis, E. and Hatch, M., The Chernobyl accident—An epidemiological perspective, *Clin. Oncol.*, 23, 251–260, 2011.

Cardis, E., Krewski, D., Bobiol, M., Drozdovitch, V., Darby, S.C., Gilbert, E.S., Akiba, S. et al., Estimates of the cancer burden in Europe from radioactive fallout from the Chernobyl accident, *Int. J. Cancer*, 119, 1224–1235, 2006.

Cole, L., Much ado about radon, *The Sciences*, 30, 19–23, 1990.

Corle, C., Makale, M., and Kesari, S., Cell phones and glioma risk: A review of the evidence, *J. Neurooncol.*, 106, 1–13, 2012.

Deadly toll of Chernobyl, http://news.bbc.co.uk/2/hi/europe/722533.stm (accessed on June 14, 2012).

Death toll from Chernobyl was overestimated: report, www.timesonline.co.uk/tol/news/world/eyrope/article563041.ece (accessed on June 14, 2012).

Diehl, J.F., *Safety of Irradiated Foods*, Marcel Dekker, New York, 1990.

Douple, E.B., Mabuchi, K., Cullings, H.M., Preston, D.L., Kodama, K., Shimizu, Y., Fujiwara, S., and Shore, R.E., Long-term radiation-related health effects in a unique human population: Lessons learned from the atomic bomb survivors of Hiroshima and Nagasaki, *Disaster Med. Public Health Prep.*, 5(Suppl. 1), S122–S133, 2011.

Dowson, D.I. and Lewith, G.T., Overhead high voltage cables and recurrent headache and depressions, *The Practitioner*, 232, 435–436, 1988.

Draper, G., Vincent, T., Kroll, M.E., and Swanson, J., Childhood cancer in relationship to distance from high voltage power lines in England and Wales: A case-control study, *BMJ*, 220, 1290, 2005.

Eijgenraam, F., Chernobyl's cloud: A lighter shade of gray, *Science*, 250, 1245–1246, 1991.

Food and Drug Administration, Radiation of lettuce and spinach, *Fed. Regist.*, 73, 49593–49603, 2008.

Hardell, L., Carlberg, M., and Hansson, M.K., Pooled analysis of case-control studies on malignant brain tumours and the use of mobile and cordless phones including living and deceased subjects, *Int. J. Oncol.*, 38, 1465–1474, 2011.

Henshaw, D.L., Eatough, J.P., and Richardson, R.B., Radon as a causative factor in induction of myeloid leukemia and other cancers, *Lancet*, 335, 1008–1012, 1990.

Jerrard, H.G. and McNeill, D.B. (eds.), *Dictionary of Scientific Units*, 6th edn., Chapman & Hall, London, U.K., 1992.

Kroll, M.E., Swanson, J., Vincent, T.J., and Draper, G.J., Childhood cancer and magnetic fields from high-voltage power lines in England and Wales: A case-control study, *Br. J. Cancer*, 103, 1122–1127, 2010.

Little, M.P., Rajaraman, P., Curtis, R.E., Devesa, S.S., Inskip, P.D., Check, D.P., and Linet, M.S., Mobile phone use and glioma risk: Comparison of epidemiological study results with incidence trends in the United States, *BMJ*, 344, e1147.

Malagoli, C., Fabbi, S., Teggi, S., Calzari, M., Poli, M., Ballotti, E., Notari, B., Bruni, M., Palazzo, G., and Paolucci, P., Risk of hematological malignancies associated with magnetic fields exposure from power lines: a case-control study in two municipalities of northern Italy, *Environ. Health*, 9, 16, 2010.

McGregor, R.G., Vasudev, P., Letourneau, E.G., McCullough, R.S., Prenti, F.A., and Taniguchi, H., Background concentrations of radon and radon daughters in Canadian homes, *Health Phys.*, 39, 285–289, 1980.

Minami, I., Nakamura, Y., Todiriki, S., and Murata, Y., Effect of γ irradiation on the fatty acid composition of soybean and soybean oil, *Biosci. Biotechnol. Biochem.*, 76, 900–905, 2012.

Morgan, L.L., Herberman, R.B., Philips, A., and Davis, D.L., Re: Mobile phone use and brain tumors in children and adolescents: A multicenter case-control study, *J. Natl. Cancer Inst.*, 104, 635–637, 2012.

Munshi, A., Cellular phones: To talk or not to talk, *J. Cancer Res. Ther.*, 7, 476–477, 2011.

Parnes, R.B. and Lichtenstein, A.H., Food irradiation: A safe and useful technology, *Nutr. Clin. Care*, 7, 149–155, 2005.

Perrson, B.R. and Holm, E., Polonium-210 and lead-210 in the terrestrial environment: A historical review, *J. Environ. Radioact.*, 102, 420–429, 2011.

Pool, R., Electromagnetic fields: The biological evidence, *Science*, 249, 1378–1381, 1990.

Report of the Working Group on Electric and Magnetic ELF Fields, *Electric and Magnetic Fields and Your Health*, Publ. # H46-2/89-140E, Min. Supply and Services Canada, 1990.

Rutkowski, C.A. and Del Bigio, M.R., UFOs and cancer? *CMA J.*, 140, 1258–1259, 1989.

Samet, J.M., Radiation and cancer risk: A continuing challenge for epidemiologists, *Environ. Health*, 10 (Suppl. 1), S4.

Savitz, D.A., Wachtel, H., Barnes, F.A., John, E.M., and Tvrdik, J.G., Case control study of childhood cancer and exposure to 60-Hz magnetic fields, *Am. J. Epidem.*, 128, 21–39, 1988.

Severson, R.K., Acute nonlymphocytic leukemia and residential exposure to power frequency magnetic fields, *Am. J. Epidem.*, 128, 10–20, 1988.

Shore, R.E., Electromagnetic radiations and cancer. Cause and prevention, *Cancer*, 62, 1747–1754, 1988.

Shea, K.M., Technical report: Irradiation of food. Committee on environmental health, *Pediatrics*, 106, 105–110, 2000.

Smith, H., ICRP publ. 50. Lung cancer risk from indoor exposures to radon daughters, *J. Can. Assoc. Radiol.*, 39, 144–147, 1988.

Sohrabi, M.R., Tarjoman, T., Abadi, A., and Yavari, P., Living near overhead high voltage transmission power lines as a risk factor for childhood acute lymphoblastic leukemia: A case-control study, *Asian Pac. J. Cancer Prev.*, 11, 423–427, 2010.

Stone, R., Can a father's exposure lead to illness in his children? *Science*, 258, 31, 1992a.

Stone, R., Polarized debate: EMFs and cancer, *Science*, 258, 1724–1725, 1992b.

Swanson, J., Vincent, T., Kroll, M., and Draper, G., Power-frequency electrical and magnetic fields in the light of Draper et al. 2005, *Ann. NY Acad. Sci.*, 1076, 318–330, 2006.

Tar-Ching, A.W., Living under pylons, *BMJ*, 297, 1469–1470, 1988.

Thériault, G., Goldberg, M., Miller, A.B., Armstrong, B. et al., Cancer risks associated with occupational exposure to magnetic fields among electric utility workers in Ontario and Quebec, Canada and France: 1970–1989, *Am. J. Epidemiol.*, 139, 550–572, 1994.

Tronnes, D.H. and Seip, H.M., Health risks caused by indoor radon exposure. In *Risk Assessment of Chemicals in the Environment*, Richardson, M.L. (ed.), Royal Society of Chemistry, London, U.K., 1988, chap. 16.

Wertheimer, N. and Leeper, E., Magnetic field exposure related to cancer subtypes. In *Environmental Sciences*, Sterrett, F.S. (ed.), New York Academy of Sciences, New York, Vol. 502, pp. 43–54, 1987.

U.S. coal mining deaths 1990–2009, http://frankwarner.typead.com/free_frank_warner/2006/01/us_coal_mining_html (accessed on June 14, 2012).

Yamaguchi, M., Japan nuclear disaster "man-made" says report, Toronto Star, July 5, 2012, http://www.thestar.com/news/world/article/1221748—japan-nuclear-disaster—man-made-says-report

14

Gaia and Chaos: How Things Are Connected

Life itself is a religious experience.

James Lovelock

When we try to pick out anything by itself, we find it hitched to everything else in the universe.

John Muir (1911)

Gaia Hypothesis

Formulated in 1965 by the independent British biologist James E. Lovelock and elaborated by Lynn Margulis, distinguished biology professor at the University of Massachusetts, it proposes that certain kinds of life on the planet grow, change, and die in ways that lead to the persistence of other life forms. In some circles, this has been interpreted as meaning that life on Earth forms a single, complex continuum, one ecosystem throughout time and space. The Earth, according to this view, can thus be considered as a single organism and its various components as cells in that organism. The name is taken from the Greek Earth goddess Gaea. Although Lovelock intended his first book to be taken as a scientific treatise, there was a considerable amount of mysticism and spiritual significance attached to it by segments of the public, and this tended to turn serious scientists away from the theory for a long time. As information accumulated about the role of rain forests in consuming CO_2, and producing O_2, and of the role of wetlands in purifying water, and of ocean phenomena such as *El Nino* in affecting climate, the idea of Earth as an integrated biosystem gained credibility. This was strengthened as it became evident that human disruption of components of it, such as the ozone layer, could have serious consequences for life on Earth.

To a microbiologist like Margulis, the Gaia hypothesis made perfect sense as is seemed, when stripped of its Earth goddess mystique, to be simply another, perhaps more complex, example of symbiosis, so commonly encountered in the world of microorganisms. As she pointed out during an address to other microbiologists, "Without the few pounds of bacteria in each of our guts, no one would ever digest food, and without the nitrogen-fixing bacteria in the soil, no food would ever grow in the first place." Lynn Margulis, aged 73, died November 22, 2011.

Both Lovelock and Margulis took considerable pains to point out that the Earth-as-single-organism view is not what Gaia is all about. To quote Margulis, "the surface temperature, chemistry of the reactive gaseous components, the oxidation-reduction state and the acidity-alkalinity of the Earth's atmosphere and surface sediments are actively (homeorrhetically) maintained by the metabolism, behavior, growth and reproduction of organisms (organized into communities) on its surface. Gaia is not an individual, it is an ecosystem."

One example of a self-regulating system involves dimethylsulfide (DMS) produced by algae in the ocean. Bubbles produced by wave action burst, injecting their DMS into the air. Solar heat causes rising convection currents to carry it into the stratosphere, where it oxidizes, providing nuclei for the formation of water droplets. The resulting cloud formation has a cooling effect on the ocean surface, reducing the transfer of DMS to the stratosphere by convection. Algae use DMS as a propellant gas. Its formation thus increases when the population of algae explodes, and the cooling effect can help to regulate algal growth. Algae inhabit coral reefs and other benthic invertebrates, often determining the color of the host organism. If too great an increase in temperature occurs, coral polyps may expel the exploding algae in the phenomenon known as coral bleaching. If the problem is short-lived, the algae will reestablish in the coral. If the situation persists, however, the coral will die. Since coral reefs cover only 10% of the ocean floor but account for 25% of its fish life, the consequences for the whole reef ecosystem are serious. Global warming, regardless of the cause, can thus have catastrophic effects on coral reef systems.

Another example of how one organism can affect others is two billion years old. At that time a family of organisms, called Cyanobacteria, were dominant on Earth. They were photosynthetic, and in the process of consuming CO_2 they produced large quantities of O_2 which was poisonous to most other species. Those with resistance to O_2 survived and gradually evolved into the aerobic organisms that came to dominate the planet's species. There has never been as radical a change in the Earth's population since.

The resurgence of interest in the Gaia hypothesis as a result of environmental concerns has had both good and bad consequences. On the plus side, there is increased awareness of the interconnectedness of life on Earth and the possibility that a disruption in one part of an ecosystem can have far-reaching consequences. Less desirable is the proliferation of fuzzy-minded philosophies (we are all one with the universe, etc.) that have led to such new age phenomena as Shirley McLain, with her crystals and prior lives. (Why was everyone a princess or a warrior in a previous life, but never a toilet cleaner?) Beliefs such as these tend to erode interest in, and trust of, science and this may be reflected in declining enrollments in science programs at a time when society needs to be improving and increasing its science and technology.

Of course, the science of ecology is based on knowledge of the interconnectedness of the components of an ecosystem. An example of the practical importance of such knowledge was noted in the May/June 1999 environmental issue of *Canadian Geographic*. A decline in pollock stocks in the Bering Sea, attributed to overfishing, has led to a decline in the population of Steller's sea lions, which are now considered a threatened species. These sea lions depend on pollock as a food source. Their decline has forced orcas to prey on sea otters instead of sea lion pups. As a result, sea urchins have proliferated, and these have decimated the kelp beds that constitute a nursery for many other species.

Chaos Theory

In the introduction to his book *Chaos: From Theory to Applications*, Tsonis points out that simplicity and regularity are associated with predictability, whereas complexity and irregularity are virtually synonymous with unpredictability. Chaos, in the language of mathematics, is defined as "randomness" generated by simple deterministic systems. The word randomness is presented in inverted comas to suggest that the random nature may be apparent and that the determinism may persist, albeit in forms difficult to recognize.

Chaos theory has its roots in bifurcation theory first formulated by Poincaré but it was developed by Edward Lorenz who showed that non-linear differential equations exhibited final states that were nonperiodic, i.e., apparently random. During investigations using computer simulations of computer networks it was discovered that, under some conditions, routings became random and chaotic instead of following the orderly sequence that the system had been designed to use. For example, if a request was placed to use a specialized computer in location A (e.g., designed for theoretical mathematics) during a slack period (lunch hour), the request would be honored and several minutes of use might be available. If, however, the request clashed with several other simultaneous ones, it might be rerouted to location B along with other requests and disrupt use of this facility with resulting further rerouting to tertiary locations and the subsequent production of ripples throughout the system. Graphic plots of usage reveal oscillating patterns that are neither organized nor completely random. They have been described as "organized complexity."

An important feature of chaos theory is the existence of the so-called low-level attractors. Using the example of a free-swinging pendulum, where x_1 = position and x_2 = velocity, the tendency to return to the equilibrium state, where both x_1 and $x_2 = 0$ is defined as the trajectory and the equilibrium state is the attractor. In a system in which the effect of friction is offset by a mainspring, a disturbance in the motion of the pendulum will eventually be overcome and it will return to its periodic state. In this case, the *cycle* is the attractor. When chaotic processes are plotted on phase-space (three-dimensional) graphs, patterns are produced that are not truly random, as they would be if the process were completely disorganized. These patterns are called "strange attractors."

Chaos theory is now being applied to such diverse fields as the physics of fluid mixing and weather forecasting. In the latter regard, computer models literally suggest that a butterfly beating its wings in China can influence weather patterns in North America. While this may seem farfetched in the real world, it illustrates once again the interconnected nature of nature, and it helps explain why accurate weather forecasting is a difficult goal to achieve over a time span of more than a few days. There are a number of models used for weather forecasting. They differ in the physics and other parameters. Chaos theory may provide a means of exploring which model is most appropriate under given conditions.

Other Examples of Interconnected Systems

Students of ecology should be highly familiar with the concept of symbiosis and how important it is in the maintenance of an ecosystem, but for those whose education has been centered largely on human health, a few examples of how biological events can be interconnected might be illustrative.

Vicious Circle

In Chapter 11, the subject of fungal infections of cereal grains was discussed, and how these caused economic losses measured in billions of dollars in Canada alone. Such economic loss to farmers means fewer dollars to spend on consumer goods. This leads to a slowdown in the manufacturing sector, leading to higher unemployment, leading to further declines in the purchase of consumer goods (a vicious circle) and, eventually, contributes to a recession. The Great Depression that began in 1929 is generally blamed on the crash of the stock market, precipitated by margin buying and uncontrolled borrowing. There is another theory, however.

Following the Great War (1914–1918) there was a tremendous upsurge in mechanization of farm machinery fueled by technological developments such as the internal combustion engine used in aircraft and tanks. Prior to the war, one in every five Americans earned their living in agriculture. After mechanization it fell to one in twenty. The dust bowl of the Midwest further exacerbated the situation. Thousands of unemployed farmers and farm hands migrated to the big cities where heavy industry was already in decline due to the loss of war contracts. Forced onto the dole they bankrupted city coffers, causing yet more borrowing until the financial collapse occurred. Farmers could not borrow on their farms anymore and foreclosures were the common consequence. This foretold the subprime mortgage crisis of recent times.

That was a combined economic and natural vicious circle, but a biological one may also occur. Figure 14.1 illustrates how this might work. In the

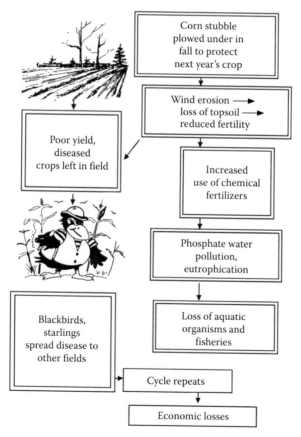

FIGURE 14.1
A vicious circle affecting agriculture.

past it has been a practice to plough under corn stubble in the fall so as to expose the roots and allow frost to kill fungal spores. If the following spring was dry, wind would erode the topsoil, reducing its fertility. This may require the increased use of chemical fertilizers that can, in turn, increase water pollution and possibly cause eutrophication. Other possible consequences are also illustrated.

Domino Effects of Global Warming

The impact of climate change on the natural world was discussed in Chapter 4. No other occurrence in recent history better illustrates how things are connected. Walruses have been dying in onshore stampedes because ice floes, their natural haul-out sites, are disappearing. Polar bears may be threatened because sea ice is disappearing and that is where they hunt seals at their blowholes. Insects, including venomous ones are migrating north as summer temperatures rise. Crop yields and livestock may be threatened by drought. When a habitat changes dramatically, every species within it is affected, including our own.

Even a small increment in mean annual temperature, including a cyclical one, could have profound effects on the biosphere. The current "hot spot" for fungal infestations of grain in Canada is southwestern Ontario. Warming would move the demarcation line northward, so that *Aspergillus flavus*, the source of the carcinogenic mycotoxin aflatoxin, which cannot survive cold winters, could change its distribution.

Certain insects could also move north. The Africanized honeybee is already in the southwestern United States and it would follow a warming trend northward. Mediterranean fruit flies, screwworm flies, even malarial mosquitoes, could follow. Malaria-carrying mosquitoes have been detected as far north as New York State. Insects are vectors of many diseases of animals and humans. These would spread along with their hosts. Venomous insects also could move north. The brown recluse spider has moved from Florida to Pennsylvania and will undoubtedly cross the Great Lakes at some time if it has not already done so.

There is some evidence that ecological disturbances caused by anthropogenic activity may result in viral infections jumping species barriers, especially from animals to humans. Recently, the Marburg virus and the Ebola virus emerged as life-threatening infections of humans and are believed to have originated in monkeys. Every variety of influenza antigen has been identified in ducks and other waterfowl, and an outbreak of flu in Hong Kong was traced to chickens and resulted in the slaughter of thousands of birds in an effort to contain the outbreak.

Swine have long been known to harbor flu viruses and are thought to have been the reservoir of the Spanish Flu epidemic during World War I. The plague bacillus, *Yersinia pestis*, periodically jumps from rats to humans, carried by fleas. Slash-and-burn agriculture, practiced in Africa for eons, creates an environment favorable to the Anopheles mosquito that carries malaria. These mosquitoes are present in the Great Lakes Basin, where malaria was endemic in the 1800s, and the possibility that there may be a resurgence of malaria associated with a warming trend in the climate cannot be dismissed. Lyme disease has already reached the Southern Ontario from its original identification site in New England. An outbreak of a viral infection in 1993 killed 12 people in New Mexico. The virus (Hantavirus) was identified as belonging to the Hantaan group, which is spread in the feces and urine of rodents. It was first identified during the Korean War as the cause of hemorrhagic fever in soldiers. The condition observed in the American southwest is called Hantavirus pulmonary syndrome (HPS). An exceptionally high yield of piñón nuts, a staple diet for rodents, led to an explosion in the rodent population and a resulting increase in the exposure of humans to their droppings. It has since been identified as a cause of human infection in many states, including Florida. The deer mouse appears to be the most common carrier, but other rodents such as the cotton rat have also been identified as carriers.

Many of these problems are exacerbated by the ease with which people can now move from one locale to another, and some medical experts are warning that their profession must be on the alert for diseases not normally seen in northern climes. Global warming may allow easier movement of birds. In 1999, cases of West Nile encephalitis in New York were traced to contaminated wildfowl that apparently brought the disease from North Africa. Also attributed to the movement of birds was an outbreak of Salmonella type DT104 in Vermont. This strain of bacteria, prevalent in cattle in Great Britain, has been responsible for an epidemic there. The organism was isolated from the milk tank on a dairy farm. One family member became critically ill but recovered.

On a more cheerful note, a longer growing season could mean increased crop yields and more arable land.

Feedback Loop

An elegant example of how an ecosystem can self-regulate is the manner in which water temperature is regulated in small bodies of freshwater. It has been shown recently that bioregulation of water temperature occurs in a manner reminiscent of negative feedback control systems in mammals and similar to the DMS feedback system in oceans. The system is shown in Figure 14.2. Again, algae play an important role. In an algal bloom, water turbidity is increased and penetration of sunlight lessens. The cooling effect inhibits algal growth. The same effect can occur if the pond is stocked with predatory fish. These will eat smaller fish that prey on algae eaters (small crustaceans, etc.), allowing uninhibited growth of the algae.

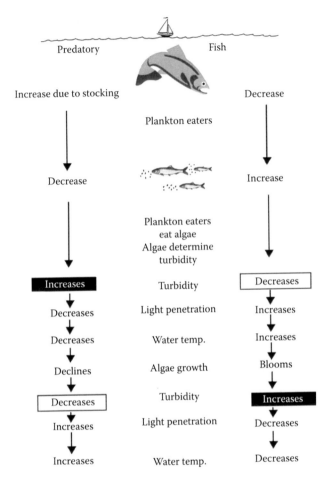

FIGURE 14.2
A feedback loop regulating temperature in a small body of freshwater.

Food Production and the Environment

This text is primarily about the relationship between the environment and human health. It is undeniable that starvation and malnutrition are the greatest killers of humankind and that they relate to the ability of the Earth to feed its population. Some consideration of this question is thus not inappropriate.

Meat versus Grain

It is often stated that a vegetarian diet is environmentally friendlier than a diet containing meat, because one can produce more food by growing plants than by grazing animals. Besides, animals produce methane, which is a greenhouse gas. This argument is frequently put forth by animal rights activists to support their philosophical position, which also draws heavily on the mystical side of Gaia. This conventional "wisdom" has even appeared in the popular press in articles written by nutritionists. But does it stand up to scrutiny? Notwithstanding the fact that a diet rich in fruits and vegetables is undeniably healthier than one which is heavy with meat, consider the following:

1. Livestock can be, and are, grazed on grasslands unsuitable for cultivation and in colder climates with a very short growing season. This occurs in several locations in North America and northern Europe. In Australia and New Zealand, sheep are grazed extensively on land that is harsh and inhospitable to cultivation.

2. Before the North American prairies were ploughed up to plant grain, they supported an estimated 50,000,000 bison and millions of pronghorn antelope, elk, and caribou (at least seasonally) without damage to the soil. Cultivation coupled with drought brought the dustbowl of the 1930s. In Southern Ontario, a dry spring and high winds may result in thick clouds of brown dust coating autos, houses, and grassland as topsoil is blown from surrounding farms that were fall-ploughed to eliminate the spores of the *Fusarium* mould.

3. Even in semiarid areas such as the American southwest, studies have shown that the footprints of wild ungulates and cattle form little traps for seeds and water. Grass tufts develop in these that help stabilize the soil. It must be kept in mind that there is a vast difference between intelligent grazing and overgrazing. Much of the evidence against livestock grazing is taken from underdeveloped countries where overgrazing occurs extensively and agricultural technology lags. Another bit of evidence is taken from the Amazon basin, where deforestation for cattle ranching destroys the soil and the rain forest. This is quite true, but deforestation also occurs for paper production (some years ago, Japanese entrepreneurs floated an entire pulp and paper mill up the Amazon) and for crops such as sugarcane. All are equally destructive because the culprit is the deforestation. In Florida vast areas of the Everglades have been drained to provide land for growing rice and other crops. The loss of such wetlands is an ecological disaster.

4. What about that methane? Best estimates are that the rice paddies of the world produce as much methane as all of the animals combined, both domestic and wild (about 100 megatons per year). Since rice is a staple grain for much of the world, shifting reliance to it from meat would not prevent as much methane pollution as one might think. Besides, if we increase the wildlife population by protective measures (a laudable goal) we will also be increasing

methane production. Is there good methane and bad methane? How much methane can an elephant make?

5. The average daily caloric intake of 40 countries from the poorest (Bangladesh) to the richest (United States) is 2571 cal of which meat provides 205 and vegetables only 41. Even in Guatemala meat provides 49 of 2020 cal/day, vegetables only 18. The bulk is from maize (corn), which probably is high in carcinogenic mycotoxins.

6. Water pollution by animal wastes is often cited as evidence that livestock is less desirable environmentally than crops, but we have already seen how the latter can pollute through chemical fertilizers.

Decisions concerning what is best for the environment should be made on the basis of sound scientific knowledge rather than on a philosophical position. The same applies to decisions regarding the healthiest diet.

There is mounting evidence that world food supplies are declining at a time when the population is exploding. These two factors appear to be on a collision course. The production of cereal grains, notably corn, is being diverted to the production of ethanol for fuel, leading to concerns that this might impact on the world's food supply. Fish stocks s are being depleted. A prime example of this latter point is the decline in North Atlantic cod stocks. Overfishing has been singled out as the major culprit, but competition from an expanding seal herd for caplin, a herring-like fish that is the principal food source for both species, has also been incriminated. Surveys indicate that *all* ground species, including noncommercial ones like monkfish and eelpouts, are in decline from waters off northern Newfoundland all the way up the coast of Labrador, suggesting that natural phenomena may be contributing to the decline. Levels of pollutants such as dioxins and heavy metals are negligible in Atlantic cod, making it unlikely that pollution is the culprit. A possible explanation may be the 0.5°C–1.0°C cooling of northern spawning waters, which may reduce dramatically the survival of cod fingerlings. *El Nino*, which involves a massive pooling of warm water in the mid-Pacific, has caused marked reductions in the fish catches off South America.

Genetically Modified Plant Foods

With regard to cultivated foodstuffs, there is now great concern that the development of special strains of food grains like rice and wheat, with increased resistance to specific diseases, may render them more susceptible to other diseases, some of which may not have emerged as yet. There is now an effort to collect and preserve the wild strains of important food sources such as rice, corn, potatoes, and fruits to constitute a genetic library that can be called upon in the future when it is needed.

A more recent concern over the environmental impact of human-created species relates to the development of genetically engineered species of macro- and microorganisms. Herbicide resistance has been conferred genetically on cultivated plants, resulting in concern that their survival advantage might lead to their invasion of inappropriate habitats. The use of *Bacillus thuringiensis* (Bt) to control insect pests such as the gypsy moth was discussed in Chapter 4. While safe for mammals, the nonspecificity of Bt means that desirable species of insects can be adversely affected. The insecticidal gene from Bt has been transfected into plants using a species of *Agrobacterium* as a "gene taxi." The purpose is to impart resistance to some insect pest for that plant. There is concern that rDNA-modified species might interpollenate with wild ones (in the case of plants), attack benign insects (in the case of predatory or parasitic pest control species), or protect undesirable species (as in the case of the antifrost bacteria *Pseudomonas syringae*).

There is also evidence that Bt might not be as benign for people as previously thought. It has been implicated in ophthalmic infections, gastrointestinal infections, and hospital-acquired infections.

There have been calls for improved methods of risk assessment of genetically engineered species before they are turned loose in the world. A further concern is that allergens might be introduced to another species, resulting in an allergic reaction in an unsuspecting individual who consumes the altered food item. This has been demonstrated to be a real possibility when a gene from the Brazil nut was introduced into a strain of potato. The presence of the allergen was detected during safety testing and the product was not introduced to the market. Nevertheless, the public is becoming increasingly vociferous in its demand that genetically altered foods should be labeled as such. While there may be little reason for the public to be concerned about what is on the shelves of the supermarket, there is good reason to demand a high level of vigilance on the part of regulatory agencies. In the past, many corporations have submitted proposals to market bacterial and insect toxins for pest control without due regard for their allergenic potential. Fortunately they were not approved, but given the ease with which transgenic plants and animals can be produced (there are over 10,000 already), the possibility of a hazardous species falling through the cracks must be guarded against.

Ethical and legal concerns have been raised by the fact that it is now possible to patent a living organism. This has already occurred. Monsanto, holder of the patent on the herbicide Roundup (the isopropylamine salt of glyphosate), has also patented two genetically modified plants species of corn and soybean that carry a gene that imparts resistance to Roundup. This allows control of broadleaf weeds in crop fields without harming the crop. Already there have been lawsuits because a neighboring farmer's field showed growth of the resistant strain. Charges of patent infringement have led to countercharges of cross-pollination or seed contamination.

A further concern has arisen more recently. A microfungus has been identified in Roundup Ready corn and soy that have been implicated in spontaneous abortion in livestock, sudden death syndrome in the soy, and wilt in the corn. Emeritus professor D.M. Huber, a respected plant scientist, sent an open letter (February, 2011) to U.S. Secretary of Agriculture T. Vilsack warning of this threat. It was discovered by colleagues by electron microscopy.

Not all authorities are in agreement regarding the degree of crisis in food production. J. Ausubel, writing *The Sciences*, predicted that the Earth will hold 8 billion people by the year 2020. In 2011 it had already reached 7 billion. His premise is that advanced technology will cope with the larger numbers. He believes that we are currently in a period of "creative destruction" brought on by the flagrant abuse of credit, both public and private, in the 1980s. The cycle of bankruptcies, fiscal crises, and unemployment that followed will in turn be followed by a period of sustained growth that will foster the emergence of new technologies to solve the food crisis. The financial crisis of 2008 clearly indicated that the Promised Land has not yet been reached. His recommendations, however, focus mainly on the need to foster cooperation among nations to promote sustainable development and reduce the competitive nature that dominates most current international affairs. Apart from stressing the need to reduce our dependence on polluting energy sources, as by substituting natural gas for coal and oil, no specific innovative measures are offered. Not surprisingly, his article generated a flurry of letters in support of the conventional wisdom that zero population growth must be achieved within the span of the present generation if we are to have any chance of reversing the pollution and starvation generated by overpopulation. There was no discussion, however, of the fact that much of the world's food-producing capacity is underutilized because of market forces that prevent its redistribution to areas of need. In 1992, farmers in Prince Edward Island buried millions of kilos of potatoes because there was no market for them and no agency could be found to convert them to potato flakes for shipment to war-torn areas such as Bosnia, even if they were donated at no cost. American-Canadian trade disputes over wheat exports further illustrate the fact that cost is often the limiting factor in getting food to those who need it.

In his reply to his critics, Ausubel uses a quotation that bears repeating. It goes as follows:

> The most convincing examinations of the phenomenon of overpopulation hold that we humans have by this time become a weight on the earth, that the fruits of nature are hardly sufficient for our needs, and that a general scarcity of provisions exists, which carries with it dissatisfaction and protests, given that the earth is no more able to guarantee the sustenance of all. We thus ought not to be astonished that plagues and famines, wars and earthquakes come to be considered as remedies, with the task, held necessary, of reordering and limiting the excess population."

These words were written by Tertullian, a priest, around AD 200. Were they prophetic or merely alarmist?

The Environment and Cancer

The public fear that anthropogenic chemicals in the environment may be contributing to cancer incidence has already been noted, as has the existence of natural carcinogens such as certain mycotoxins and radon gas. Statements in the press that a high percentage of cancers are "environmentally produced" are usually taken to mean "produced as a result of anthropogenic activity," with no reference to the existence of natural carcinogens. Indeed, anthropogenic carcinogens may be of greater significance to aquatic organisms than they are to human beings. Cancer statistics for North America indicate that, in fact, the incidence of most cancers is declining. There are some noteworthy exceptions, such as smoking-related lung cancer. In contrast, the apparent "epidemic" of breast cancer in North America has been discounted by both the American Cancer Society and the National Cancer Institute as being a statistical aberration due to better detection and a population bulge in the over-40-year age group of women. Despite the generally encouraging statistics regarding cancer incidence, the (U.S.) National Academy of Sciences recently issued a report "Pesticides in the Diets of Infants and Children" in which it is stated that allowable levels of pesticides may be several hundred times too high for these age groups because of their age-related susceptibility (they consume more food per unit weight, and may not be efficient detoxifiers) and because their eating habits may lead them to consume many times more of a particular food than the amounts used to calculate allowable levels. The economic need for pesticides in agriculture may not be as great as previously thought, and the pressure is increasing to limit their use. Organic farming, while no panacea for world food supplies, is gaining in popularity largely due to public concerns over the use of pesticides, and perhaps chemical fertilizers, in agriculture. In Chapter 1, considerable space was devoted to the subject of the genetics of chemical carcinogenesis. While panic is not warranted, vigilance certainly is and any reasonable effort to reduce our intake of carcinogens is worthwhile.

Evidence has emerged in Utah of an epidemic of bone cancer, including chondrosarcoma (cartilaginous tumors), which is believed to be associated with an environmental carcinogen. The victims were dinosaurs. They lived 135 million years ago. The problem of cancer obviously is not a new one.

Further Reading

Ausubel, J.H., 2020 vision, *The Sciences*, 33, 14–19, 1993.

Comments on above and response, *The Sciences*, 34, 7 and 51–52, 1994.

Damgaard, P.H., Diarrhoeal enterotoxin production by strains of *Bacillus thuringiensis* isolated from commercial *Bacillus thuringiensis*-based insecticides, *FEMS Immunol. Med. Microbiol.*, 2, 245–250, 1995.

Fisher, D., *Fire and Ice: The Greenhouse Effect, Ozone Depletion and Nuclear Winter*, Harper & Row, New York, 1990.

Ginsberg, C., Aerial spraying of *Bacillus thuringiensis kurstaki* (Btk), *J. Pest. Reform.*, 26, 2613–2616, 2006.

Hantavirus infection: Southwestern United States: Interim recommendations for risk reduction, *Morbid. Mortal. Week. Rep.*, 42, RR-11, 1993.

Hantavirus pulmonary syndrome: United States, *Morbid. Mortal. Week. Rep.*, 42, 816–820, 1993.

HO, M.-W., Emergency! Pathogen new to science found in Roundup Ready GM Crops? Inst. Sci. Soc., ISIS Rep. February 21, 2011, http://www.i-sis.org.uk/newPathogenRoundupReadyGMCrops.php

Huberman, B.A., An ecology of machines: how chaos arises in computer networks, *The Sciences*, July/August, 39–44, 1989.

Jackson, S.G., Goodbrand, R.B., Ahmed, R., and Kasatiya, S., *Bacillus cereus* and *Bacillus thuringiensis* isolated in a gastrointestinal outbreak investigation, *Lett. Appl. Microbiol.*, 21, 103–105, 1995.

Joseph, L.E., *Gaia: The Growth of an Idea*, St. Martin's Press, New York, 1990.

Lloyd, S., The calculus of intricacy; can the complexity of a forest be compared with that of Finnegan's wake? *The Sciences*, September/October, 38–44, 1990.

Lovelock, J., *Gaia: A New Look at Life*, Oxford University Press, Oxford, U.K., 1979.

Lovelock, J., *The Ages of Gaia*, WW Norton & Co., London, U.K., 1988.

Margulis, L., Gaia in science (letter), *Science*, 259, 745, 1993.

Margulis, L. and Dobb, E., Untimely requiem, *The Sciences*, January/February, 44–49, 1990.

Marshall, E., Hantavirus outbreak yields to PCR, *Science*, 262, 832–836, 1993.

Mazumder, A., Ripple effects: How lake dwellers control the temperature and clarity of their habitat, *The Sciences*, November/December, 39–42, 1990.

Miller, H.I. and Gunary, D., Serious flaws in the horizontal approach to biotechnology risk, *Science*, 261, 1500–1501, 1993.

Morse, S.S., Stirring up Trouble; environmental disruption can divert animal viruses into people, *The Sciences*, September/October, 16–21, 1990.

News and comment: Experts clash over cancer data, *Science*, 250, 900–902, 1990.

Regush, N., Border crossing pathogens: microbes on the march, *Can. Geograph.*, 120, 62–66, 2000.

Tsonis, A.A., *Chaos: From Theory to Applications*, Plenum Press, New York, 1992.

Yanko, D., Are animal disease patterns changing because of global warming? *Vet. Mag.*, 2(3), 18–21, 1990.

15

Case Study Reviews

All case studies are based on actual occurrences.

Case Study 1

This population of laboring miners has an incidence of a rare cancer many times that of the general population, and those who live near the mine, as well as family members of miners, have a lower, but still elevated, incidence of the same cancer. The incidence of lung cancer in smoking miners is 60 times that in nonsmoking miners, and several times higher than in smokers who are not miners.

This is actually the situation in asbestos miners who were first employed 30 years ago or more. The cancer is mesothelioma. The miners carried home the asbestos fibers on their clothing. Family members, especially the wives who did the family laundry, inhaled these. Living in close proximity to the mine also was a risk factor, as particles were airborne over short distances. Cigarette smoke and asbestos fibers were acting as co-carcinogens and/or promoters to greatly increase the risk of lung cancer.

Socioeconomic factors that might increase cancer incidence, in addition to smoking, would include diet (high saturated fats and low fiber), high alcohol consumption, lack of exercise, and other workplace hazards (e.g., benzene). Miners would likely be 18–65 years of age and predominantly male, so that the elderly and the very young, as well as women, would be excluded from the group at risk.

Case Study 2

Working in a confined space always constitutes a risk whenever volatile solvents or gases, explosive or otherwise, are in use. In this particular case, explosion-proof electrical devices (lights, ventilating fans) should have been available. Failing that, the ventilation fan could have been placed at a distance from the site and flexible large-bore conduit used to conduct air into the tank. The tank should have been pumped dry before attempting to work in it. A safety person should have been left at the surface to summon help (not to enter the tank alone), and safety lines should have been attached to the workers in the tank so that they could be recovered if they lost consciousness. Breathing apparatus could have been used if available. Ideally, air quality should have been tested before any workers were allowed into the tank ($O_2 > 19.5\%$, flammable substances <10%, toxic chemicals to meet published standards).

This accident took place in the Philippines, in a remote area. Epoxy paint solvents may contain a variety of volatile substances that are sedating and heavier than air. In this case, volatile glyceridyl ether was identified.

Case Study 3

This 4-year-old boy had acrodynia, a rare form of childhood mercury poisoning. His 24 h urine mercury level was 65 µg/L as were those of his mother and siblings. The house was air-conditioned and therefore sealed. The interior latex paint used on the walls of the house was identified as containing about 950 ppm mercury. Seventeen gallons (68 l) had been applied to the interior walls. The U.S. EPA allows the addition of 300 ppm mercury as phenylmercuric acetate to interior latex paint as a fungicide and bactericide to prolong shelf life. This limit was exceeded by more than threefold.

Case Study 4

The manure pit is another example of the dangers of working in a confined space. In this case, methane gas from the decomposition of the manure displaced O_2 so that the workers died of asphyxiation. All of the safety measures noted for CS #2 apply here as well. It is almost unbelievable that four men would consecutively enter the confined space, but it is a testimony to how panic can overcome training and common sense. Over 25 people have died in this type of accident in the last 5 years in the United States.

Case Study 5

Yet a further example of a "confined space" problem, these workers was using a chlorofluorocarbon (Freon-113) as a degreasing agent in a pit beneath a large piece of machinery when they were overcome and one died en route to hospital. The portal of entry was obviously by inhalation, and the target organs were the brain and the heart. CFCs are CNS depressants, causing narcosis, stupor, and loss of consciousness. Asphyxiation may occur from respiratory depression as well as from displacement of O_2. These agents, like the related anesthetics halothane and chloroform, sensitize the heart to adrenaline, with resulting arrhythmias and ventricular fibrillation. A number of years ago, sniffing CFCs in aerosol propellants became a cheap way of becoming "high" in some adolescent cultures. A number of deaths resulted from cardiac arrest. From 1983 to 1990, 12 deaths were recorded in the United States from the industrial use of CFCs.

Case Study 6

These Inuit soapstone carvers are exposed to the dust of their carving stones. This can cause pneumoconiosis similar to silicosis. Like the asbestos miners, they could carry the dust home on their clothing and thus put their families at risk as well. The use of respirators capable of filtering out the dust and of special work clothes left at the site, are necessary safeguards.

Case Study 7

The variety of signs and symptoms, not all of which are infectious, in these agricultural workers does not suggest that the primary cause is a microorganism but rather an inhaled pollutant. Potential airborne causative agents in this environment would include animal dandruff, dried feces, grain dust, dust from feed additives (antibiotics, sulfa drugs, minerals, etc.), parasites, bacteria, bacterial endotoxins, fungi, ammonia (adsorbed to dust particles), and methane. Many of these agents are allergens and an allergic component may be present in some individuals.

Corrective measures could include improved ventilation, the use of filter masks or respirators, damping down floors to control dust, and removal from this work environment of anyone suffering from allergies or respiratory disease. In many jurisdictions, agricultural workers are not protected by workplace legislation.

Case Study 8

The array of respiratory symptoms is strongly suggestive of an inhaled toxicant. Since ice-surfacing machines use internal combustion engines, CO poisoning should be suspected. Carboxyhemoglobin levels in blood should be measured. This was done and values ranging from 10% to 20% were found (normal levels are <2% for nonsmokers, 5%–9% for smokers). CO would account for the headache, nausea, and dizziness, but not for the difficulty in breathing nor the coughing up of blood. This is a strong indication of nitrogen dioxide (NO_2) poisoning.

The following day, tests were conducted with an ice-resurfacing machine equipped with an internal combustion engine. The machine had not been serviced for some time. Levels of NO_2 reached 1.5 ppm. The standard set by the U.S. National Safety and Health Administration (OSHA) is 1 ppm for the Short Term Exposure Level (STEL), which has a 15 min limit. The CO reached 150 ppm and may have been higher the previous evening. The maximum recommended for ice arenas is 30 ppm.

Case Study 9

These family members became dizzy and nauseated after eating a snack food purchased at a local convenience store. Two of them suffered convulsive seizures. The brain is obviously the primary organ of toxicity. The rapid onset (<1 h) indicates a preexisting toxin rather than an infectious agent. Since common bacterial toxins do not produce this type of reaction, and since the preparation did not involve seafood or mushrooms, the likely offender is a chemical contaminant.

It is difficult to identify the source of the contaminant since all cases involved the same retail outlet and the same manufacturer. The fact that the snacks were prepackaged would suggest the manufacturer as the source, but the nature of the package was not identified, and it is not impossible that some chemicals could penetrate paper or polyethylene bags.

Endrin is an organochlorine, cyclodiene insecticide. No traces were found in the store or the manufacturing plant. Another possibility is that the flour from which the taquito snacks were made was contaminated. There have been previous incidents of such contamination.

Signs of acute organochlorine intoxication include headache, nausea, vomiting, dizziness, clonic jerking, and epileptiform seizures.

Case Study 10

This is another case of a worker dying from exposure to fumes from a volatile chlorinated hydrocarbon solvent being used as a degreaser. Trichloroethane acts like CFCs to sensitize the heart and depress the CNS.

Inadequate ventilation is the critical factor in all these cases (refer to Case Study 2). Chronic intoxication would have probably involved central necrosis of the hepatic lobules.

Case Study 11

These demolition workers are cutting up an old iron bridge. The organ systems involved in the toxic reaction are the musculoskeletal system (joint and muscle pain), the CNS (headache), and the gastrointestinal system (nausea). This last symptom could also be central in origin, through stimulation of the chemoreceptor trigger zone. The most likely portal of entry is the lungs. These torch cutters would be wearing coveralls and gauntlets to protect against sparks. The source of the toxicant is thus likely to be vapors from the cutting process.

An old bridge of this nature is unquestionably going to be coated in many layers of lead-based paint. Blood lead levels (BLLs) were performed and ranged from 60 to 160 µg/dL. U.S. regulations require that workers having levels >60 µg/dL be removed from the work site. The highest level was detected in the barge worker who did not benefit from the nearly constant breeze encountered up on the superstructure of the bridge. The paint from the bridge was found to contain 30% lead by weight.

Treatment was initiated with chelation therapy (EDTA) and substantial amounts of lead were excreted in the urine with an accompanying reduction in symptoms.

The employer was fined for not providing appropriate respirators, clean work clothing, and facilities for washing up before lunch and at the end of the day.

Case Study 12

This is another case of lead poisoning. BLL was 70 µg/dL in the primary patient and elevated in other family members. In chronic gastric pain that cannot be attributed to ulcer or cancer, a blood lead determination is useful, as gastric distress is a common symptom of lead poisoning.

The portal of entry in this case is almost certain to be oral since no activities were in place that could have caused lead vapors. The ceramic jug was the suspect because it was imported from Mexico where lead glazing still is used. The jug had been used to store a fermenting beverage, so that considerable time was available for leaching to occur. The entire family partook of this beverage over the course of the summer.

Case Study 13

An investigation of the home environment revealed that the house had been built after the banning of lead paints. Water was obtained from a well by means of a galvanized pipe system, eliminating solder joints as a potential source of the lead. No suspicious ceramic or pewter utensils were used for food storage or preparation. All of this pointed to the work environment as the source of contamination.

A 19 day course of treatment was begun with dimercaptosuccinic acid (DMSA), an orally administered lead chelating agent. The patient was instructed to remain off work during the treatment period. At the end of this period, his BLL fell to 13 µg/dL. One month after returning to work it was back up to 53 µg/dL, confirming this environment as the source of the lead.

Further careful questioning elicited the information that he habitually chewed on bits of insulation cut from the ends of electrical wires. Analysis of this colored plastic (white, blue, and yellow) indicated that it contained 10,000–39,000 µg of lead/G of plastic. He was instructed to desist from this habit and within 4 months his BLL fell to 24 µg/dL and he reported a subjective improvement in his symptoms.

Lead, usually as lead chromate, is used in pigments employed in the manufacture of colored plastics. Lead salts are used in the manufacture of polyvinyl chloride (PVC) plastics. Thus, any colored plastic may contain lead and should not be chewed or ingested. Cadmium is also present in a number of pigments.

Lead sometimes appears as an unexpected contaminant in unusual circumstances. Ethnic health remedies are such an example. Lead, in amounts up to 90% by weight, has been detected in "Azarcon" from Mexico (used as a digestive aid), "Greta," also from Mexico (same use), "Paylooah" from Southeast Asia (applied to inner lower eyelid to improve vision) and a substance from Tibet given to improve development. All of these have resulted in raised BLLs in children (20–80 µg/dL) and symptoms of lead poisoning (Centers for Disease Control and Prevention 1993).

Case Study 14

These five steam press operators became ill with a variety of signs and symptoms suggesting involvement of the CNS, the blood, and the heart. The use of a solvent-borne adhesive raises questions about ventilation of the workplace, the nature of the solvent and its toxicity, whether respirators were in use and, if so, were they approved for the task, and whether protective gloves and clothing were in use. Since the steam presses had been in use for many years without incident, it is important to probe carefully to identify any changes in work habits or materials that had taken place recently.

A wide variety of aromatic nitro compounds is capable of causing methemoglobinemia. Remember that nitrites are administered deliberately in cases of cyanide poisoning to form methemoglobin, which has a higher affinity for CN than does reduced hemoglobin. The top 10 aromatic nitro compounds implicated in methemoglobin formation include, in order of decreasing potency, *ortho*-chloraniline, dinitrobenzene, *meta*-nitroaniline, *para*-toluidine, nitrobenzene, *meta*-toluidine, *ortho*-nitrochlorobenzene, aniline, *para*-dinitrosobenzene, and *ortho*-toluidine.

Since analysis of air samples was not helpful, analysis of the actual solvent should be undertaken if possible. Analysis of the adhesive in use at the time of the toxic event was undertaken and the results compared with a newly delivered lot. Samples were extracted with carbon disulfide-methanol and analyzed by gas chromatography with flame ionization detection. The "old" sample was found to contain 1% by weight para-dinitrobenzene (pDNB) versus 0.03% in the new sample. Tracing the product back to the manufacturer, it was found that a proprietary solvent used in the preparation of the adhesive was contaminated with pBNB. MetHb levels were monitored after the removal of the contaminated lot and found to be normal. Periodic monitoring was instituted, and workers were required to wear butyl rubber gloves as it was felt that significant skin absorption may have been occurring.

Case Studies 15 and 16

The rapid onset of symptoms in both cases, the hemodialysis patient and the people who attended the picnic, makes it unlikely that an infectious agent is involved. There were no waterborne bacterial toxins that could account for the symptoms. The CNS appears to be the organ system to which most symptoms are related (sleepiness, dizziness, etc.).

In both of these cases, it emerged that the potable water system was cross-connected with a chilled water system used for air-conditioning. Ethylene glycol was used as an antifreeze. The ethylene glycol diffused into the tap water and was consumed by those at the picnic and diffused across the dialysis membrane into the bloodstream of the patient with renal disease. Ethylene glycol acts first as an intoxicant similar to alcohol. Subsequently, the formation of oxalate occurs. This chelates calcium to form calcium oxalate, which precipitates in the kidneys and other tissues to cause renal failure and other tissue damage. Blood calcium levels are low also because of the binding of calcium to oxalate. Ethanol is administered because alcohol dehydrogenase is the metabolic enzyme for ethylene glycol but ethanol is the preferred substrate.

Case Study 17

These patrons of a Chinese restaurant most likely took in something orally. Again, the rapid onset of symptoms suggests a preexisting substance rather than an infectious agent. The signs and symptoms relate to the CNS. The symptoms are not compatible with known pesticide toxicity.

The physician inquired about fish and mushroom consumption because a type of scale-fish poisoning produces similar symptoms. Toxic mushrooms (toadstools) also might have been responsible.

The fact that only these three patrons consumed a particular dish is helpful. Inquiry revealed that this particular dish, pork chow yuk, had been prepared by three different chefs. Each had added monosodium glutamate to the full amount, resulting in concentrations three times normal. This is a classical case of "Chinese restaurant syndrome."

Case Study 18

This "crop-duster" pilot is displaying characteristic signs of excessive cholinergic activity. The signs and symptoms are opposite to those of atropine overdose and include pupillary constriction, visual disturbances, mental confusion, and other mental disturbances, profuse sweating, dizziness, weakness, and diarrhea. Parathion is an organophosphorus insecticide. Questioning revealed that the pilot had been careless in handling the concentrated stock solution, had not worn protective clothing, including gloves, and had not worn a respirator. The signs and symptoms of organophosphorus poisoning result from its irreversible inhibition of cholinesterase and they are frequently delayed. Treatment would consist of atropine to block the effects of acetylcholine overload, and pralidoxime to reactivate the acetylcholinesterase enzyme by removing the phosphate group.

Pralidoxime would be contraindicated if a nonorganophosphorus inhibitor of acetylcholinesterase had been used, such as a carbamate like Sevin. Since no phosphate is involved, no antidotal response occurs, and the situation could be worsened because pralidoxime has some anticholinesterase activity in its own right.

Plasma cholinesterase levels and red-cell acetylcholinesterase would be depressed in this individual.

Case Study 19

The signs and symptoms of this outbreak are characteristic of poisoning with a cholinesterase inhibitor. Questioning revealed that 23 of the men had been involved in the mixing, loading, or application of mevinphos, an organophosphorus insecticide being used for aphid control. It has high toxicity (EPA Class 1). The remaining men had entered and worked in orchards within 24 h of spraying.

Questioning of the orchard operators revealed that protective clothing or equipment either was not available or had not been used. Respirators, gloves, goggles, coveralls, and rubber footwear are recommended and, in some jurisdictions, required by law.

Seven individuals required hospitalization. Their plasma and red blood cell cholinesterase levels were depressed 75%–95%. Other workers were treated as outpatients and had levels depressed by 15%–25%. Atropine was given to a total of 11 patients.

Case Study 20

This restaurant-related syndrome involved a small number of patrons (6) out of four dozen who had eaten lunch there. Careful inquiry should be undertaken to determine whether these six had eaten anything different from the others. It emerged, after questioning these and other available patrons, that the six had consumed yellow-fin tuna. On the surface, this resembles a monosodium glutamate reaction, but the chefs adamantly denied using this flavor enhancer in the fish dish. The persistence of the symptoms, up to 9 h, also argues against MSG as the cause of the reaction.

Three of the affected individuals reported that the tuna had a distinctly peppery or "Cajun" taste, but again, the chefs claimed not to have used any such spicing.

This is a typical case of scombroid fish poisoning. Analysis of the yellow-fin tuna revealed histamine levels of 50–160 mg/100 g (normal < 1 mg/100 g). A telephone survey of hospital emergency departments in the city uncovered nine more cases over a period of 2 days. All had eaten yellow-fin tuna either in restaurants or in the home. Investigation did not elicit any evidence of a serious lapse in refrigeration or handling. The fish were cleaned and packed in ice onboard the fishing boat, delivered by refrigerated vehicle to a distributor, where they were repacked in ice in smaller lots for delivery to retailers and restaurants. It was extremely hot during the period of the outbreak, however.

All cases were treated successfully with oral antihistamines. In asthmatics and cardiac patients, the condition can be life threatening, requiring more intensive emergency treatment.

Case Study 21

One would want to enquire carefully as to recent ingestion of food, especially seafood, given the environment. These six sport fishermen consumed blue mussels that had been collected in deep water far offshore by commercial fishing boats. This makes bacterial contamination from sewage effluent or other human sources unlikely. The toxin is obviously a neurotoxin. Shellfish toxins that must be considered include domoic acid (amnesic shellfish poisoning), okadaic acid (diarrhetic shellfish poisoning), and saxitoxin (paralytic shellfish poisoning). Tingling and loss of sensation are strongly indicative of saxitoxin poisoning. Significantly, the episode occurred in June, when red tides would be more common in the northern hemisphere.

The symptomatology and mechanism of action of saxitoxin are identical to those of tetrodotoxin from fugu (puffer fish).

Analysis of the remaining blue mussels revealed saxitoxin levels of 25,000 µg/100 g in uncooked mussels and 4,300 µg/100 g in cooked mussels. The boiling destroyed much of the saxitoxin, otherwise the attacks would have been fatal, as they were in a similar incident in which clams were steamed but not boiled. The raw clams contained up to 13,000 µg/100 g of saxitoxin, most of which survived the steaming.

Case Study 22

In this case, the differential diagnosis must include food poisoning (some of the symptomatology could be due to *Salmonella* infection or *Staphylococcus* toxin), a chemical contaminant such as a pesticide, or some other cause. The time to onset is not suggestive of poisoning by a preexisting toxin as in Staph food poisoning. Careful questioning might reveal whether a possible source of *Salmonella* existed in the diet, e.g., undercooked eggs, chicken, or beef. The symptomatology is reminiscent of poisoning by an organophosphorus insecticide (see Case Study 18), but it is unlikely that spraying to control insects would occur so close to harvesting the tobacco.

This is a condition known as Green Tobacco Disease. It is due to the absorption across the skin of nicotine and it characteristically occurs in periods of wet weather. The signs and symptoms are those of stimulation of nicotinic receptors in the ganglia of the autonomic nervous system and of the neuromuscular junction. Some of the symptoms will relate to parasympathetic stimulation, as do those of organophosphorus poisoning.

Case Study 23

This poisoning obviously involves the gastrointestinal tract and the peripheral nervous system with different times-to-onset of signs and symptoms. Without knowing the common denominator (amberjack) one would initially have to consider an infectious agent, a contamination occurring in the restaurant, and a preexisting toxin. Stool and vomitus cultures were negative for all common bacterial causes of gastroenteritis (*Salmonella*, *Shigella*, *Campylobacter*, and *Yersinia*). Monosodium glutamate has a different array of symptoms and a more rapid time-to-onset. Seafoods that could be responsible include shellfish (diarrhetic shellfish poisoning, paralytic shellfish poisoning) and scale fish (scombroid and ciguatera poisoning). Shellfish poisoning is ruled out by the commonality of amberjack consumption. The rapid onset of histamine-related symptoms of scombroid poisoning is lacking. The combination of gastrointestinal and neurological symptoms is characteristic of ciguatera poisoning. Analysis of samples of the amberjack by mouse bioassay was positive for ciguatera-type biotoxins.

Case Study 24

The rapidity of onset of the symptoms after ingesting the root points clearly to this as the source of the toxin. There is also a clear dose dependency as the man who consumed three bites died, whereas his brother, who ate only one bite, survived. This is a classical case of water hemlock poisoning, with cicutoxin as the causative agent. The involvement of the gastrointestinal tract and the central nervous system are typical.

Case Study 25

These teenagers had collected some jimsonweed from an empty lot taken over by weeds. They had consumed some of the seeds from the seed pod. These are rich in scopolamine and atropine, muscarinic blocking agents with central effects. This accounts for the hallucinations (typically involving insects), confusion, combative behavior, disorientation, seizures, and coma. Peripheral effects include dry mouth, blurred vision and photophobia, and urinary retention.

Treatment is supportive and may include induced emesis and activated charcoal to decontaminate the gastrointestinal tract plus physostigmine, a cholinesterase inhibitor, to increase the availability of acetylcholine.

Case Study 26

This case is based on an actual occurrence. This boy was bitten by a Massasauga rattlesnake. The Eastern Massasauga (*Sistrurus catenatus*) is the largest of the pygmy rattlesnakes and is usually around 60–75 cm (24–30 in.) in length. It is gray-brown in color with darker blotches. It is the only venomous snake in Ontario and is restricted to some fairly circumscribed areas including the Bruce Peninsula and the Niagara escarpment, an area along the north shore of Lake Erie near Wainfleet (the Wainfleet Bog), and another in southwest Ontario near Windsor. Sporadic pockets exist along the shore of Georgian Bay and on Manitoulin Island.

If the trip to the hospital was going to take more than a few minutes, a tensor bandage could have been applied to the arm from the elbow to the fingers with about the same tension as for a sprain. Supporting the arm in a sling would also have been helpful. The hospital would have administered polyvalent equine crotalid antivenin. The bite of a Massasauga is rarely fatal. A bite to the face would be more dangerous and preexisting medical conditions like cardiac problems increase the risk, especially in the elderly. For further information see http://www.publichealthgreybruce.on.ca/Fact_Sheets/Eastern_Massasauga_rattlesnake.htm

Case Study 27

In a case of food poisoning in hot weather and in the likely absence of electric refrigeration, Staphylococcal food poisoning is immediately suspected. Anything with a creamy salad dressing could be the source. In this case it would be the potato salad. The other symptoms, tremors and delirium, suggest a neurological involvement not typical of Staph food poisoning. Home-smoked meat can be a source of botulinum toxin that can involve vomiting but this is typified by flaccid muscle paralysis and shallow, labored breathing due to progressive paralysis of the intercostal muscles, and the diaphragm.

The onset to symptoms is 8 h to 8 days. The milk is the most likely source of the toxin since dozens of toxic plants may contaminate cow's milk. The signs of poisoning in the nursing calves confirm this as the likely source. This is a case of milk sickness from cattle eating white snakeroot. On a walk through the pasture the veterinarian discovered a patch of the weed. He would be more familiar with the signs and symptoms since livestock of almost all kinds are more likely to suffer from this poisoning.

Case Study 28

This case is based on an actual incident (see Mulloy et al. 2001). In the late 1960s and 1970s a uranium rush occurred that rivaled any gold rush of the previous century. It was complete with claim jumping, murder, and honky-tonk boom towns. Rich deposits were discovered in the four corners area of the American southwest. It is the only place where one can be in four different states, Arizona, Colorado, New Mexico, and Utah, by taking a few steps. It is also the traditional lands of the Navajo people. Many of them worked for years in the mines. Because of the pressures of the Cold War, safety measures were almost nonexistent. Workers told of blowing on Geiger counters after their shift just to set them off. Lung cancers were still showing up decades after the mines were closed in the early 1980s. Many of the older generation on the Navajo reservation have already died. Carcinogenic substances that could have been encountered by the miners include radon daughters (the most serious risk), silica, and diesel exhaust fumes.

Safety measures that were usually lacking include adequate ventilation, respirators, special clothing, and showers.

References

Centers for Disease Control and Prevention (CDC), *Morbid. Mortal. Week. Rep.*, 47, 522, 1993.

Mulloy, K.B., James, D.S., Mohs, K., and Kornfeld, M., Lung cancer in a nonsmoking underground uranium miner, *Environ. Health Perspect.*, 109, 305–309, 2001.

Index

A

Acetone (C_3H_6O), 216
Actinia equina, 312
Active transport, 5
Aflatoxin B_1 (AFB$_1$), 285–286
African sleeping sickness, 263
Agonist, 12
Airborne toxicity, 395
Air pollution
  absorption, 129
  aerial spraying
    adverse reactions, 139–140
    biological pesticides, 136–137
    conventional insecticides, 140–141
    DDT, 136
    LBAM, 137–138
    painted apple moth and Asian
      gypsy moth, 138–139
  climate change
    carbon dioxide, 154–155
    global cooling, 157–158
    global warming, 152–154
    greenhouse effects, 157
    methane, 155–156
    motor vehicle exhaust, 156–157
    natural factors, 158–159
    sulfur dioxide, 156
    water vapor, 154
  health effects, 134–135
  pollutants
    chlorine, 151–152
    gaseous pollutants, 133–134
    ozone depletion, 150–151
    particulate pollutants, 134
    stratosphere, 132
    sulfur dioxide and acid rain,
      149–150
    troposphere, 132
    water and soil transport, 132–133
  remedies, 158–159
  sources, 130–131
  types, 129–130
  workplace
    asbestos, 142
    CO and NO$_2$, 144
    dust, 143
    MCS (*see* Multiple Chemical
      Sensitivity (MCS))
    metal-fume fever, 141
    methane, 143–144
    plastic pyrolysis, 143
    silicosis, 143
    systemic poisoning, 141
Alkylbenzenes, 218
Amanita muscaria, 322
Anaplastic lymphoma kinase (ALK), 135
Animal poisons
  freshwater algae, 312–313
  land animals
    arthropods, 319–320
    snake (*see* Snake poisons)
  marine animals
    coelenterates, 311
    echinoderm venoms, 312
    mollusk, 310
    red tide dinoflagellate toxicity, 307
    scale fish (*see* Scale-fish toxins)
    shellfish toxins, 307–309
    stinging fish venoms, 309–310
    venoms and toxins, 304
Antagonist, 13
Anthropogenic carcinogens, 391
Arsenic (As)
  concentration, 200
  environmental effects, 202
  organic, 201
  Red River delta area, 200
  tobacco, 201
  toxicity, 201–202
  toxicokinetics, 201
  treatment of, 202
Aryl hydrocarbon hydroxylase (AHH),
  109, 177

Aryl hydrocarbon receptor (Ahr), 177–179
Asbestos, 142
Attention deficit hyperactivity disorder (ADHD), 231–232
Autism spectrum disorder (ASD), 231
Azaspiracid toxin (AZP), 308–309

B

Bacillus thuringiensis, 136, 138, 140, 389
Bacillus thuringiensis var. *kurstaki* (Btk), 271
Benzene leukemia, 219
1,2-Benzisothiozolin-3-one (BIT), 138
Biodegradation, 181
Biological control methods, 271–272
Bis(chloromethyl) ether (BCME), 219
Blood lead levels (BLLs), 397
Botanical insecticides, 268
Botulinum toxin, 404–405
Boysen company, 222
Breast cancer, 391
Brevetoxins, 308
Butylated hydroxyanisole (BHA), 233–234
Butylated hydroxytoluene (BHT), 233–234

C

Cadmium (Cd)
    application, 198
    toxicity, 199–200
    toxicokinetics, 198–199
    treatment, 200
California Department of Food and Agriculture (CDFA), 137
Canada Environmental Protection Act (CEPA), 58
Cancer risk prediction
    acute exposure, 55
    carcinogenesis
        distribution model, 56
        dose-response curve, 57
        extrapolation, 56
        inhaling volcanic ash, 56
        linearized multistage assessment technique, 57–59
        mechanistic model, 56

chronic exposure, 55
sources of error
    age effects, 61
    co-carcinogens and promoters, 61
    extrapolation, animal data to humans, 63
    hormesis, 63–64
    natural *vs.* anthropogenic carcinogens, 64
    portal-of-entry effects, 59–61
    species differences, 61–63
    test reliability, 65
    very-low-level/long-term exposure, 55
Carbamate insecticides, 267–268
Carbon monoxide (CO)
    carboxyhemoglobin, 133, 395
    motor vehicle exhaust, 156–157
    and NO_2, 144
    O_2 transport, 28, 133
    poisoning, 27, 183, 395–396
Cardiopulmonary resuscitation (CPR), 311
Ceiling exposure value (CEV), 68
Centers for Disease Control (CDC), 306
Chaos theory
    definition, 381
    graphic plots, 381–382
    low-level attractors, 382
    random and chaotic routings, 381
    randomness, 381
    strange attractors, 382
    weather forecasting, 382
Childhood mercury poisoning, 394
Chinese restaurant syndrome, 400
Chloracne, 172, 174–175
Chlorofluorocarbon (CFC), 394–395
Chloroform, 183
Cholinesterase inhibitor, 401
Chronic fatigue syndrome (CFS), 145
Chronic lymphocytic leukemia (CLL), 360
Ciguatera fish poisoning (CFP), 304
Ciguatoxin (CTX), 304–305
Claviceps purpurea, 284
Cone shells, 310
Conotoxins, 310
Cyanide poisoning, 398–399
Cyanobacteria, 296
Cytochrome P450 (Cyp) 3A4, 286

D

Deoxynivalenol (DON), 291–293
DES, *see* Diethylstilbestrol (DES)
Dicarboximides, 271
Dichlorodiphenyltrichloroethane
    (DDT), 264, 339
Dieffenbachia, 321
Diethylstilbestrol (DES)
    abnormal sexual development, 247
    active/reactive metabolite, 247
    beef and poultry production, 246
    breast cancer, risk of, 247–248
    cervical malformation, 246
    feminization sign, 246
    fetal risks of exposure, 247
    homosexual preference, 248
    pellet implant, 246
    psychological costs, 248
    statistical evidence, 248
    synthetic estrogens, 245
    uterine carcinoma, 247
Dimethylsulfide (DMS), 380
Dinitrobenzene, 218
Dinoflagellates
    CTX, 304–305
    ichthyotoxin, 306–307
    red tide, 307
    shellfish toxins, 307–309
Dioxin (TCDD) toxicity
    Ahr and enzyme induction, 177–179
    carcinogenicity, 175–176
    chloracne, 174–175
    in Great Lakes, 109–110
    hepatotoxicity, 174
    industrial accident, 74–75
    metabolic disturbances, 177
    neurotoxicity, 176
    porphyria, 174
    reproductive toxicity, 177
Dissociation constant (pK$_a$), 5
Domoic acid, 308
Drug residues
    allergic sensitization, 239
    antibiotics
        allergy, 245
        IDR (*see* Infectious drug resistance
            (IDR))
        infectious disease, 244–245
        MDR, 240
        multiple drug resistance, 240
        Shigella dysentery, 240
        tetracycline, 239–240
    anti-infective agent, 239
    hormonal growth promoter
        anabolic female hormone, 249
        bovine growth hormone,
            248–249
        DES (*see* Diethylstilbestrol
            (DES))
        ZEAs, 249–250
    toxic manifestations, 239
Dust, 134, 143

E

Electromagnetic fields (EMFs),
    369, 371
ELF electromagnetic radiation,
        see Extra-low frequency (ELF)
        electromagnetic radiation
Elixir of Sulfanilamide, 53
Endrin, 396
Environmental Protection Act, 221
Environmental Protection Agency
    (EPA), 100
Eosinophilia-myalgia syndrome
    (EMS), 252
Equine leukoencephalomalacia
    (ELEM), 288
Escherichia coli, 118, 121
Ethoxyresorufin-*O*-deethylase
    (EROD), 117
Ethylene glycol, 216–217, 399–400
Ethylene oxide (CH$_2$CH$_2$O), 220
European Childhood Leukemia-
    Lymphoma Study
    (ECLIS), 360
Expanded chick edema disease, 109
Extra-low frequency (ELF)
        electromagnetic radiation
    cancer, 370–371
    deleterious effects, 369–370
    electric blankets, 369
    EMF, 369, 371
    lymphomas and sarcomas, 370
    oddball theory, 371
    types, 369

F

Favism, 250–251
Food additives
    artificial food colors
        ADHD, 231–232
        ancient tartans, 231
        ASD, 231
        banned/restricted dyes, 232
        double-blind crossover study, 231
        Feingold diet, 231
        GRAS list, 230
        synthetic food dyes, 230–231
    artificial sweetener
        aspartame and cyclamate, 236
        GRAS list, 236
        phenylalanine metabolism, 236
        sodium saccharin, 235
        sorbitol, 235–236
        Stevia rebaudiana, 237
        "sugar-free" gum, 235
        xylitol, 235
    emulsifiers, 232–233
    flavor enhancer
        cardiovascular disease, 237
        childhood obesity, 237
        food stuff, 238
        MSG, 237
        Northern Manhattan Study, 238
        SSB consumption, 238–239
    food and drug regulation, 227–228
    preservatives and antioxidants
        Belgian study, 235
        BHA/BHT, 233–234
        curing agent, 234
        FAO/WHO recommends, 234–235
    types of, 228–230
Food contamination
    carcinogens, 253–254
    EMS, 252
    favism, 250–251
    herbal remedies, 252–253
    toxic oil syndrome, 251
Food production
    Bacillus thuringiensis, 389
    creative destruction, 390
    living organism patent, 389
    meat *vs.* grain, 386–388
    microfungus, 389–390

    overpopulation, 390
    rDNA-modification, 389
    sources, 388
Fusaria, 290–291
Fusarium graminearum, 290

G

Gaia hypothesis
    Canadian Geographic, 381
    coral bleaching, 380
    DMS, 380
    Earth, 379
    rain forests, 379
Gaseous pollutants, 129–130, 133–134
Generally regarded as safe (GRAS) list,
        230, 236
Genetic factors
    acetylation, 23
    hemolytic attack, 25
    hypothetical dose-response
        distribution curves, 22–23
    single gene locus, 22
    strains of organisms, 24
    TPMT, 24
Glue sniffing, 221
Glutathione-*S*-transferase (GST), 109
The Grassy Narrows Story
    mercury contamination, 196, 198
    mercury discharges, 196–197
    social costs, 197
Green fluorescent protein (GFP), 338
Green Revolution, 265

H

Halogenated aliphatic hydrocarbons,
        214–215
Halogenated aromatic hydrocarbons
        (HAHs), *see* Halogenated
        hydrocarbon
Halogenated hydrocarbon
    chloracne, 172
    physicochemical characteristics
        accidental human exposures,
            181–182
        Ahr and enzyme induction,
            177–179

antibacterial disinfectants,
172–173
biodegradation, 181
biphenyls, 180
disposal site, 182
herbicides, 173
insecticides, 179
paraquat, 179
pharmacokinetics and
metabolism, 180
solvent, 182–183
TCDD toxicity (*see* Dioxin (TCDD)
toxicity)
trihalomethanes, 183–184
structural formulae, 171–172
toxicity, 180
Hantavirus pulmonary syndrome
(HPS), 385
Hepatotoxins, 312
Herbal remedies, 252–253
Hexachlorophene, 172–173
Hormone disrupters
BPA and phthalates, 341
DES *in utero*, 335
endocrine function
(*see* Xenoestrogens)
estrogens and androgens, 336–337
interpreting and extrapolating,
346–347
Lake Apopka incident, 336
phytoestrogens (*see* Phytoestrogens)
Hydrogen cyanide (HCN), 324–325

I

Idiopathic environmental intolerance
(IEI), 145–146
Infectious drug resistance (IDR)
coliform resistance pattern,
243–244
conjugation, 241–242
E. coli O157:H7 strain, 243
gram-negative enteric bacteria, 241
hinges mechanism, 241
livestock industry, 242
MDR strain, 243
NARMS, 244
NTA, 243

resistance pattern, 240
transduction and
transformation, 241
transposon, 241
Insulin-like growth factor-1 (IGF-1), 249
Interconnected systems
feedback loop, 385–386
global warming, 384–385
vicious circle, 382–384
International Tanker Owners Pollution
Federation (ITOPF), 117
Inuit soapstone carvers, 395
Irukandji, 311
"Itai-Itai" (ouch-ouch) disease, 199

J

Jellyfish, 311
Jimsonweed, 403–404

K

Karenia brevis, 307
Karlodinium veneficum, 306–307

L

Lead (Pb)
cellular toxicity, 191–192
chronic lead poisoning, 190–191
CNS toxicity, 191
fetal toxicity, 192
occupational exposure, 190
poisoning, 397–398
toxicokinetics, 191
treatment of, 192–193
Leiurus quinquestriatus, 319
Lepidopteran pheromones, 136
Life Span Study (LSS), 358
Light-brown apple moth (LBAM),
137–138
Lindane
(1,2,3,4,5,6-hexachlorocyclohexane),
266
Linearized multistage assessment
technique, 57–59
Lock-and-key model, 12
Lung cancer, 393

M

Magnaporthe oryzae, 294
Malaria and yellow fever, 263
Manure decomposition, 394
Marine pollution
  nonpoint sources
    acidification, 116
    deleterious effects, 114
    freshwater species, 115
    metal content, 113
    Microciona prolifera, 114
    plastics, 116
    sea-surface microlayer, 116
    specimens, 113
    sponge cells, 114–115
  point sources
    Exxon Valdez spills, 117–118
    French Frigate Shoals, 116–117
    German submarine U-864, 116
    oil spills, 117
Material data safety sheet (MDSS), 137
Maximum allowable concentration
    (MAC), 54
Mercury (Hg)
  elemental toxicity, 194
  inorganic salt, 194
  mechanism of, 195–196
  organic mercurials, 194–195
  poisoning treatment, 196–198
Metal toxicity
  aluminum, 203
  antimony, 204
  arsenic
    concentration, 200
    environmental effects, 202
    organic, 201
    Red River delta area, 200
    tobacco, 201
    toxicity, 201–202
    toxicokinetics, 201
    treatment of, 202
  cadmium
    application, 198
    toxicity, 199–200
    toxicokinetics, 198–199
    treatment, 200
  carcinogenicity, 205
  chromium (Cr), 202–203

CNS disturbances, 189
  double-digit specific gravity, 189
  heavy metal exposure, 205–206
  hippocrates, 189
  lead
    cellular toxicity, 191–192
    chronic lead poisoning, 190–191
    CNS toxicity, 191
    fetal toxicity, 192
    occupational exposure, 190
    toxicokinetics, 191
    treatment of, 192–193
  manganese, 203–204
  mercury
    elemental toxicity, 194
    inorganic salt, 194
    mechanism of, 195–196
    organic mercurials, 194–195
    poisoning treatment, 196–198
  metallothioneins (MTs), 204
  nutritional elements, 204
  uranium, 204
Methane, 143–144
Methanol, 215–216
Methyl isocyanate (MIC), 102
Microwaves
  cell phone and brain tumor
    CEFALO study, 368
    cranial exposure, 366
    gliomas, 366–367
    risk factor, 367
    tissue exposure, 366
  frequency, 365
Minnesota multiphasic personality
    index (MMPI-2), 147
Monosodium glutamate (MSG), 237
Multiple chemical sensitivity (MCS)
  CATS, 146
  CFS, 145
  definition, 144–145
  depression, 147
  IEI, 145–146
  odors, 148
  psychological component, 148
  SHR evidence, 147
  somatization, 146
  SPECT, 149
  TRPV1/VR1 receptor, 147
Multiple drug resistance (MDR), 240

Mutagenesis and carcinogenesis
  aplastic anemia, 30
  chromothripsis, 39
  Darwinian model, 38
  genetics of
    antisense gene, 34
    chemoprotection, 36
    drug resistance gene, 33–34
    epigenetic mechanism, 34–35
    growth factor receptors, 32
    hormone receptors, 33
    immunomodulation, 37
    induced cell death, 36
    oncogene, 31, 36
    predisposition to cancer, 34
    proto-oncogene, 31
    Rous sarcoma virus, 35
    TSGs augmentation, 36
    tumor suppressor gene, 31–32
    viral-mediated oncolysis, 36
  genome instability, 38
  mutational model, 37–38
  nongenotoxic, 38
  risk/benefit analysis, 30
  squamous cell carcinoma, 29
  stages of
    initiation, 39
    progression, 40–41
    promotion, 39–40
  teratogenesis, 30
Mycotoxins
  aflatoxins
    AFB_1, 285–286
    chemical structure, 287
    corn, 286–287
    Cyp 1A2, 286
    exo-and endo-epoxide, 286
    peanuts, 286
    "X" disease, 286
  aleukia, 285
  biological function, 283
  DON, 291–293
  economic impact, 294
  ergotism, 284–285
  fumonisins, 287–288
  fusarium species, 290–291
  grain detoxification
    binding agents, 295
    canines and felines, 295
    carnivores, 295
    chemical treatments, 295
    harvesting and milling, 294
    herbivores, 295
    human food sources, 296
    omnivores, 295
  ochratoxins, 288–289
  patulin, 289–290
  trichothecene, 293
  zearalenone, 291

N

National antimicrobial resistance
    monitoring system
    (NARMS), 244
National Precipitation Assessment
    Panel (NAPAP), 103
Natural toxicants
  EMS, 252
  favism, 250–251
  toxic oil syndrome, 251
Nerve growth factor (NGF), 147
Nitrogen dioxide (NO_2), 133, 144, 396
Nonsmall cell lung carcinomas
    (NSCLC), 135
Nontherapeutic antibiotics (NTAs), 243
No observable effect level (NOEL), 68
Normal equivalent deviations (NEDs), 17

O

Occupational Health and Safety Act, 66
Occuptional health, 393–394, 397
Okadaic acid, 308
Ophidiophobia, 313
Organic farming, 391
Organic solvents
  application, 213
  cause of cancer
    BCME, 219
    benzene, 219
    dimethylformamide and glycol
      ethers, 219–220
    ethylene oxide, 220
  chemical industry, 213
  classes
    aliphatic alcohols, 215–216
    aliphatic hydrocarbons, 213–214

aromatic hydrocarbons, 217–218
glycols and glycol ethers, 216–217
halogenated aliphatic
    hydrocarbons, 214–215
food and pharmaceutical
    industry, 213
household poisoning, 213
nonoccupational exposure, 221–222
toxic reaction, 221
Organophosphates, 100–101, 267–268
Organophosphorus insecticides
acetylcholinesterase, 100
advantage, 267
botanical, 268
carbamate, 267–268
cardiac arrhythmias, 101
cholinesterase, 101
irreversible inhibitors, 267
parathion, 267
Organophosphorus poisoning, 400

P

Palytoxin (PTX), 309
Partial agonist, 12–13
Particulates, 130, 134
Penicillium expansum, 290
Peroxisome proliferator-activated
    receptor (PPARγ), 341
Persistent organic pollutants (POPs), 335
Pesticides, 99, 391
African sleeping sickness, 263
agricultural pest control, 263
chemical classification, 99
DDT, 264
definition, 263
environmental contamination, 275
farming mechanization, 264
fungicides, 271–272
government regulation of, 272–273
Green Revolution, 265
health hazards
    carbamates, 101–102
    chlorinated hydrocarbons, 99–100
    chlorophenoxy acid herbicides, 100
    organophosphates, 100–101
herbicides
    bipyridyls, 270
    carbamate herbicides, 270

chlorphenoxy compounds, 269
dinitrophenols, 269–270
triazines, 271
insecticides, classes
    organochlorines (chlorinated
        hydrocarbons), 265–266
    organophosphorus insecticides,
        267–268
malaria and yellow fever, 263
multiple pesticide resistance, 274
nonspecificity, 275
resistance development, 273–274
risks and benefits, 276
syphilis, 263
toxicity of, human, 277–278
Pfiesteria piscicida, 307
Pharmacodynamics
biological variation and data
    manipulation, 14
cumulative effects, 19–21
definition, 12
dose response
    exponential curve, 15–16
    Gaussian distribution curve, 14–15
    minimum lethal dose, 15
    NOEL/NOAEL, 17
    quantal response, 15
    S-shaped curve, 15–16
    therapeutic index, 15, 17
factor influences, xenobiotics
    age, 21–22
    body composition, 22
    genetic factors (*see* Genetic factors)
    pathology presence, 25
    sex, 22
    xenobiotic interactions, 25–27
ligand binding and receptors, 12–13
probit analysis
    arithmetic curve, 18
    hypothetical toxicity data, 18
    NED, 17
    percent responder, 17
    probit curve, 18–19
    semilogarithmic curve, 18–19
Pharmacokinetics
absorption, 4–5
biotransformation
    phase I reaction, 7–9
    phase II reaction, 9–10

distribution
blood–brain barrier, 7
blood flow rate, 6
DDT, 6
lipid-soluble chemical
disposition, 6
nonspecific binding site, 5
elimination/excretion, 10–12
law of mass action, 4
partition coefficient, 4, 7
Pharmacology
acute exposure, 2
definition, 3
extrapolation process, 1–2
pharmacodynamics
(*see* Pharmacodynamics)
pharmacokinetics
(*see* Pharmacokinetics)
reproductive and fetal toxicity, 2
Physalia physalis, 311
Phytoestrogens
coumestrol, 342, 345, 347
estrogenic activity, 341
genistein and daidzein, 342
livestock and wildlife, 345–346
zearalenone, 342
Pinocytosis, 5
Plants poisons
autonomic agents, 322–323
cardiac glycosides, 321
cyanogenic glycosides, 323–325
microtubule dissolvers, 323
phorbol esters, 323–324
pyrogallol tannins, 322
research and treatment, 325–326
vesicants, 321
water hemlock, 324, 326
Plastic pyrolysis, 143
Polybrominated biphenyls (PBBs), 180–181
Polychlorinated biphenyls (PCBs),
180–181
Polymerase chain reaction (PCR)
method, 24
Portal of entry, 4
Proportional Mortality Rate (PMR), 370
Prospectors and Developers Association
of Canada (PDAC), 102
Protoperidinium crassipes, 308
Prymnesium parvum, 313

R

Radiation hazards
electromagnetic spectrum
components, 353–354
ELF exposure (*see* Extra-
low frequency (ELF)
electromagnetic radiation)
energy and damage measurements,
355–356
foodstuffs irradiation, 372–373
insect pests, 373
microwaves (*see* Microwaves)
nuclear disasters
Chernobyl, 359–361
Fukushima, Japan, 361–362
Hanford release, 359
Hiroshima, 357–358
TMI, 358
radioactive energy, nuclear decay, 355
radiophobia, 353
radon gas
alpha-particle, 364
Canadian homes, 362–363
lung cancer, 363
myeloid leukemia, 363
reindeer, 364
uranium deposits, 362
sources, 354–355
tissue sensitivity, 364–365
UV radiation, 368–369
Risk analysis and public perceptions
Bill 208 features, 67
definition, 67–68
designated substances, 66
Elixir of Sulfanilamide, 53
environmental monitoring, 65–66
environmental risk
real and potential, 69–70
risk avoidance, cost of, 71–72
toxic substance level, 68–69
voluntary risk acceptance *vs.*
imposed risk, 70–71
human health implication
dioxin (TCDD), 74–75
formaldehyde, 73–74
nuclear accidents, 72–73
legal aspects
Delaney Amendment, 76–77
De Minimis concept, 75–76

precautionary principle, 79–80
risk assessment, 54–55
risk management, 78–79
risk prediction (*see* Cancer risk
     prediction)
setting acceptable limits, 66
statistical problems, 77–78
WHMIS regulation, 66–67
Rous sarcoma virus, 35

S

Safe Chemicals Act, 2011, 53
Saxitoxins, 307, 402
Scale-fish toxins
     Chinese restaurant syndrome, 400
     ciguatera poisoning, 403
     ciguatoxin, 304–305
     ichthyotoxin, 306–307
     scombroid, 306, 403
     tetrodotoxin, 305–306
Scombridae, 306
Scombroid fish poisoning, 401–402
Sea anemones, 312
Sea urchins, 312
Sensory hyperreactivity reaction (SHR), 147
Shellfish poisoning, 403
Short-term exposure limit (STEL), 221
Short-term exposure value (STEV), 68
Silicosis, 143
Single photon emission computed
     tomography (SPECT), 149
Small cell lung carcinoma (SCLC), 135
Smog, 130
Snake poisons, 404
     bite, 313–314
     components, 317–318
     Crotalidae, 315
     Elapidae, 315–316
     erbutoxins, 318
     first aid, 318
     Heloderma spp., 316–317
     Hydrophiidae, 316
     neurotoxins, 317–318
     ophidiophobia, 313
     phospholipase A2 complex, 317
     Trimorpohodon spp., 316
     viper and crotalid bites, 317
     Viperidae, 313, 315

Soil pollution, *see* Water pollution
Somatoform disorder (SFD), 146
Specific absorption rate (SAR), 366
Stachybotrys chartarum, 291
Staph food poisoning, 402–403
Staphylococcus aureus, 296
Sugar-sweetened beverages (SSBs),
     238–239
Syphilis, 263

T

2,3,7,8-Tetrachloro-dibenzo-*p*-dioxin
     (TCDD), *see* Dioxin (TCDD)
     toxicity
Tetrodotoxin (TTX), 305–306
Therapeutic index (TI), 15, 17
Thiopurine *S*-methyltransferase
     (TPMT), 24
Three Mile Island (TMI), 358
Threshold limit values (TLV), 68
Time-weighted average exposure value
     (TWAEV), 67
Toxicity equivalency factors (TEFs), 109
Toxic oil syndrome, 251
Toxicology
     acute toxicity
          central neurotoxin, 28
          intermediary metabolism, 29
          oxidative phosphorylation
               inhibitor, 28
          peripheral neurotoxin, 28
          uncoupling agent, 28
     carcinogenesis (*see* Mutagenesis and
          carcinogenesis)
     chronic toxicity, 27, 29
     definition, 3
     DNA repair and cell repair, 41–42
     environmental and economic
          toxicology, 3
     fetal exposure
          teratogenesis, 42–44
          transplacental carcinogenesis,
               44–45
     forensic, 3
     population and pollution, 45–46
Toxic Substances Control Act, 53
Transferrable drug resistance, *see* Infectious
     drug resistance (IDR)

Trichloroethane, 396–397
Trichothecene, 293

U

Ultraviolet (UV) radiation, 368–369
Unexplained medical symptoms
    (UMS), 146
Unicellular members
   antibacterial and antifungal
      antibiotics, 296
   Cyanobacteria, 296
   mycotoxins (*see* Mycotoxins)
   protein toxin, 296
   Staphylococcus aureus, 296
Union Carbide Corporation (UCC), 102
Uranium poisoning, 405

V

Veratrum californicum, 346
Volatile organic compounds (VOCs),
    221–222

W

Waste disposal hazards
   burning, 104
   landfill sites, 104
   Love Canal
     cancer, 106
     EPA, 107
     lindane and dioxins, 107
     long-term health effects, 105
     methane accumulation, 108
     preterm birth child, 106
     toxic *vs.* natural disasters, 106–107
   rehabilitation, 105
   surface impoundments, 104
   tunnel construction, 104–105
Water hemlock poisoning, 403
Water pollution
   abiotic modifiers
     dissolved organic carbon, 91
     light stress, 91
     oxygen, 91
     pH, 89–90
     temperature, 91
     water hardness, 90

acclimation, 94
acclimatization, 94
acidity and toxic metals, 102–104
anthropogenic, 94
aquifer, 95
bioaccumulation, 94
bioconcentration, 94
biological hazards, drinking water,
    118–119
biomagnification, 87, 94
biosphere, 87
biotic modifiers, 91–92
chemical hazards, waste disposal
    (*see* Waste disposal hazards)
dead zone, Gulf of Mexico, 112
detergents, 98
eutrophication, 88
Great Lakes
   adverse effects, 110–111
   global warming and water levels,
     111–112
   GST enzymes, 109
   potential reproductive effects,
     108–109
   TEF method, 109
groundwater, 95
invasive species
   alewife, 92
   Asian carp, 93
   coho salmon, 92
   in Great Lakes, 94
   lamprey eel, 92
   quagga mussel, 93
   round goby, 93
   Zebra mussels, 92–93
liquid freshwater, 95
marine pollution (*see* Marine
    pollution)
pesticides (*see* Pesticides)
sources
   agricultural runoff, 96
   drainage, 96
   municipal sewage discharge, 97
   nitrates, 97–98
   rain, 96
   storm drains, 97
   surface runoff, 96
toxicant dissemination, 88–89
toxicity testing, 94–95

unicellular organisms, 87
Walkerton
    animals hog operation, 120
    boil water, 120
    botulinum food poisoning, 121–122
    E. coli, 120–121
    gastrointestinal infection, 119
Workplace Hazardous Information
        System (WHMIS), 66

X

Xenoestrogens
    Ah receptor, 340–341
    breast cancer, 343–344
    DDE exposure, 344
    endometriosis, 343
    E2 receptors, 338–339
    hormone-modulating activity,
        337–338
    interpretation and extrapolation,
        346–347

livestock and wildlife, 345–346
males
    cryptorchidism, 342
    DES *in utero*, 339
    organochlorines, 343
    prostate cancer, 343
    testicular cancer, 343
    vinclozoline, 339–340
    xenobiotics, 339, 342
mechanisms, 337
ovoviviparous species, 338
Vietnam veterans, 342
wildlife species, 338

Y

Yessotoxin (YTX), 309

Z

α-Zearalenol (ZEAs), 249–250
Zooids, 311